In Peril

ALL PEOPLE, ALL LIFE, OUR EARTH

In Prospect:

BETTER HEALTHCARE & MEDICINE

EXPLORING MASSAGE, CANCER, QIGONG, & CLIMATE CHANGE

By

ALBERT HOWARD CARTER, III

Perspectives in Medical Humanities

Perspectives in Medical Humanities publishes peer reviewed scholarship produced or reviewed under the auspices of the University of California Medical Humanities Consortium, a multi-campus collaborative of faculty, students, and trainees in the humanities, medicine, and health sciences. Our series invites scholars from the humanities and health care professions to share narratives and analysis on health, healing, and the contexts of our beliefs and practices that impact biomedical inquiry.

General Editor

Brian Dolan, PhD, Professor of Social Medicine and Medical Humanities, University of California, San Francisco (UCSF)

Other Titles in this Series

Clowns and Jokers Can Heal Us: Comedy and Medicine
Albert Howard Carter III (Fall 2011)

Bioethics and Medical Issues in Literature
Mahala Yates Stripling (Fall 2013) (Pedagogy in Medical Humanities series)

From Bench to Bedside, to Track & Field: The Context of Enhancement and its Ethical Relevance
Silvia Camporesi (Fall 2014)

Heart Murmurs: What Patients Teach Their Doctors
Edited by Sharon Dobie, MD (Fall 2014)

Soul Stories: Voices from the Margins
Josephine Ensign (Fall 2018)

www.UCMedicalHumanitiesPress.com

This series is made possible by the generous support of the Dean of the School of Medicine at UCSF, the Center for Humanities and Health Sciences at UCSF, and a Multicampus Research Program Grant from the University of California Office of the President. Grant ID MR-15-328363.

In Peril:

ALL PEOPLE, ALL LIFE, OUR EARTH

In Prospect:

BETTER HEALTHCARE & MEDICINE

EXPLORING MASSAGE, CANCER, QIGONG, & CLIMATE CHANGE

By

ALBERT HOWARD CARTER, III

First published in 2019
by Virtuoso Press
In partnership with the UC Medical Humanities Press

Medical Humanities Consortium
3333 California Street, Suite 485
San Francisco, CA 94143-0850

Cover design Ian MacNeil

Book design by Virtuoso Press

ISBN: 978-0-9963242-8-1

Printed in USA

DEDICATION

This book is dedicated to all living creatures,
including humans, trees, and bacteria
that support the health of all the inhabitants of the earth,
even the earth itself that is our home…and is now endangered.

CONTENTS

LIST OF FIGURES

INTRODUCTION

THE HEALTH OF EVERY HUMAN, ALL LIFE, AND OUR PLANET EARTH

In 2001, at the age of 58, I went to massage school. I had been interested in the field for a long time, and early retirement from college teaching was near. Concurrently, however, I had been diagnosed with a leukemia, and I started chemotherapy for that cancer—even while still working. These two contrasting paths provided a wellness story and a sickness story, and material for my thinking, feeling, and writing over the next several years.

All of us live between peril and safety, danger and security, sickness and wellness, death and life. The threats range from annoyance to traumatic violence, from a head cold to the Climate Change that threatens all the life on earth. How are we to consider such topics and create strategies and positive outlooks? These are the subjects of this book.

Everyone loves stories. Comic stories tell us of happy times—a goal achieved, lovers united, a conflict resolved—while tragic stories present sad times—a betrayal, trauma, illness, exile, or death. There are also combinations of the two types. As readers (or listeners or viewers) we learn of characters' experiences and compare them to our own. What brings us pleasure, pain, or hope?

The first half of this book presents two stories of wellness: (1) I become a licensed massage therapist and (2) I am certified as a Qigong healer (a Chinese healing art). There is also the story of illness, as I experience cancer and a difficult aftermath of treatment. Next there is a story of illness and healing, as I provide massage and Qigong for cancer patients and their families.

The second half of the book moves to a wider scope: (1) the Climate Change that threatens to disrupt the health of all humans and all life on our planet Earth and (2) possible changes we can make locally and globally, including changes in healthcare and medicine.

Throughout, I test these topics with a notion borrowed from MIT physicist Max Tegmark that *matter, energy*, and *mind* are all strongly related. I realize that modern society has, sadly, reduced the potentials of each one of these. I find that his three basic notions and the ancient art of Qigong seem to line up, and all three can help us improve healthcare and medicine.

This book explores the health of all humans, sick or well, including all caregivers who support our health. Modern medicine largely focuses on disease and trauma. We need these resources, certainly, but what if healthcare were primarily based on assets, not deficits? A sick or injured person still has resources. How may these be supported, recruited, and honored? When a patient is considered only "a patient" or "*a case*," he or she is reduced to the *material* body.

Health is the condition of wellness, we usually say, often thinking of individual people. We think of wholeness, vigor of mind and body, and the absence of illness. We may extend these ideas to include the healing assets of persons, communities, and the entire earth—all contexts for health.

As wonderful as modern medicine is, it is still largely *materials*-based. It could be extended, improved, and be more effective if it drew on the resources of *mind* and *energy* in both caregivers and patients, indeed in all persons sick or well. Modern physics, furthermore, sees close connections between matter, mind, and energy.

Drawing on *mind* and *energy* would have benefits: people would stay well longer, heal faster when injured or ill, have shorter hospital stays, and, when their lives come to an end, die in a more healthy manner. Furthermore, energy and mind have roles in wider contexts for health: social values, communities, public health, and the environment. Our reduction of "energy" to carbon sources (coal, gas, oil) have been a driver for Climate Change that will dramatically impact the health of all humans, all living creatures, and the earth itself. Personal energy is often reduced to efficient performance at work, in sports, etc.

In a massage workshop, cell biologist James L. Oschman suggested that "all medicine is energy medicine," but we specialize in surgery (mechanical energy) and drugs (chemical energy) and ignore wider concepts of energy. His two books describe and analyze various uses of energy healing (see SOURCES FOR INTRODUCTION below).

A cancer survivor, I am grateful to the medical care that put me into remission from my disease, but I also think all medicine can be improved.

Indeed there have been increasingly wider perspectives for medicine: bioethics, literature and medicine—all included in the Medical Humanities and now in the even wider field of Health Humanities. Such approaches have less emphasis of surgery, drugs, disease, and linear thinking and more emphasis on health, assets of patients and families, and multi-dimensional thinking. At some hospitals there are no more "visiting hours" because family members are now considered part of the healthcare team and welcome at any time.

MY INTERDISCIPLINARY BACKGROUND INCLUDES STUDYING HUMAN ANATOMY AND A HOSPITAL CLOWN

I'm an interdisciplinary scholar with a Ph.D. in Comparative Literature; I taught in college for many years. In 1983 I studied bioethics with James F. Childress at Virginia then spent a year at the Center for Bioethics, Kennedy Institute of Ethics at Georgetown. Edmund Pellegrino, the director at the time, mentioned a new journal *Literature and Medicine*, and this field became my main area of research. For the books that followed, I was primarily an observer, an analyst of the stories, images, and values. For *First Cut: A Season in the Human Anatomy Lab* (1997), I was neither medical student (nor cadaver!). For *Rising from the Flames: The Experience of the Severely Burned* (1998), I was neither a caregiver nor a burn patient; for *Our Human Hearts: A Medical and Cultural Journey* (2006), neither heart patient nor medical provider.

In other areas, I was more directly involved: I went through emergency medical technician training, became certified as an EMT and served as a pastoral-care volunteer in an ER/Trauma Center for a dozen years. There I could not only observe but also interact with patients suddenly torn from their ordinary lives. My job was to support patients with conversation, prayer (if appropriate), phone calls to family, and so on. I remember a woman upset with worry about her cat's dinner at home, also man with slashed wrists who told me that his girlfriend was unfaithful. Many were frightened, angry, sad, or guilty about accidents they had caused. Physicians and nurses gave excellent technical care but had little time to visit with patients. It was clear that listening, small talk, and human-to-human presence had power to calm and—to some extent—reintegrate patients into a more normal social world. My next book *Clowns and Jokers Can Heal Us: Comedy and Medicine* (2011) took these ideas further, including description of a hospital clown at work with patients.

HOW I CAME TO THIS PROJECT: MASSAGE, CANCER, QIGONG, PHYSICS, AND CLIMATE CHANGE

Several strange developments in my life made this book possible. Retiring early from college teaching, I went to massage school and became a massage therapist. While that adventure was planned, the next one was not: I was diagnosed with a leukemia and treated with chemotherapy for six months. Healed (and with further training in hospital massage) I provided, part-time, massage to cancer patients and their families for ten years. It was clear that massage helped them in body and in mind.

My own massage therapist mentioned an afternoon presentation about Qigong, a 20th-century revival of ancient Chinese healing practices. We went to the event, and I began five years of study. Although initially skeptical, I grew to appreciate the power of this art, and how dominant ideas of today—rational and Newtonian—were unable to understand or explain it. Qigong touches body and mind and also energy ("Qi" or "chi"). After certification, I provided Qigong to cancer patients, their families, and others.

I found work of MIT physicist Max Tegmark; he suggests that matter, mind, and energy are all strongly related; I use this as a speculative hypothesis for understanding health, illness, even death. Western thought deals largely with *matter*, relegates *mind* largely rationalism, and has limited senses of *energy*. I speculate that matter, energy, and mind transform back and forth and are important resources for healthcare and medicine. I am not a scientist; I hope that physicists and other scientists will provide technical proof for (or modifications of) my speculation.

Another unwelcome "adventure" has come to *all of us*: Climate Change, formerly known as the pleasant-sounding "global warming." I capitalize the phrase in the style of "World War" to emphasize its wide and destructive power. Among the many impacts—now and projected to become worse—

are many health impacts. These include loss of food and water, increases in tropical diseases, violent and extreme weather, heat exhaustion, lung problems, mental stress, and more. A book on human health must discuss Climate Change. Much of this discussion owes to my wife, Nancy Corson Carter, raised as a farm girl, later a Ph.D. in American Studies. Active nationally in ecological issues, she's provided me with most of the sources that I cite.

A MUSE: ESSAYIST MICHEL DE MONTAIGNE

My approach and writing owe a lot to Montaigne, the 16th century French essayist. "What do I know?" ("*Que sçay-je?*") he asked. What do we learn from our experience, experts, the ideas of the day, and our careful testing of such ideas? In his view, the literary form of the essay, or trial, plumbs both the topic at hand and also in the mind of the writer considering the topic. Further, our minds have many levels: rational, emotional, instinctual, as well as the influences (often tacit) of contemporary social values. Accordingly, an essay is multidimensional and synthetic.

When I was teaching a course in the essay, we read Montaigne's "Of Practice." One student who had been in a serious car wreck said it meant a lot to her because Montaigne not only described in detail a collision of horses that severely injured him but also his thoughts during his lengthy healing, even fears of death. She felt affirmation that his experience was like hers, and that she also found new meanings to life as she healed. Scholar Sarah Bakewell sees Montaigne's accident as a pivotal event in his life: he decided to stop worrying about death and to focus on life (*How to Live*, or *A Life of Montaigne*, 2010, pp. 15-22). When I was sick with cancer, I thought about medical care, my current debility and possible death, and, as I healed, implications of health for me, for my caregivers, and for all life around us. I began to wonder why modern medicine seemed more based in deficits rather than in assets.

MORE SYNTHESES, SPECULATIONS, AND WORLDS, LESS OF LINEAR PATHS

In a noontime conference I heard a doctor say, "Well, I got the PE [physical exam], the history, and the labs, and after that, you know, you just ramify your way on down." He was referring to a diagnostic grid like a decision tree that would presumably come to the right answer for this—or any other—patient. Possibly he had a great bedside manner and knew the patient well, but his comment appeared to focus only on algorithms, and the patient as a particular person had disappeared, along with any reference to nature as suggested by the words "ramify" or "tree."

The modern West is awash with habits of linear thought. One reason for the prevalence is that they are very useful for focusing on particular goals. Examples include flowcharts, decision trees, Monte Carlo simulations, trajectories, and itineraries. As helpful as linear models are, they are often reductive, based on limited evidence, and aimed at a predetermined goal, often ignoring contexts, novelty, and other dimensions and approaches with wider ranges of speculation and synthesis. Linear approaches can do a lot, but so can the wider contexts of conceptual worlds.

Syntheses...including emotions and new perspectives

My approach owes to Montaigne in drawing from personal experience, research and concepts from experts, ideas of the day (some criticized), and attention to the wider ranges of our mental lives, imagination, emotions, intuitions, and instincts.

I believe that emotions are one of the ways we perceive and evaluate the world; to ignore them is to be vulnerable to them, even controlled by them. Accordingly, each Section includes a rant, clearly labeled as "Rant." These are emotional outbursts. They are devoid of academic tone or careful argument. Examples are cherry-picked and counter-examples recklessly ignored. These passages reflect my emotions, some negative and emphasizing losses, some positive and emphasizing gain, values, and aspirations.

Unlike Montaigne, I do not draw primarily from Latin or Greek authors, the basis of his education; instead I refer to contemporary experts and writers from a variety of time and places.

Unlike Montaigne, I have more structure to my Sections and to this book as a whole.

Each Section has a closing Essay that seeks to pull the ideas together and consider implications the health of all living creatures, humans included, and for improving contemporary medicine and healthcare.

Each Part has one more "New Perspectives," ideas or approaches from the last decade or so that allow for new understandings, particularly for medicine and health care.

Part I   Affiliation and reaffiliation (from Rita Charon et al.)

Part II   Universals of matter, mind, and energy (expanded from Max Tegmark)

Part III   Integrative Medicine and Health Humanities (from Andrew Weil and Therese Jones et al.)

I acknowledge, as well, a fictional character, Capt. Kirk from TV's *Star Trek* who always accepted the mission "to boldly go" into new areas and make connections, even if it meant splitting an infinitive.

In the spirit of all these muses, this book is *an essay*, an exploration of topics and also of the many minds perceiving them past and present. It is *speculative*—a wonderful word, with origins in Latin for "watching."

Speculations and Prospects

Each Part has a speculation. Speculations are imaginative explanations, often without the data or facts that would provide firm proof. Some speculations may be proved later, some disproved, some otherwise modified to better formulations.

The medical word "speculum" has the same Latin origin (to "view," to "see") and stands for a mirror, a reflector for seeing an internal structure. The implied values include a crafted, material instrument, a workable strategy to probe a human body, and a human viewer, such as a doctor, who looks with diagnostic intent. "Speculation" became specialized in a financial sense (to look at trends and take risks) but also suggests a wide range of thought, from thinking, reflecting, surmising, hypothesizing, even taking "shots in the dark." A deeper root for all of these is a "watchtower," a raised viewpoint from which the surrounding landscape may be seen and assessed for approaching enemies (perils), friends, or anything new. Such outlooks and high viewpoints are also called "prospects." While prospects for the future may be dismal or very good, we tend to think positively about them, like prospectors looking for gold.

The speculation explored in this book is this: as we understand that matter, energy, and mind all influence each other, we can make applications to health. The interchangeability of matter and energy is well known in physics, especially through Einstein's equation: $E = mc^2$. More speculative are relations between mind and matter (spoon bending for example; see Section IV) and mind and energy (a basic principle of Qigong; also Section IV).

Such interactions can have powerful applications to health and medical treatment, but Western medicine focuses largely on the matter of bodies and rational minds that make assessments and treatments. A vision that includes matter, mind, and energy may see these universals acting in dynamic harmony, a systemic health. While standard Western medicine is largely oriented to treatment of disease and injury, integrative medicine and the emerging Health Humanities (see Sections VIII and IX) both emphasize health maintenance and a wider sense of mind that is part of healing agency.

In closing essays for each Section, summarize main points, raise further questions, and refer to the three "universals" of matter, mind and energy.

Worlds, not linear paths

We speak of *worlds* metaphorically as realms of disparate material, ideas, and data, but with a central notion of theme: the world of sports, the fashion world, the modern world.

A digression on globes: how do we perceive reality and live there?

My parents had a globe perhaps a foot high. A semicircular arm gripped it at the poles for easy turning. As a child, I enjoyed seeing the borders of countries, the names, the pretty colors, and so on; these seemed to give order and control. The globe was well designed—orderly, clear, even purposeful—suggesting that the world made sense, but I also enjoyed twirling it quickly to make all the categories blur.

I had questions. Why were the seas colored black? What was a large mathematical construct in the Pacific shaped like a figure 8? What might the metal disks at the poles be hiding? As an adult, I know that many countries are now different: some new, others entirely gone. Like other models, any globe is a temporary and approximate. Indeed a perfectly round globe distorts the somewhat flattened poles, not mention mountain ranges and marine trenches. Globes don't show the shell of the atmosphere—with its lovely clouds—that makes possible life on earth by shielding us from ultraviolent and infrared radiation and by providing cycles of rain. Globes don't show how the sun makes the earth half light, half dark. Over geologic time, all continents are not permanently contained by shorelines but temporary landmasses that expand or shrink according to sea levels. Furthermore, the earth's tectonic plates float on a sea of magma, and landmasses change in location and size. With the earthquake of 2011, Japan moved four feet closer to the United States.

Nonetheless, this globe gave me my first notions of the planet we all live on, and its variety of land masses to oceans.

The earth, our home

The concept of our world expanded for everyone in the late 1960s with the new set of images of the earth from the NASA Apollo Missions 4, 8, and 10. These detailed, colorful photos have fascinated and inspired us: we suddenly had a new vision of our earth, its beauty, its thin atmosphere with swirling clouds, the lit and the unlit halves, and how small it looks surrounded by space. We saw that we all lived the same beautiful planet, the "big blue marble" that was/is/will be our home. Perhaps we should take care of this, our world, and even each other! This is a major theme for this book: *we all live on this earth, this home to many interacting and balanced systems. We should fit in, doing no harm and also doing good.*

Multiple worlds

The metaphors of "world" in this book are several: worlds of massage, cancer, Qigong, energy, Quantum physics, Climate Change, as well as strategies and prospects for improving health. These worlds take us beyond linear paths because of the wide dimensions of those worlds and the wonderful human minds that have many capacities.

We may contemplate the matter, mind, and energy as universal qualities that are dynamic parts of each one of us, and we may expand our notion of home to the earth and to the universe.

WORLDS OF LANGUAGE AND LITERATURE

Words have multiple dimensions, including denotations (standard diction-ary meanings) and connotations (associative meanings that vary with subcul-tures, families, even each person). Thus meanings of words are synthetic and variable, as are our own minds. Words can function logically in argument or scientific usage; they can function lyrically in poems, spirited conversation, or a dramatic monologue. Words have qualities of rhythm, rhyme, and sound, exploited especially in poetry; these qualities give us pleasures of recognition and variation. Words are an extraordinary artistic medium; they can delight and influence our minds.

The history of literature shows a worldwide interest in images, characters, stories, and many themes: adventure, conflict, families, love, death, faithfulness, betrayal, and so on. One basic theme is tragedy, the realm of separation, pain, and death. Its polar twin is comedy, the realm of integration, pleasure, and life. Much modern medicine partakes of tragedy when it focuses on deficits, asymmetrical power relations of doctor and patient, and death as a failure. We need more comic features of social integration, supportive communities, and joy, even in the face of injury and illness, even in the face of death. Life is inherently a mixture of both: it is trag-comic, and we live in the liminal space between then. Further, if we withdraw from the grand comedy of large communities, we risk creating uncooperative tribes that are, sadly, tragic.

Like any modeling system, language and literature have strengths and weaknesses. Meanings of words shift over time. Stories do not solve eternal human problems, but they do present and illuminate them so that we can feel their depth and think about what they mean.

Sarah Bakewell describes Montaigne's style as flowing as association in a stream of consciousness (p. 35), but with a central question of "how to live." She finds twenty answers, one for each of her chapters. The answers include "Survive love and loss," "Use little tricks," "Question everything," and "Let life be its own answer" (Contents, pp. vii-ix). At base, he was an optimist, even while testing and criticizing contemporary knowledge, and so am I.

SOCIAL CRITICISM

In questioning everything, Montaigne wonders about the knowledge of his day, especially medicine. In addition to the pivotal serious trauma from his collision of horses, he suffered over the years from kidney stones. In "Of Experience," he describes the attacks in distressing, anatomical detail, but he also affirms, "We must run through the bad and settle on the good" (pp. 427, 445).

Medicine offers him no relief for the attacks. He writes about doctors,

"My quarrel is not with them but with their art." For him, "The very promises of medicine are incredible" (pp. 293, 301, 303). Indeed he rants that medicine "up to this moment…has been good for nothing but killing men" (p. 241). In "Apology for Raymond Sebond," he calls the attacks of kidney stones, "the worst of all maladies, the most sudden, the most mortal, and the most irremediable." Nonetheless, "the more my illness oppresses and bothers me, the less will death be something for me to fear. I had already accomplished this: to hold to life only by life alone" (p. 293).

Like Montaigne, I consider current models of reality that guide our thinking and behavior, implicit or explicit, helpful or not. These include values of *materialism, positivism, Neodarwinism*, and *unlimited progress*. Some of these ideas are, regrettably, the deep roots for Climate Change, an enormous threat—today, and worse in the future—to the health of all humans, all living creatures, and the earth itself. In the realm of health and medicine, we typically hold values that death is an enemy and that we should fight death at all costs.

OVERVIEW OF THE BOOK

PART ONE: HEALTH IS PERSONAL: CONTRASTS OF THE WORLDS OF RELAXING MASSAGE AND FRIGHTENING CANCER

While attending massage school, I think about our material bodies, social values about touch, and the minds of both massage therapist and the recipient of massage. I pass a national exam and become licensed to provide massage.

Concurrently, I become ill with infections, low white cells, and swollen lymph glands. A diagnosis, slow in coming, is a leukemia, cancer in my blood and bone marrow. Formerly robustly healthy, I am now a patient with low energy. Massage helps me feel better. I experience a hospital stay. Chemotherapy cures my cancer but leaves me weak and vulnerable to infections for nine months. I think about the limits of my material body, the wide range of my thoughts and emotions while sick, and also the other patients at the clinic where I am treated.

Well again, I take further training in massage for cancer patients. For a decade I provide, part-time, massage for cancer patients, their families, and, on occasion, hospital personnel.

I see first hand that massage helps patients relax, makes them happy, and supports them in their illness and therapy. Massage helps bowels wake up after anesthesia so that patients can go home sooner. Further, massage therapists have no risk of giving bad news or providing painful treatment; they can make small talk and listen. Massage used to be routine in hospitals. Regrettably it disappeared from hospitals but is now making a slow comeback.

New perspective: affililiation, reaffiliation

I learn a valuable use for healthcare of the term "affiliation" from *The Priniciples and Practice of Narrative Medicine* (Charon et al., 2017, pp. 3, 130). Charon and her co-authors emphasize healing relationships between patients and clinicians through careful listening, reflecting back to patients, and creating a healing relationship that benefits both (or all) parties. This term also explains some of the benefits of massage, even though massage therapists aren't medical clinicians. In some instances, I expand the term to "reaffiliation" to understand several sources of healing that restore a loss: reconnecting patients to their own bodies, reconnecting them to their sense of well-being, and reconnecting them to the society of healthy persons. Massage therapists working with sick people also reaffiliate to the gifts of appropriate touch, cooperation and agreement of two people, and the gratitude of both people in the relationship. If they work in the medical world, they affiliate or reaffiliate to trauma, sickness, and mortality.

Speculation: matter and mind

While modern medicine often focuses on the material body of a patient, it's important to recognize also the patient's mind: ideas, thoughts, emotions, hopes, fears, and pleasures. Medicine can be improved by many reaffiliations through integrative medicine, integrative healthcare, including massage and mind-body approaches.

Cancer patients receiving massage often want to talk to the massage therapist, a "safe" person. Some complain that doctors are all focused on the disease, not the patient who has it.

High-tech medicine often works in linear and rational ways, but the mind can contribute as well. Robert Pirsig's novel *Zen and the Art of*

Motorcycle Maintenance: An Inquiry into Values (1974) explored values, both personal and social, and demonstrated multiple ways the mind may work.

PART TWO: HEALTH IS CONTEXTUAL: CONTRASTS OF HEALING QIGONG AND THREATENING CLIMATE CHANGE

Energy (beyond coal, oil, etc.)

While Part One focused on dyads of a sick person and a clinician or therapist, Part Two looks at wider contexts that influence of the health of all humans, indeed all living creatures.

During my time as a cancer patient, I observed levels of my energy, sometimes at rock-bottom. Working with cancer patients, I further sensed their compromised energy.

Typically Westerners understand energy as personal energy to compete in the world or various commodities of electricity, oil, gas, coal, etc., but other cultures—and often indigenous—around the world have other awareness of energy, often relating to living plants and animals. Tom Bender lists 65 such cultures (Bender, 2016).

I visit an afternoon workshop on Qigong, an ancient Chinese healing practice rediscovered in the 20th century. "Qi" (or "chi") can be translated as "energy," "bioenergy," or "universal" energy. "Gong" can be translated as "study," "practice," or "use." This energy is different from the commodities of petroleum or electricity or the forces of Newtonian physics. Qi is more like "the force the through the green fuse drives the flower" in Dylan Thomas' memorable phrase. I begin study of Qigong with Lisa B. O'Shea (Qigong Institute of Rochester).

While massage is widely known and practiced in many styles, Qigong is not well known in America, except in California and New England. Originating in China at least 3,000 years ago, it overlaps with the modern Tai Chi exercises seen today in parks and with Kung Fu and Karate, the martial arts. Qigong includes kinds of meditation, breath practices, and medical applications. A basic principle is that *the mind can move energy within the body and into the body from universal energy.* Qigong is widely used in China for many diseases, and especially cancer. My English translation of the basic medical text used in China has 637 pages (Liu, *Chinese Medical Qigong*, 2010).

With Daoist (as well as Buddhist and Confucian) roots, Qigong stresses the dynamic balance of nature; medical applications in Chinese hospitals seek rebalancing energy in patients. During five years of training that culminate in certification, I learn that Qigong works: it can balance energy, energize people, and give people tools to move their own energy.

Two earlier nonfiction works explored modern physics, energy, and Eastern philosophies: Fritjof Capra's *The Tao of Physics: An Exploration of the Parallels Between Modern Physics and Eastern Mysticism* (1976) and Gary Zukav's *The Dancing Wu Li Masters: An Overview of the New Physics* (1979). Although some 40 years old, they are still useful, applicable, and inspirational in understanding nonwestern concepts of energy. Unfortunately their wisdom has been ignored by the dominant culture.

New perspectives:

Qigong and Quantum physics provides ways of understand entrainment of energy, energy that pervades the universe, and ways that human intent can influence matter—strange notions in a society dominated by Newtonian, cause-and-effect energy.

Climate Change: energy running amok!

Another new perspective—for some people, that is—is Climate Change.

Unfortunately, during the last decade science has learned much more about global warming, and the term "Climate Change" has entered our language as a more accurate phrase. Climate Change may radically change the earth and all life of plants, animals, and humans. Even in the short term, health impacts threaten: heat exhaustion, increased vectors of tropical diseases polluted air, lack of fish, other food, and potable water. There will be (and are already) wars and migrations because of lack of resources linked to Climate Change.

We can eat salads, do Yoga, drive fuel-efficient cars all we want, but these will not change the directions of Climate Change. Only massive changes in energy policies and usage can slow warming and, with persistence and luck, allow for a peak of warming and a slow settling back to cooler decades. (See the National Geographic documentary *Before the Flood*, 2016).

The roots of Climate Change are multiple and sad to consider: material comfort, easy travel, and extremes of wealth. All these use resources extracted from the earth, principally coal, oil, and gas that are assumed to be infinite. Social values are also drivers. Wealthy people (and nations) may be "doing well," but these are already perilous times for the earth.

New perspective: our world is in deep trouble now (and more in the future) because of Climate Change. We must change our usage of energy and mitigate the harms of temperature rise, sea rise, increased unstable weather, increased disease transmission, shortages of food, and more, all affecting the health of humans and all living creatures.

Speculation: three universals: matter, mind, and energy

When I found a paper by MIT physician Max Tegmark, "Consciousness as a State of Matter" (2015), the notion fascinated me, although the mathematics were beyond my comprehension. I followed up with his book *Our Mathematical Universe* (2014), which went further in discussing how matter, energy, and mind interact. Physicists believe that the matter (the elements) of the universe has the same chemical elements and uses the same energies (even in the mysterious "dark energy") throughout. Tegmark stops short of seeing that all three are (or can be) equivalent, but we can speculate that they, in fact, are. If $E = mc^2$, and matter and mind are somehow equivalent, then energy and mind can also be considered equivalents. We may speculate further that versions of "mind" also pervade the universe and relate to minds of all creatures, especially humans; many cultures have believed (and believe now) in *universal mind*, sometimes—but not necessarily—with a religious dimension. One way to understand health is to see it as *orderly transformations of matter, energy, and mind*. Further, human health relates to the universe and to the many systems of transformation here on earth. I like the notion that the universe is, thereby, our home, the largest of many, and not an alien, hostile place that is somehow running down, flying apart, or otherwise doomed.

Western medicine focuses largely on the matter of bodies and rational minds that make assessments and treatments. While standard Western medicine is largely oriented to treatment of disease and injury, Integrative Medicine and the emerging Health Humanities (see Sections VIII and IX) also include health maintenance and a wider sense of what mind is. Indeed, using mind and energy as healing resources can improve both medical treatments and healthcare in general—for sick people, well people, and all care-givers.

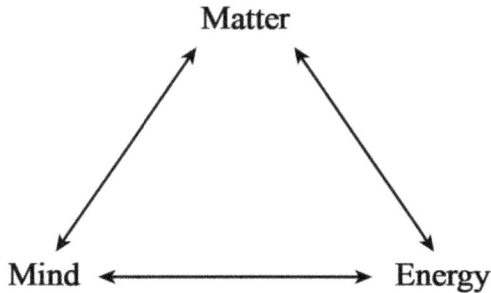

Figure 0: Interactive Universals of Matter, Mind, Energy

PART THREE: CHANGES WE CAN MAKE TO SAVE THE HEALTH OF THE EARTH AND ITS INHABITANTS HUMANS INCLUDED

Denial of Climate Change leads to a perilous stasis. The Trump administration withdraws from the Paris Agreement and seeks to expand, not limit, carbon usage by further fracking for gas, drilling for oil, mining for coal—while ignoring renewable sources of energy.

Projections of how Climate Change can make life difficult or even impossible are frightening…even overwhelming. What can we do individually or together? This part discusses small doable strategies as well as large-scale changes, including changes in healthcare and medicine, especially medicine that integrates on several levels.

New perspectives:
Health maintenance vs. intervention

Interventive medicine is especially valuable in emergencies and serious illness. Our culture loves interventive medicine, and we admire high-tech procedures that save patients, the image of the powerful physician who rescues patients, and dramatic TV shows of emergency care. Sociologist Arthur W. Frank calls the underlying story for this fascination "The Restitution Narrative" (Chapter 5, *The Wounded Healer, Body, Illness, and Ethics*, 1995). This is a "fix it" model for health care that suggests we need—and deserve—repair to become "as good as new" (p. 77). Agency belongs to caregivers, not patients. Further, Frank, a cancer survivor himself, observes that people with the same disease may have different courses (p. 90). Still further, I'll argue, we should use healthcare assets and emphasize values of social equality among caregivers, patients, family, and others.

More realistic stories include awareness of medicine's limits, variations and vulnerabilities in individuals, and compassionate understanding for trauma and illness in persons. Such qualities can be applied to nature and our earth. Both realms need continuous care and maintenance.

Healthcare in the West has several levels, including public health that promotes vaccinations, health promotion and education, surveillance for diseases, public works for water, sewer, and other sanitary measures, many of which we take for granted. More often, however, we think of healthcare as limited to what doctors provide when we are sick or injured.

Curing vs. healing

Physician Eric J. Cassell stressed the distinction between *curing* and *healing*; curing is disease-oriented, but healing is health-oriented. This insight became clear to him in a public-health seminar; he saw parallel with Chinese medicine that emphasizes, not diseases per se, but the healthy balance of energy in Yin and Yang (*The Healer's Art*, 1985, pp. 14-23). Because death is inevitable, we need healthy attitudes toward it and optimal care for the dying (Chapter 7, "Overcoming the Fear of Death").

Similarly, physician Bernard Lown argues against reductive medicine; he believes "healing is best accomplished when art and science are conjoined, when body and spirit are probed together" (*The Lost Art of Healing: Practicing Compassion in Medicine*, 1999, p. xv).

Yes, we need interventive medicine but also many forms of healthcare, including preventive medicine to keep us well. There is a long history of health movements in the U.S., including feminist, nutritional, and environmental approaches.

Integrative Medicine and Health Humanities

Integrative Medicine is now familiar to many; it takes wider view of patients, caregivers, and health maintenance.

Health Humanities, however, is but a few years old. It sees healthcare as assets-based, not deficits-based, and it assumes that the healing powers of all persons, sick or well, patient or family member, physician, nurse, allied health worker or clergymen, also other healers, animals included… even nature. Heath Humanities has evolved from Medical Humanities that included bioethics, literature and medicine, law and medicine, religion and medicine. Health Humanities includes these and other disciplines, notably the arts (music, film, etc.). Furthermore, emphasis shifts away from disease and concepts of repair and restitution to notions of health, patient assets, and narratives that include healing and also—we may even say—a natural and healthy death. While physicians in medicine and trauma care will always be important, we may expand the notion of healers further: to all persons.

Speculation: can we change?

I hope so…but it can only be with a huge shift in human minds and behaviors. Some days I doubt this can happen, but other days I see it as a prospective path we must try.

While the pressures of economics, politics, and Climate Change seem overwhelming, there are small things we can do, and larger things that are still possible.

Climate Change will become worse in the next few years. If we take committed and concerted action it may stabilize eventually, even mitigate. There may also be factors we cannot now foresee—in our favor…or not.

Whatever the future brings, we can feel gratitude for today, celebrate our bodies, energies, and minds, and seek to support the health of all humans, all living creatures, and all of the earth.

The book closes with notions of gratitude and song. Whatever the future may bring, we can, in Walt Whitman's phrase, "sing the body electric."

PRACTICAL APPLICATIONS FROM THIS BOOK

There are numerous implications for a more comprehensive view of health and medicine. Here are a few:

Who are healers?

Healers include not just doctors and nurses, important as they are, but also the 40 or more allied health professions, including social workers, phlebotomists, lactation consultants, medical interpreters, and various therapists for music and art as well as recreation, physical, and occupational therapists. Also massage therapists and practitioners of Qigong and other energy modalities. Also family members, friends, coworkers, congregants, even pets, pleasant rooms, and nature.

What do sick or injured people deserve?

Loving care, respect for the whole person, recruitment and promotion of their assets and agency, uses of the arts to comfort and inspire, and advocacy, especially for children and any with low status or low agency. All patients should be treated as whole persons, with mental and emotional lives, specific social backgrounds, philosophical, spiritual, and/or religious beliefs, and a variety of stresses and pressures as well as sources of pleasure.

All healthcare and medicine should seek to increase the agency of humans, their families, and their communities.

As we are all mortal, we need to honor the events at the end of life and offer palliative care as death approaches.

Given contexts of health, what should we do?

We should increase primary care, public health, various health screenings, good nutrition, psychological counseling, recreation and exercise that reduce stress. Furthermore: considerations of healthy neighborhoods, cities, nations, and the earth itself where all humans and other creatures live—all these are social justice concerns.

We should reduce violence, pollution of air, water, and earth, reduce access to guns, street drugs, as well as over-prescribed drugs for pain. We should increase funding for health promotion.

We need to evolve our concepts and values about matter, mind, and energy.

Because Climate Change may be the single largest threat to humans in the future, we must take action to mitigate it.

ARE THERE MISTAKES IN THIS BOOK?

Probably. I draw on a wide range of areas, and I am very sure that I am not an expert in all of them. I hope any oversimplification or outright mistakes are not fatal to my larger inquiry.

Some claims are provisional or even speculative, to be weighed and judged as further information is available; certainly some will be outmoded as science progress, or judged wrong.

A NOTE ON CONFIDENTIALITY

Laws, common sense, and respect require that patients not be identifiable. Patients are not recognizable from any of my descriptions because names and details have been changed and some persons portrayed here are combinations of patients.

DISCLAIMER REGARDING HEALTH

My descriptions and suggestions for improving health are not prescriptive for any given patients.

MY CLAIM

As wonderful as modern medicine is, it is still largely materials-based. It could be extended, improved, and be more effective if it used more carefully and consciously the resources of mind and energy in both caregivers and patients, indeed in all persons sick or well.

People would stay well longer, heal faster when injured or ill, have shorter hospital stays, and, when their lives end—as they surely will—die in a more healthy manner.

Albert Howard Carter, III, Ph.D., LMBT

Adjunct Professor
Social Medicine, School of Medicine
University of North Carolina-Chapel Hill
and
Faculty Affiliate
Trent Center for Bioethics, Humanities & History
of Medicine
Duke University & School of Medicine

SOURCES FOR INTRODUCTION

Bakewell, Sarah. *How to Live, or: A Life of Montaigne.* New York: Other Press, 2010.

Bender, Tom. "The Physics of Qi." tombender.org/energeticsarticles/qi_physics.pdf. accessed March 9, 2016. This is a "brief summary" of his DVD "The Physics of Qi (2006).

Capra, Fritjof. *The Tao of Physics: An Explanation of the Parallels Between modern Physics and Eastern Mysticism.* New York: Bantam Books, 1976.

Carter, III, Albert Howard. *First Cut: A Season in the Human Anatomy Lab.* New York: Picador, 1997.

_____. and Jane Arbuckle Petro. *Rising from the Flames: The Experience of the Severely Burned.* Philadelphia: Univ. of Pennsylvania Press, 1998.

_____. *Our Human Hearts: A Medical and Cultural Journey.* Kent, Ohio: Kent State Univ. Press, 2006.

_____. *Clowns and Jokers Can Heal Us: Comedy and Medicine.* San Francisco: Univ. of California Medical Humanities Press, 2011.

Cassell, Eric J. *The Healer's Art.* Cambridge, Massachusetts: MIT Press.

Charon, Rita, Sayantani DasGupta, Nellie Hermann, Craig Irvine, Eric R. Marcus, Edgar Rivera Colón, Danielle Spencer, and Maura Spiegel. *The Principles and Practice of Narrative Medicine.* New York: Oxford Univ. Press, 2017.

Frank, Arthur W. *The Wounded Storyteller: Body, Illness, and Ethics.* Chicago: Univ. of Chicago Press, 1995.

Jones, Therese, Delese Wear, and Lester D. Friedman, eds. *Health Humanities Reader.* New Brunswick, NJ: Rutgers Univ. Press, 2014.

Lown, Bernard. *The Lost Art of Healing: Practicing Compassion in Medicine.* New York: Ballantine, 1999.

Liu, Tianjun. *Chinese Medical Qigong*, eds. Tianjun Liu, and Kevin W. Chen, trans. Tianjun Liu, 3rd ed. (London: Singing Dragon Press, 2010).

Montaigne, Michel Eyquem. *Essays and Selected Writings; a Biligual Edition.* Trans. and ed. Donald M. Frame. New York: St. Martin's Press, 1963.

_____. "Apology for Raymond Sebond." In Frame, pp. 198-253.

_____. "Of Experience." In Frame, pp. 401-461.

_____. "Of Practice." In Frame, pp. 163-195

Oschman, James L. *Energy Medicine: The Scientific Basis*. Edinburgh: Churchill Livingstone, 2000.

_____. *Energy Medicine in Therapeutics and Human Performance*. Edinburgh: Butterworth Heinemann, 2003.

Pirsig, Robert M. *Zen and the Art of Motorcycle Maintenance: An Inquiry into Values*. New York: William Morrow & Company, 1974.

Tegmark, Max. "Consciousness as a State of Matter." March 18, 2015. https://arxiv.org/pdf/1401.1219v3.pdf, accessed May 8th, 2015.

_____. *Our Mathematical Universe: My Quest for the Ultimate Nature of Reality*. New York: Vintage, 2014.

Zukav, Gary. *The Dancing Wu Li Masters: An Overview of the New Physics*. New York: Bantam Books, 1979.

PART ONE

HEALTH IS PERSONAL: CONTRASTS OF RELAXING MASSAGE AND FRIGHTENING CANCER

I RUBBING THE RIGHT WAY: THE WORLDS OF MASSAGE

Done right, massage is enjoyable and nurturing. As a child I loved comforting touch. As an adult, I still enjoy massage, as a recipient and a provider.

Late in life, I thought about going to massage school after I retired from teaching, but an unexpected bout of leukemia raised this question: "How much time do I actually have?" My wife said, "If you want to go to massage school, how about right now?" I discussed this prospect with my academic dean and took a partial leave from teaching. I bought back some of my time—which gave me a sense of freedom and choice. Less working for the Man! More important, I had a commitment to a path, an adventure that was absurd in some ways but glorious in others. I'd be a beginner in a new field, taking instruction, doing homework, and commuting five days a week. I'd learn to *rub the right way*, deepen my knowledge of medicine and health, have a gift to offer other humans, and be able to prepare for life beyond the academy.

Following massage school, I passed the national licensing exam and enjoyed a ten-year career as a part-time massage therapist.

MY FIRST DAY AT MASSAGE SCHOOL

First day of classes: July 21, 2000. What will it be like? The teachers? My classmates? How will I fit in? I feel like a child at a new grade school. The night before, I checked my alarm clock several times, worrying that I might sleep late and arrive late—with everyone staring at me. I'd be Charles Bovary, the new boy in Flaubert's novel, laughed at and shamed on his first day of school. What will I wear? Shorts are OK but school rules say they must be "finger-tip length." Whose fingers? Also closed shoes are required, a standard in healthcare to guard against injuries. I have some loafers I like to wear without socks; I hope they pass muster. I trim an errant nose hair because people will be touching my body, even viewing it in detail. I try on my wedding ring, hoping it might fit. It doesn't. This hot, steamy summer in Florida has swollen my fingers.

There's no time for my morning jog. Heck, I can't even read the newspaper. I eat my breakfast, say goodbye to my wife Nancy, and hit the road at 8:15 a.m., just in time to join rush hour traffic. Nonetheless, I cover the 12 miles from south St. Petersburg to Pinellas Park in good time because most of my trip is on Interstate highway. I find the parking lot two blocks away, and walk to the front door of The Humanities Center Institute of Allied Health/School of Massage. Although nervous about being late, I arrive in plenty of time. I find the classroom and look in. A dozen people are already there.

"Take any open place with a bag," Lynn Otterbeck calls out to me. She's the admission officer who has helped me through the process of signing up. We shake hands, and I take a seat at one of the long tables. This is clearly the anatomy room, with life-size plastic skeletons hanging on metal stands on one side of the room. On shelves there are 15-inch high manikins with red modeling clay stuck on, and human-size models of isolated, skinned arms and legs with rust-red muscles. Hanging on the wall are very large charts of the superficial musculature. Nine long tables in three rows face the front of the room, where there is a large desk and a screen on the wall ready for slides or videotapes. I take a seat and introduce myself to a blonde woman on my right. There is a pile of material in front of me: papers, books, a big bottle of massage lotion, and a large green bag. The class fills up, totaling 20, 12 females and eight males. There are two other men in my 50-something age range; I'm glad about that because I didn't want to be Grandpa. The youngest seem to be in their 20s.

"Hello everyone, and welcome here," says Sherry Fears, the director of this school. She sits behind the large desk and smiles at us, her new class. Lynn hands out the first month's schedule, and Sherry explains it, pointing out days we need to bring our notebooks, anatomy books, and linens. "No need to haul all this stuff to every class," she says. She leads us through our materials of two notebooks, two texts, several handouts, the half gallon of lotion, a bottle of germicide and two smaller samples of massage lotion and gel. Students are responsible for bringing a clean sheet, two bath towels (for draping) and a pillowcase (for the face cradle). Men will be draped over the hips, women over hips and breasts.

At the school there are already three classes on staggered schedules. We are the newest and will always be called *July* because of our start date. With the Christmas vacation, our six-month course will end in February of 2001. Sherry explains that *July* will "lay down" for classes ahead of us so they can work on us. That's good, I think; I like being massaged...but this will also be a chance to see what lies ahead in our training.

As the newest class, we must park two blocks away and walk to the center. Sherry has some practical advice about our path on sidewalks and crossing the driveway to the center. We should move briskly there so as not to hold up cars wishing to turn in but risking being rear-ended on Park Boulevard, a state highway with four lanes and very busy at morning rush hour. I think there's an allegory here: we students take time out of our lives to learn about massage while the rest of the world whizzes on past us. Indeed massage itself is a time-out, a retreat, a specialized time of pleasure and heal-ing—in contrast to the many race-tracks, rat-mazes, and hamster wheels that circumscribe much of our "normal" American lives.

Sherry suggests how to organize our notebooks with dividers for the many handouts we will be receiving, some six or seven inches thick when all totaled up. These will cover anatomy, theory, history, responses to massage, alternative health, basic technique, advanced techniques, law, and hydro-therapy.

"Don't give a massage to anyone on drugs," she warns. "And remember some drugs can be picked up as long as six days after use, which means you could lose your license. This is a professional setting. Don't come to school influenced by drugs or alcohol." Evidently she's talking less about our future clients—still months away—than, in fact, about us. I'm guessing there have been previous students under the influence, and she certainly doesn't want that.

Sherry urges us to study together. There won't be a competitive curve for grading, so it is possible for all of us to do very well.

"Will there be any time on five-element theory?" asks one woman. Whatever does she mean, I wonder…and how does she know about that?

"We mention it for exposure in our alternative medicine unit, but we are mostly Western in approach here. Orgonal and energetic approaches are not currently covered by insurance."

Orgones? I have heard of orgone boxes…nut stuff from the 1930s, pretty much discredited and/or parodied, as in Woody Allen's *orgasmatron*.

(A decade later, I'll be instructed in detail about the five-element theory from Traditional Chinese Medicine and about Qi, a kind of energy and ancestor by a few thousand years to Wilhelm Reich's orgones.)

What's this business about insurance, I wonder? Who controls that? Licensure—sure, I've heard about that. It's one of the ways states can stop "massage parlors" from being whorehouses. According to Florida state law, if you use the word "massage" in advertising or signage, you must list your license number, and licensure depends on schooling, certifying, paying for a license, continuing education, and periodic recertification.

Bang! Sherry lifts a very large roll of toilet paper from below her desk and drops it on the desk. "It's time for the toilet paper ritual," she says with a straight face.

There are a few nervous chuckles. What can she mean?

She continues, "Tear off about the amount you would normally use for Number Two and drape it across you desk." She hands the roll to a man in the front row. I'm glad it's him and not me.

There's nervous laughter.

Number Two? What adult uses that term with other adults?

"You mean like after a big Mexican meal?" a man on the back row calls out.

I'm proud of him.

There's laughter that is more committed, more communal.

Sherry smiles and continues, "And then tear off squares. Use the edge of your desk." She demonstrates.

"Put your squares in rows and columns."

I receive the roll from my left and start to pull off a strip of paper. I look at my neighbor's strip for guidance and see a healthy three or so feet. I take a modest two feet and pass the roll to my right. What's the difference between a row and a column? Which is North-South? East-West? Who cares?

Sherry forges ahead, "We'll introduce ourselves using these. For each square, you say something about yourself. I could use one square to say I have two sons, or two squares naming each one separately."

"Suddenly I'm constipated!" bursts out another man on the back row. We all laugh wholeheartedly. I must get to know him, very likely our class clown.

"Well, let's unblock you," Sherry says. "And won't you please start? And when you're done, select the next person."

Somehow this nutty scheme works. Students flap over pieces of toilet paper and tell about themselves. There's a former mortgage broker, a chef from Trinidad, an Egyptian, a former paralegal, and a barge captain. There's a med tech who has two dogs and a three-foot boa constrictor. A woman who quit smoking two days ago (*applause*), a computer consultant, a fashion designer who left the field (too competitive), a casino dealer, a former Army Ranger who rappelled down cliffs, a bookkeeper who loves plants, also a woman who studies astrology, has four children, and will be a grandmother next week (*applause*). Next, a bar manager who worked until 3 a.m. last night and who is sometimes a pole dancer. Also a woman from Scotland who has a delicious accent; she has worked as a florist, she's worked at Disney World, and she's flown all over the world because she worked for an airline. And me? I'm a local academic with an undramatic life. I state that I teach college, sing in choral groups, write on medical topics, and I'm thrilled to be here. I feel eyes staring at me as I talk and, focusing on my words, I fail to turn over my toilet paper squares.

Today is just a half-day. Sherry reminds us to get our linens and other gear in order for Monday, our first full day. Sanitation is important, she adds. "We can't have flu bugs passed around the entire school. When you work with elderly clients who come to our clinics, you don't want to infect them either. Many of them have less than robust immune systems."

As I drive home—all the new materials stacked in the back seat—I think about the morning. I'm elated with the group. The other massage school I visited had a clutch of kids just out of high school. My class has some in their mid-twenties, and most scattered through their 30s, 40s, even 50s. I sense a high level of intelligence and motivation among them. Many do sports, work out, or have other interests in health. And, it seems, there is a pervasive yearning to create a learning community. When Sherry mentioned learning groups, some of the men said, "The sooner, the better." I'm dragging my feet a little here, the only one living 12 miles to the south of the Center and, worse, having egoistic pride in my abilities to learn quickly by myself. But bodywork is a helping profession, and I'm a beginner like everyone else, so I commit to that goal.

And what a variety of searchers, many tired of their previous lines of work or burned out. Not me: I will continue to teach literature and write about health, but I have, as an amateur, loved massage for a long time and I sense the possibility of a book. Our variety is good: different perspectives to share and different bodies to touch, and the wide range of experience suggests the universality of massage. Further, we all share the goal of learning about massage. We're giving up a chunk of our lives for six months and a chunk of cash—whether outright payment (in my case), or student loans in many others—all on a gamble to learn massage and provide it for other, yet unknown recipients. Whatever our backgrounds, we believe massage is good and want to know more about it.

On the center's walls are photos of earlier classes, all lined up and smiling. Clearly they've made it through and are, somehow, "out there."

WHY DID I GO? MY "TOUCH HISTORY" AND OTHER REASONS

In massage school we learned that people have a "touch history," that is, a cumulative sense of what touch has meant over their lives and, therefore, what it means—positively and/or negatively—in the present. For some, there has been violence, abuse, or incest; for them, touch can be threatening, even terrifying. For others, touch has been pleasant, and can be welcome. Probably for most of us, there's a mixture of good touches and bad touches in our histories.

I recall many positive memories. My family and relatives were affectionate in general, although I remember early years when I wanted to repel embraces, especially those of aggressive aunts. I've also seen pictures of me as a baby, nursing at my mother's breast; I don't remember this specifically, but I figure that's a fine way to start life. I always liked back rubs. "Drizzle" is what I termed the gentle, swirling touch of fingers. I also remember injuries and painful dental experiences.

As we age, there are progressions: from hugging a teddy bear to hugging a pet cat or dog, from childish wrestling and games to adolescent experiments in sex, from sexual naiveté to experience. Many of these are pleasurable, but some—fights, corporeal punishment, sexual exploitation, or touches by bullies or other violence—are hurtful and can be hurtful in the mind many years later.

One memory floats up. I was somewhere around eight or nine. It must have winter because there was a fire in the living room fireplace. It was late evening, time for me to go upstairs to bed. The fire was dying down to a small quilt of red embers. No other lights were on. My father was lying on a couch, facing the fire. Somehow he invited me (or I initiated) to lie next to him, against his large warm body, also facing the fire. The family dog Ben came up to us, sensing a good thing, but he was forbidden to be on any of the living room furniture.

My dad said quietly, "He can come up." I could hardly believe my ears: this was *revolutionary*! Before he could change his mind, I patted the cushion and the dog rose up onto the couch and lay with his back to my chest, just as I was nestled against my father, my arm over the dog, my dad's arm over me. We were quiet. The red coals glowed. Time stopped.

So…from childhood, I had a positive view of comforting touch.

Hot Springs, Arkansas

The first professional message I received was in Hot Springs, Arkansas, in 1960. I was 17 at the time and eager for a new experience but also nervous. I understood that people took their clothes off and were touched, perhaps by some gorilla-like, hard-chopping masseur. Would this feel good? Would there be discomfort, even pain? And what about a total stranger touching and looking at my skinny body, still gaining—all too slowly—its adulthood? I didn't mind being naked with boys in my phys. ed. class, but I had no experience with complete strangers in a new setting. And what about touching between men? There was almost none of that in high school, only the carefully codified slaps on the back and comradely punches to the upper arm. There was some outright fighting, but even for that the boys were clothed, and the intent was usually domination more than outright injury. Could I just lie on a table, totally passive, while an unknown person worked his will on me?

My Dad was the leader here; he wanted to have a massage and felt sure I would too. I thought it sounded good and was willing to take the risk.

Hot Springs is well named: water rushes out of the ground at 147 degrees. A guide showed my family the steaming water coming out of a wall of rock and said, "You can cook an egg in that." We were suitably impressed and signed up for baths and massages. Some details are still vivid to me more than half a century later. Men went one way, my mother and sister another. I was led to a small, serviceable room with a very large tub—a size anyone might always yearn for but never experience. The attendant turned two large taps and twin waterfalls flowed in. He said that I should mix them to the desired temperature and then left the room. Massive amounts of hot water in an enormous tub gave me a feeling of luxury, and this was the beginning of a life-long interest in hot tubs, spas, and volcanic hot springs from North Carolina to the Pacific Northwest, even the Japanese *ofuro* ("honorable bath"). I luxuriated in the tub and wondered about my massage.

When summoned, I put on an enormous white terry cloth robe and walked to a warm room that smelled of menthol. I was told to lie down on the massage table. I laid my robe aside and lay face down, entirely naked. I was concerned that my body didn't look right, but no one seemed to care, and soon I didn't either. I don't remember much else about the room; there were a few other tables with massage therapists and male bodies. Could I really join this society? Be accepted as a real *mensch*? The massage therapist greeted me and went right to work, using lotion on my back and shoulders. I was tense at first but soon gave myself over to pleasure and relaxation. I worried that I might have an erection all too evident when I turned over, but I was increasingly relaxed, and the sensations were sensuous, not sensual, let alone sexual. I remember men discussing a scar that was still pink and, therefore, a sign of a fairly recent wound. I was impressed by this knowledge of how the body works but soon realized that that had been my experience with injuries to my own skin.

So, that was my introduction to the art, and I liked it.

Esalen Institute

Another turning point was a visit to Esalen institute, Big Sur, California. The year was 1974; the times were post-Viet Nam, post-hippie, and—at least for some Americans—everything seemed possible. My wife Nancy and I drove up from Claremont, where I was part of an NEH Summer Seminar. We had made reservations for various weekend workshops because we wanted to see this place famous for its contributions to the human potential movement. Another motivator, at least for me, was one of the signature features, the hot tubs. These were, we had heard, halfway down the tall cliffs where land overlooked the Pacific Ocean. Many years before—one website says 4,500 years—the Esselen Indians had bathed here, although not, of course, in the modern bath houses. I imagined something primal about the centuries involved, the hot water heated by the magma of the earth, and whatever ceremonies the Indians had for cleansing, pleasure, and experience of this dramatic place where land, sky, and sea all united.

Soon after arrival, I walked down the path that slanted along the cliff. I quickly learned that the signs MEN and WOMEN over the two rooms were routinely ignored, so that people of both sexes and various ages all entered, stripped off clothes, and soaked in the hot water together. On the adjoining deck, naked people sunned or stretched. One woman was in a shoulder stand with her feet pointing to the sky, her breasts hanging improbably toward her ears.

Esalen-style massage was well known by the time of our visit. Indeed my wife and I were working our way through the *Massage Book* by George Downing (1972), the renowned Esalen-style body worker. I still have my copy and use some of his strokes.

At the office I paid my money and made a reservation. I was told to be in the hot tubs some 15 minutes before my appointment time—hardly a problem because I spent most of my waking hours outside of my workshop in the hot tubs. At the appointed time, a man with a slip of paper entered the bathing space and called my name. "Right here!" I called out. I climbed out of the hot tub, grabbed a towel, and followed him to a room between the two bathing areas. He lifted up a piece of plywood away from a window open to the Pacific so that we could hear the surf pounding below, also the sounds of seagulls and seals. I climbed onto the table and was soon lost to bliss as my therapist went to work. For a short while I tried to follow his routine to see how it matched up with Downing's book, but soon my left brain shut down. I don't remember details, just the great pleasure I felt.

Later adventures

Nancy and I continued to study the book and learned how to do a full-body massage. Later we sometimes would find another couple to exchange with, our chaste version of the movie *Bob & Carol & Ted & Alice*. In fact, anyone can do massage…but trained therapists do it better, with more techniques, more knowledge, more acceptance of whoever comes for treatment. As our careers in teaching progressed, we decided to be regularly massaged by professionals, for escape from the grind, for stress-reduction, for health, and for pleasure.

WHY DID MASSAGE STUDENTS SWEAT A LOT?

Our classes in strokes usually went like this: the teacher, usually Bonnie, would explain and illustrate the stroke upon one of us lying down while all other students watched intently. Then we'd fan out to tables in pairs and decide who was to lie down first. That person would shuck his or her outer clothes, placing them under the table. A towel covered the hips of a person lying down. For a woman, the breast towel, folded lengthways, lay across the table, and she would lie face down with her breasts on the towel. Bra etiquette was that the therapist (we were called that already!) would undo the back of the bra, lay the horizontal straps to either side on the towel portions sticking out below, and roll up the towel ends with the bra ends to meet her body…all neat and tidy…and modesty preserved. When a woman lay face up, we laid the towel across her bra and tucked in the ends at each side of her chest.

Bonnie would observe our first attempts at a new stroke, make corrections, and—when all pairs were checked—ask therapists and recipients to change places.

We all looked forward to our first chances to learn strokes. What we didn't expect is that we *sweated*…not just a little, but copiously. I'm not talking about "sweating a test," as student slang had it. Sure, we used that phrase, but the sweat I'm talking about was the obvious perspiration that soaked our shirts during our first hands-on sessions.

We were required to wear shirts with the name of our school and with our own first name printed on them just below our left clavicle. The names made it easy for the teachers and, of course, for all the students. In the early weeks of school I was surprised to see huge blotches of sweat in my armpits and below my pecs, even on my back and stomach…on other students as well. In two weeks or so, this strange sweating faded away. Why did we sweat? It wasn't because of heat in the building or our physical exertion. It was something else. Our excessive sweating ("hyperhidrosis") was caused by some combination of our apprehension, our worry, our anxiety—in short, our stress. Somehow we sensed danger…threat…peril.

Typically, massage relieves stress! Why are we so stressed? Our sweating was never discussed in class; this omission was unfortunate because that could have been an opportunity to talk about personal and social values involved in touching. Instead, the emphases were steadfastly on the mechanics of strokes and the body as a *material system*.

Years later, I believe that there were six reasons for the sweating.

1. Breaking taboos of touch in general, specifically in a touch-aversive society

Every culture has its own rules to manage touch, since there are many issues of privacy, sexuality, pleasure, and power. American culture is touch-aversive: in general we avoid touching each other. Many cultures are more permissive, allowing hugs, kisses, walking arm in arm. In Morocco, you can see men talking to each other with an arm around the other man's shoulder. In the U.S., there would immediately be interpretation that they were gay and guilty of "PDA" or "public display of affection," a judgmental phrase from college campuses.

Anthropologists speak of "personal space" even a "personal space bubble," but this is not a constant; various cultures understand differing sizes needed for privacy and safety. American culture assumes larger bubbles than other cultures. Perhaps our Puritan heritage, perhaps our heightened sense of personal autonomy, perhaps our sense of wide-open prairies, as in the song "Don't Fence Me In." Our touch aversion draws on elements of Christianity, Puritanism, and individualism. In particular, there's a fear of touching between men.

Whatever the social causes—conscious or unconscious—bodies touching bodies are largely taboo in this culture except for traditional ways: touching between lovers (and even here there can be limits), touching between parents and children (again, only certain kinds), and touching according to social norms. Currently acceptable are social greetings of hugs, kisses on the cheek, even male athletes patting each other on the ass—on national TV!

In the toilet paper exercise, Sherry suggested one other taboo area of urine and feces, and our laughter suggested our new group norms of acceptance of piss, shit, even assholes and genitalia. After all, we will appear before each other in our bras, underpants, boxers, and briefs. Any one of my classmates will lay down for me, and I will lay down for any of them. *Laying down* might suggest *getting laid* in some contexts, but not here. When we lay down for a fellow student we have another purpose agreed upon by society at large, by this school, and by this class, with no peeking or groping.

Touching our classmates was stressful at first, not just because of any sexual resonances, but because touching in general is taboo in this culture.

2. Massage as a euphemism for prostitution; proper and improper intimacies

Oddly enough the verbs "lie" and "lay" were regularly (but understandably) confused at the school. Strictly speaking, "lay" is a transitive verb: you actively lay down a burden or your own body to take a nap. When you "lie" down to receive massage, however, the verb is intransitive with a passive sense. Both make sense in recognizing the activity and the passivity in receiving a massage. Your brain makes it possible, choosing to lie down; you actively put your body on the table. Then you are laid down to relax and passively receive a massage from the active therapist.

The sexual phrase "getting laid" ambiguously combines both activity and passivity, but I never heard "getting massaged." Instead we said, "being massaged," "receiving a massage," or other variants that stressed the activity of the therapist and the passivity of the recipient.

Sexual activity has often been part of massage at brothels, whorehouses, "massage parlors," and the like for centuries, probably millennia. In the porno novel *Emmanuelle* (1967) the title character goes to a Thai massage parlor where she receives not only relaxation but also sexual stimulation by hand and by vibrator. For many cultures, massage and sexual satisfaction are still a norm, even in some places in the U.S.

In this discussion, massage will mean professional, legally defined massage that is non-sexual; this is the realm of therapeutic massage, relaxation massage, sports massage, Swedish massage, and the like. Such massage is about relaxation in a safe context, different from erotic massage for sexual stimulation. Erotic or sexual massage can be marvelous, but you don't go to massage school to learn that.

Massage does, however, allow for constructive intimacy. The intimacies of massage are many. As a recipient, you've disclosed information about your health on a medical form. You're alone in a room. You take off many or all of your clothes. You lie down and close your eyes. You abandon yourself to the experience. Another person sees you and touches you. During the massage, the therapist may ask you questions about your body: *Is this area sore?* or *How does this stroke feel?* or *Did you have surgery here?* If you were injured, you explain how it happened and reveal a weakness, a failure, or something to be protected. Sometimes a sore muscle is a mystery, and therapist and client may have an exploratory conversation. "I noticed you carry a large purse; do you always carry it on the same shoulder?" Or maybe there's a referral needed to a physical therapist. Therapist and client have a close relationship that is trusting, defined, and nonsexual.

When people relax, their bodies can rest and heal. Like sleep, massage is a time when a non-stressed body may take care of itself.

Therapeutic massage was a strange new world and entering it was oddly stressful, even on an unconscious level. In touching our classmates, we had to free ourselves of any ambiguities about the social histories of "massage parlors" in general.

3. Our class sorts out touching for sexual arousal and touching for relaxation

Besides the social contexts of "massage parlors," students needed to figure out their own personal values about touching semi-naked bodies, especially of the opposite sex.

One day, a woman sat on the table so that the teacher could illustrate a stroke. The woman was attractive. On her upper body she wore only a bra. I've forgotten what the stroke was but I remember that men gathered in front of her (by instinct, I'd guess, since I was there too). The woman put her hand in the center of her upper chest, as if to block our view. I understood that she was signaling that she didn't want her chest with thinly clothed breasts observed and was well aware of the male-female dynamics in the room. Another man understood the same; he handed her a towel so that she could drape her chest. By this action, he was actually the more important teacher that hour. The woman was the woman who worked at a bar and was a pole dancer. In that other setting, modesty was not an issue. At the massage school, it clearly was, and she felt a threat.

In the draping of male and female, we understood the importance of care for modesty, and we learned about ethical and legal boundaries for massage. In practice later, we read of cases of therapists who were reported for improprieties and lost their licenses. We were specifically taught about inappropriate touch.

Massage means that one human body touches another, and the touch can be influenced by intent. A person's attitude (whether giving or receiving) can give various interpretations. We students had an unconscious sense of this, but it would have been helpful for us to talk about it in class.

Strangely, American cultural values concerning touch often emphasize sex and violence. Violence was ruled out at the massage school by mutual, universal (and unspoken) consent. No one dreamed of punching anyone else, let alone stroking to cause pain. With gentle hands on bare flesh, however, the disconnection of touch from sex was not so automatic. Our class was unusual in having almost equal men and women (the others were dominated by women), and most students were young and all of us were vital: everyone seemed alert and intelligent. We worked closely five days a week over seven months. Sexual attraction and interest would be natural, but, as far as I could tell, it was not a focus for any of us. Our job was to understand massage, ethically and legally defined, and, as necessary, to leave behind sexual or erotic thoughts.

Dividing technical perception from sexual perception occurs in other fields as well, from drawing nude models in art classes to medical training. In Samuel Shem's classic novel *The House of God*, (1978) the narrator warns himself not to have "undoctorly thoughts" when observing an attractive and naked female patient.

Being semi-naked with strangers and touching them was stressful at first. As we got to know each other, however, we became more like brothers and sisters for whom there was an incest taboo. (See "Sibs Not Sluts" below.)

4. Fear of messing up, hurting someone, even acts of violence or battery
This is a simpler concept: we sweated because we didn't want to do anything wrong or endanger our fellow students. *Touching* between humans can be injurious. Boxers, wrestlers, and football players routinely injure each other. The fencer's cry "touché" indicates, symbolically, a wound. A colloquial word describing a crazy person is "tetched," meaning "touched" and now controlled by some malevolent spirit. The accompanying gesture is a tap on the speaker's head. We can be "touched" emotionally, for joy or sorrow. If you "put the touch" on someone, you are asking for a loan, sapping their strength, "hitting up" that person. A "touchy" person is irritable, easily disturbed. Clearly many uses of the word "touch" suggest threats to body and mind. In contrast, each massage stroke was a new way of touching a person to please and comfort. Nonetheless, we beginners worried about doing it correctly and how it felt to our partner, who might feel discomfort or outright pain.

We feared being judged by our partner: "No, no, that's not what she said; you are doing it wrong!" We were usually more tactful than that but we worried about doing something incorrectly. With little experience on how, for example, to massage a calf, we might fear causing a cramp or a bruise. As we gained experience, this fear diminished.

As for being judged by the teacher, this never diminished, because we continually had new things to master and he or she was always looking at us and correcting, acknowledging, or praising what we did. When it was a correction, we welcomed it as part of the training, but being judged can make a person anxious, especially as a beginner.

Furthermore, we all had touch histories that no one else knew…and maybe that we were only partially aware of. As a kid, I heard about a torture technique called "the cigarette cough." You put someone supine on the ground and rap on the sternum with your knuckles, always in the same spot. "Doesn't sound so tough," I foolishly told my buddies. They immediately pushed me down, sat on me, and one of them rapped on my chest. At first it didn't bother me. Before long, however, it was quite painful, and I yelled and bucked and either got free or they let me up. There was also some bullying in gym class. I got knocked out "accidentally on purpose" by an elbow to my throat during an end-around football play. More in the realm of give-

and-take, high-schoolers snapped wet towels at each other when instructors weren't looking.

I have received only one bad massage. It was a chair massage at an airport. I had plenty of time between planes and thought I'd treat myself to 30 minutes of massage. The man started out with gentle strokes but got rougher and rougher. "Hey," I said, "that's too hard!" He slacked off somewhat, but reverted to his overly macho approach. I was astounded. What was he trying to prove...besides that he didn't know the most basic things about massage? As a passive client, I was slow to complain, nor did I say that I was licensed and he was doing it wrong, etc., nor did I ask for his supervisor or later formally complain. Today, I would complain.

There are cultural limits in our touching norms, including many prohibitions. Some touches are completely out of bounds. Legally, touching without permission is *battery*, as in the phrase "assault and battery," the latter word evoking the phrase "battering ram." Even a doctor, generally approved by society to touch bodies, needs a patient's consent for medical touching, or he or she is liable for battery.

5. Our ignorance of the healing capacity of caring touch

This point is different from the first four dealing with doing something wrong. Instead, it's about doing something right. We students lacked a deeper understanding of what we were doing and of the long traditions of meaning and symbolism of touch as healing. As noted above, American culture avoids touch in general. More specifically, modern medicine typically does not promote touch as a healing resource.

As massage students, we were training for social acceptance and a legal license. Parallel to James Bond, "licensed to kill," we would be "licensed to touch" in carefully circumscribed ways, to accept money for this service, and fit into various niches of society: resorts, spas, health clubs, YMCAs, hospitals, and so on. (The word "license" has a Latin root meaning "to be allowed.") In the U.S. (as of this writing), 47 states plus the District of Columbia require licensure or certification. Most (if not all) states require continuing education units (CEU hours) for relicensing. To renew my license in North Carolina for each two-year period, I need 24 hours of instruction including three hours specifically in ethics.

Other cultures, both contemporary and historical, have used touch to heal, but we have largely lost this knowledge and/or instinct.

The Bible describes Jesus' healing by touch. In England, there was the tradition of the "King's Touch" that was presumed to heal; the modern version of this is a hospital visit by royalty or some other governmental leader. Faith healers touch a sick person to bring healing. In baptism, we touch an infant with water. A knight is dubbed with a sword. Blessings are often "hands on." If a doctor or nurse or other healer touches us with loving care, we often feel better. If a friend or family member gives a loving touch, we feel better.

When two people touch, there is an energy exchange. (More on this in Section IV.) Jesus asked, "Who was it that touched?" when a woman covertly touched him so that she might be healed. He said, "I perceive that power has gone forth from me" as if she had tapped and taken some of his energy (Luke 8.40-48).

While we didn't know these deeper traditions—nor were they mentioned—we had an intuitive sense, probably part of our deep mammalian history, that caring and careful touch was important.

6. Disconnections with mothering, nature, the earth.

Women do not usually start wars. Indeed, there's a book by Deena Metzger entitled *Women Who Slept with Men in Order to Take the War out of Them* (1983).

Mothering (typically by woman, but also by caring men) nourishes families, homes, communities, and the earth. Our male-dominated culture promotes action, rationalism, and urbanism; it tends to ignore being, feeling, and the natural world. By and large, we live with buildings, paved roads, and cars, as well as with calculation, and activity. (We'll further discuss modern American life-style when we look at the mythic figure of Atlas.)

Massage, for provider and receiver, is another world of calm, pleasurable sensation, the natural resources of human bodies and human consciousness. It provides powerful reconnections.

In summary

As beginning massage therapists, we were unaware of these reasons for our strange sweating, and there was no discussion of them to relieve our stress. We did, however, deepen our intuition that massage was good for people and that we intended to learn the skills to deliver competent and caring massage.

In a few weeks, as we got used to touching and being touched, our bizarre sweating came to an end.

In a few months, we saw a new class of students in the hallways with large sweat marks on their clothes. We didn't make fun of them.

RUBBING THE RIGHT WAY: WHAT FURRY PETS (AND OTHER CREATURES) KNOW

As I child, I heard my mother say that the best way to stroke a dog was with the grain of its fur. I had no idea that fur had grain, but when she showed me on our dog, my hand felt how "with" was much better than "against." Rubbing a person "the right way" with massage, I slowly learned in school and elsewhere, involved gentleness, caring, and shared values about pleasure and relaxation. The concept of *intention* is pivotal, both the intent of the person given the touch and the intent of the receiver of it.

When people look forward to a massage, they set their intention toward pleasure, and that helps them enjoy it. In *Touch: The Science of Hand, Heart, and Mind* (2015), neuroscientist David J. Linden discusses how the emotional state of a person strongly influences how a touch is perceived. The very same touch to a calm person and to a fearful person will be perceived differently because of the interactions of affective processes and sensory nerves in their brains. A light but undefined touch might be frightening—an unknown threat such as a spider, a molester, an attacker. The same touch in a massage setting could be pleasurable.

And it's not just humans. I was sitting next to a woman and holding her hand in both of mine. She had a fracture in her thumb that was slow to heal, and I was doing some energy work. Her dog came up to me and laid its head in my lap, clearly signaling *me next*. In the Reiki world, there are many stories of dogs and cats sensing energy work and wanting to be part of it.

We like to pet our dogs, cats, rabbits, and other furry pets. We like to pamper, coddle, or cosset these animals and feel them relax to our touch. ("Cosset" originally meant a pet lamb; some dictionaries relate the word to "kiss.") There is an ancient, even primeval history for our fellow mammals that includes comforting touch and mothering in general, with activities such as grooming, nestling, and cuddling. Apes groom each other to clean their fur but also to affirm social bonds. In the classic book *Touching: The Human Significance of the Skin* (1958), anthropologist Ashley Montagu describes "Harlow's monkeys," a series of famous (and controversial) experiments of the 1950s. Harlow separated baby rhesus monkeys from their mothers shortly after their birth, then arranged for them to be "raised" by surrogate mothers—robot-like structures that could dispense milk. One "mother" was made from bare wire mesh, while another was covered with soft terry cloth. The babies preferred the "contact comfort" of the mother with cloth, even to the point of ignoring the bare wire mother that provided milk. Montagu writes, "By far the most important of Harlow's observations was the finding that his infant monkeys valued tactile stimulation more than they did nourishment" (p. 30). From this and other studies, Montagu concludes, "Tactile communication forms an elaborate medium of communication among primates" (p. 34) and "As an order, primates are…contact animals" (p. 35). This is an important book, almost 400 pages long, but "massage" does not appear in the 14-page index, although it does list "laying on of hands and fingers."

Montagu discusses the importance of touch for dogs, cats, rats, dolphins, hens, sheep, and, of course, humans; throughout, he stresses that touches can be erotic, threatening, or comforting. Animals, including humans, know the difference.

While Harlow's inanimate "mothers" could not have intent in their touch, humans clearly have intent. This can be revealed by accompanying language, facial expression, and so on, but—as we'll discuss later—the recipient can feel intent by touch alone and even the energy field of the person touching.

Comforting touch is part of our deep history and still powerful today. When my massage therapist arranges sheets around me, I feel pleasure akin to being "tucked in" as a child.

Comforting touch is important for many animals, especially mammals, and humans in particular, but Americans have partially forgotten this heritage. Teachers are warned not to touch their students. The phrase "touch-starved" has entered our vocabulary as a common condition. There are now "cuddling services" in cities across the U.S.; for pay, a cuddler comes to your address and cuddles you!

Rubbing the right way—when a recipient agrees to be rubbed—is the right thing to do.

SENSATE, SENSUOUS, SENSUAL, SEXUAL: WORDS AS APPROXIMATE LENSES OF INTIMACY

The English language attempts to discriminate levels of touch.

"Sensate" is neutral…perceiving through touch; it applies both to the object sensed as well as the sensor, be that an amoeba or a massage therapist. How do we to interpret what we feel? "Sensuous" heads toward more pleasurable human activities, while "sensual" is on the slippery slope of gratification and indulgence plummeting toward the risk-laden realm of "sexual."

Even "erotic" suggests a super-charged, naughty world and not the wider sense of "loving," the realm of the Greek god of love Eros. Could it be that we hear the stressed word "rot" in the middle of "erotic," a cesspool to be not only avoided but roundly condemned?

Qigong describes how the ancient Chinese understood that rot, muck, earth were all part of the necessary and fruitful Yin or earth energy. Contemporary farmers and gardeners know this as well.

SIBS NOT SLUTS

One of the phrases I heard at school was "table slut." This inelegant phrase referred to—and made fun of—any student eager to jump on a table so the instructor could demonstrate a stroke. I learned later that it had been passed down from class to class for several years. Evidently it deftly affirmed the physical pleasure afforded by massage as well as the sexual resonances so carefully avoided.

"Slut" is, of course, a rude word for a dissolute, immoral, dirty person, even a prostitute. Our usage of the word was comical in its wildly inappropriateness; imagined sluts of "massage parlors" were far, far away. And "table" (not "bed") is the place for surgery or giving and receiving massage. Taken together, the two words represent an ironic tension between humans' basic, id-like desires for pleasure in contrast with the school's rigid structures of formal, ethical behavior. We found the phrase especially droll when applied to a man volunteering to lie down for the instructor. Although said as a joke, the phrase indirectly acknowledged our desire to give and receive massage. All of us loved being worked on, and all of us had, one way or another, a persisting love for touching flesh.

Gossip among men about attractive women—at least in my hearing—referred to women in other classes, not in our class. A temporary family of sibs, we were brothers and sisters with an unspoken incest taboo. We considered ourselves as helpers to each other, not potential boyfriends and girlfriends. On one plane, we were all equals, working hard to learn the craft. On another, we took turns at the table as giver or receiver. Early on, I resolved to partner with every single class member, "exchanging" as we called it because I wanted to feel a variety of bodies and to enjoy our growing sibship as temporary brothers and sisters. In the Introduction, we saw the notion of "affiliation" from the narrative medicine approach of the Columbia group (Charon et al., 2017); this term explains some of the benefits of massage, even though massage therapists aren't medical clinicians. The word "affiliation" traces back to Latin *filius* (or *filia*), the adopting of a son (or daughter), which suggests a commitment by one person to provide faithful care to another person over time: to affiliate, in this sense, is to join in a caring, familial relationship.

Sibs take care of each other

One day, during that dangerous hour just following lunch, I started to doze off. I was in the half of the class lying down while the other half watched the instructor demonstrate a stroke. One of my buddies shook my ankle to wake me up. Good thing, because the Center decreed that if you go to sleep in class you are *marked absent for the entire day.*

In the lecture classes I sat on the back row, where I, a literature professor, could help other students with the spelling of unusual words.

Near the end of my training a class behind us lay down for my class. By chance I was working with an attractive women. It was easy to see she was athletic, trim, and muscular. Her black bra and underpants seemed to me more like a swimsuit than lingerie. As I worked with her, I found she was quite flexible and moved easily. I asked about this; she said she was a dancer, formally trained. When she worked on me, she asked about my legs, muscular from long distance running. Was there sexual energy between us? Sure, but within a context of larger energies: athletic, intellectual, and, surrounding all, the massage energy of our shared intent to learn massage and to touch for relaxation and sensual (not sexual) pleasure.

RANT 1: YANKEE MALAISE: THE PERILS OF INDIVIDUALISM AND NEODARWINISM; STRESS AND LONELINESS VERSUS COOPERATION, SHARED GOALS, AND MASSAGE

Students often compete for the best grades, the quickest mastery, the smartest questions, or just plain "looking good," but I didn't see this in my class. Instead, we helped each other with techniques, vocabulary, missing linens and so on. As we learned the first day, there was no limitation for high grades and group learning was encouraged. As sibs, we bared our bodies for touch and gave touching with care, tenderness, and increasing skill.

We were also pack animals in the sense that we helped each other on the hunt for the common goals and that we nurtured each other and the group as a whole. But packs often have orders of domination, some of them vicious, including fights to the death. In some packs only the alpha male can breed with the females. In our class there might have been jockeying for *alpha student*, but I wasn't seeing that. It's now some 18 years after massage school, so I may have forgotten some frustrations, pettinesses, and the like, but I'm still in the massage world of practice and required seminars where we—as strangers—work on each other, and I see that, in general, massage therapists care for each other.

As we noted above, the massage world values connection, care, mothering, and the natural resources of bodies.

Sadly, the surrounding society awash is in values and behaviors that are destructive: competition, envy, loneliness, meanness, fear, and prejudice.

While massage symbolizes cooperation and shared aims, I see much of contemporary American culture as violent, unfriendly, stressed, and self-deluded.

The result? Loneliness, unhappiness, anger, and stress.

What seem to be the roots of our Yankee malaise? I think of four areas of values. Such values are typically *implicit*: we are not aware of them, but we act them out daily.

1. Individualism: to hell with others and the society as a whole

Consider these the following phrases: "Make it on your own," "top dog," "climbing the ladder," and "rugged individualism." The last is attributed to Herbert Hoover, meaning that individuals should not seek help from the government…or perhaps from anyone else. This philosophy is a boon to production and sales, because every single household must own a lawn-mower, a ladder, etc., items that can easily be shared. Barbed wired closed the West, white picket fences marked off homes, and today many people do not know their neighbors….or even care to know them.

A real American does not need help from anyone, especially is he's a man, a martyr, all stressed and lonely!

Massage allows for relationship, cooperation, and shared goals such as giving/receiving help. Both giver and receiver are nourished.

2. Tribalism: stick with folks like you and avoid "others"

Many Americans associate with similar people, avoiding people who are different. Examples abound: so-called "gated communities," charter schools, some clubs, sports fans all wearing the same color at the big games, preferences for "*nice people, people like us.*" Survivalists, doomsday preppers, street gangs, organized crime, even some capitalists have a tight focus for their own welfare that does not include the welfare of a neighborhood, a town or city, their entire country, or the world. The Internet has sites for parents who will not vaccinate their children because they "know better" and don't mind if their children are carriers who may infect other children or even if their children may become quite sick (or die!) from, say, measles.

Outsiders might be terrorists! Avoid them!

It's important to divide rich from poor people!

Professional massage allows trained people to give comforting touch to a wide range of people—potentially anyone. Caring touch between family and friends can comfort and nourish.

As for intent: neighborhoods, affiliations, collaboratives, and wider communities can share in a sense of the common good, nationally, even globally.

3. Indolence is bad; "net worth" is good

We are a work-addicted culture. Laziness is sloth! Idle hands are the devil's workshop! Follow Benjamin Franklin's example of applied Enlightenment ambition!

For many, a "full calendar" is a sign of virtue, a mark of a proper adrenaline-driven lifestyle. We love "high performance" and clear definitions of self through a work role. *Be on task! Focus!*

The central meanings of life include the following: (1) make money, (2) advance in rank, and (3) buy stuff. He who dies with the most toys wins! Materialism! Consumerism!

Pleasure is for wimps. Lazy = bad. Do not smell the roses. Take your work home. Always be available by cell phone. Allow no time for exercise, but plenty of time for alcohol, drugs, and plenty of food.

Be a stress junkie.

Unemployed people should know better and be on career paths, but corporate pyramids narrow, so that there's pervasive disappointment in not advancing up the ladder. Even service jobs may have no advancement whatsoever, few fringe benefits, and no pensions.

The numbers of salary, wealth, bonuses, etc. cannot measure a host of more important qualities and resources: faithfulness, character, kindness, imagination, good humor, calmness, love, etc. Also: wealth is rarely "enough," usually an ever-receding goal.

Massage allows rest, relaxation, recovery, healing. Massage is pleasurable because of nonsexual touch, the caring intent of the therapist, and the protected time and space, and more.

Massage celebrates the body and the present, valuable time with no past, no future, no social or monetary goals.

4. Power, violence, Neodarwinism: survival of the fittest!

Nature is red in tooth and claw, indeed the world (universe?) is violent, vicious, and uncertain. Might makes right; only the fittest survive; eat the other guy's lunch!

Lock up your home; carry a gun because people are after you and your possessions. A *siege mentality* makes sense because the peril lurks everywhere!

The best sports are violent: football, car racing, so-called mixed martial arts, boxing, hockey, and, increasingly, basketball. Watch them in person or on TV. Male values of strength and violence are good.

When thwarted, nurse a grudge and anger in general!

Other people want to keep you *unrewarded*. Defined or not, they are your enemy.

Massage provides cooperation, sharing, and touch is gentle and slow. The intents of massage are caring, love, and kindness.

In summary

I've heard it said: "if everyone were massaged every day there would be no more wars," and I believe that.

SOME MASSAGE TERMS AND HISTORY

Most dictionaries give an etymology for massage from French (*masser*) or Portuguese (*amasser*), meaning to stroke or to knead (as in dough) and, further back, to an Arabic word *massa*, meaning to stroke or anoint. I like the multiple and multicultural meanings because massage is widespread, even universal among humans.

According to Braun and Simonson, Swedish massage is attributed to Henrik Ling, the formulator of Swedish gymnastics, although he may have been one of many developers (*Introduction to Massage Therapy, 3rd ed.*, pp. 11-14). The French words for strokes were formulated in the 19th century by a Dutchman, Dr. Johan Georg Mezger, who popularized massage as a field separate from gymnastics. The words "massage," "masseur," and "masseuse" were already in vogue, and he coined frenchified terms for four of the five basic strokes still used today in Swedish massage. These are as follows.

Effleurage—to stroke lightly. These are light strokes, often progressively heavier and deeper to warm up skin and muscles. The word's history may include the notion of stripping flowers (*fleur*) or leaves from a stem or moving on a level ("floor").

Petrissage—to knead. French word means "molding" or "kneading," as if flesh were dough. These are heavier strokes to loosen muscles and increase circulation. Working on flesh or on dough have much in common: the material is soft but resilient, and the kneader takes pleasure in the feel of the material. Even expectation plays a role, if we imagine a warm loaf of bread or a happy, relaxed client.

Tapotement—to tap or drum, with fingers, side of hand, or cupped hand. This stroke can be gentle or more rigorous. This is the technique sometimes inaccurately shown in movies as rough karate chops.

Friction—to move top layers of tissue over deeper layers, creating heat; it can be fast or slow.

One more term completes the classic five strokes:

Vibration—to shake or rock with the hand, not sliding on the skin (therefore without lubrication).

There are other terms: compression, wringing, static pressure, nerve strokes (very light, sometimes called "feathering"), bending or stretching joints, and so on.

Such terms emphasize the *mechanics* of massage and the *material* bodies of a massage therapist who delivers strokes with the "tools" of hands and forearms to the recipient as target.

One day a teacher said, "we can *distract* the skin and fat to give better access to this muscle," meaning "to draw it aside" in the literal meaning of the word, not the metaphoric way we usually use it.

Later I learned that a therapist can ask a woman to move (or distract) her breast with her hand or ask a man to move his genitals with his hand; a sheet or pillowslip covers such flesh for modesty and the client's sense of security. In these cases, the mechanical and cultural values are both at work, helping to set the *intentions* of therapist and recipient.

OTHER MODELING SYSTEMS

Words are great. I love them. They are a mainstay of my life, from reading to writing, from crossword puzzles to jokes, from texts for music to emails with friends. Like any modeling system, however, words focus selectively, clarifying some things while distorting or ignoring others. To be sure, our instruction used spoken lecture and printed material, both with many new terms. The single-spaced handouts totaled some 700 pages. While words were central and important for our training, there were others ways of instructing us about muscles, bones, and strokes.

Two-dimensional drawings provided insights. There were sketches in our workbooks that showed muscles and their relationships with landmarks, such as bony prominences. Some students liked the *Gray's Anatomy Coloring Book*, and used various colors to represent muscles. I liked (and still use) Andrew Biel's *Trail Guide to the Body*; at 300 pages, it covers a lot (and he gives etymologies for anatomical terms!). The clear drawings simplify bodily structures by leaving out the obscuring fascia and fat and by showing an "average" presentation that ignores individual variation. My second edition is from 2001; there is now a fifth edition of this popular and useful text.

One day, 2-D moves to 3-D. We work as partners in the large massage room. Each table has a china-marking pen, something I've never seen before. Our job is to draw a triangle on the back of our partner to represent the shoulder blade. This bone floats on the back held by a web of fascia and muscles. It's basically triangular, looking a little bit like a spade (the meaning of the Latin word *scapula*). It might not hold up for digging, however, because one part of it is so thin as to be translucent.

Today we have a trio at my table, two men and one woman. To make our marks, we need to know where the scapula is and feel its edges through skin, fat, and muscle. The other man takes off his shirt and lies down. He is easy; the woman and I find the bone and confidently draw its shape on his skin on both sides of his spine. I lay down next; I feel both of them poking and rummaging, then marking two triangles. When the woman lies down, I feel one side of her back while Calvin, the other man, feels the other side. Our eyes meet and our eyebrows go up in despair. She's very fat and we can't feel any landmarks; we can't draw anything but a guess. We've also been told that some clients will have strange or absent landmarks, and I sense that our instructor passing by understands our dilemma. Although today's exercise is meant for visualizing structures below the skin in three dimensions, perhaps we have learned something more valuable about the variety of bodies.

Then something important happens: Calvin rubs her shoulders—not in our instructions. He free-lances this as, I guess, a kind act. She probably senses that we are not finding her scapulae, have made a quick sketch, and are just standing at the head of the table. She surely knows she's quite fat. Beyond that, she's brave enough to come to massage school, take off her clothes, and lay down for anyone. I'm proud for Calvin for the way he took the time to honor her body.

How can you move still deeper into the body? There is surgery on live people and autopsies on dead ones. Also various imaging systems: X-ray, CT, ultrasound, MRI, and endoscopes or speculums for views from within. Clearly we aren't doing any of those—although it would be interesting to have some of those images available.

Instead, we use manikins and modeling clay. The manikins are made of white plastic. About 15 inches high, they stand upright on a stand. They are one half of a skeleton, head to toe. I've seen these on shelves in the anatomy room, decorated with red-brown clay for muscles and, sometimes, silly clay hats on their heads, whimsical additions by previous, possibly bored students. One day we study schematic drawings of muscles shown in our syllabi. Entirely flat, they remind me of clothes for my sister's paper dolls. Our job is to recreate these in clay and apply them to a manikin. Today I'm partnered with Ruth, the Scottish woman; she describes anything small as "wee." We make a good team: I'm good at the shapes, so I draw and cut out the paper muscles. Ruth is dexterous with clay, owing to work she did in a florist's shop. Guided by the paper muscles, we make clay muscles and apply them to the manikin. Thus we move from 2-D to 3-D and get a feel for miniature, internal muscles. In short order we have our half manikin clothed in red-brown muscles and join the ranks of bored students. I ask Ruth about the Scottish highlands, from which my Stewart ancestors came. She says that they, like all highland Scots, were savages.

Another day we have a whole-body exercise of viewing the body from a distance. Barely clad, one class member stands in front of a grid of lines mounted on the wall. The rest of us, some fifteen feet away, are urged to look attentively. How is the head held? Are the shoulders level? The arms and hands? The points of the pelvis (the anterior superior iliac spines or the ASIS)? A large grid of lines on the wall behind the person helps us see when a shoulder is dropped—usually on side of the dominant hand. When the subject turns around: are the buttocks level, the legs parallel? If there are distortions, what muscles would be involved? What activities might the person be doing...or not doing?

What about sound? Medicine uses percussion to "sound" the chest, but not us. Percussion of the chest originated, it is said, from the French wine industry: barrels were tapped to find the fluid level. Our version is *tapotement*, tapping on flesh; done with cupped hands, it makes a *clop-clop-clop* noise. During our first clinic, we were all in the same room to massage our first clients. As we neared the halfway of 30 minutes, we used tapotement on the back. When the first student did this, all the others looked at the clock and quickly followed suit so that a bevy of claps filled the room. These are rapid, even strokes with a quick crescendo and a slower decrescendo. Besides the timing during the massage, we mimicked each other in the speed of the strokes.

The music played during massage was also a modeling system: it provided a slow or moderate tempo, a pleasant, nondramatic mood, and a non-intrusive background of harmonies in Western traditions. Music with massage was taken for granted…with no explanation or discussion. The usual assumption is that soft music helps clients relax, and this is correct. What I slowly also understand is that it also helps me as the therapist to smooth out my strokes often matching them with tempos and rhythms of the music. I've been involved in music since my youth; I'm strongly affected by it. When I get a new therapist, I always discuss the music because it can't have words, any singers, or a chorus, because it'd focus on that, not the massage. Also, I don't want classical music because I played violin in youth orchestras. Instead, I need slush music or elevator music that is calming and not engaging my mind. Fortunately, there's a whole industry for massage music that suits my need as client and as therapist. I have half a dozen such CDs, although one is my favorite; I play it while I prepare the massage room to get me in the mood and again when my client is on the table. I can pace my time without using a clock because I know when various melodies and instrument mixes occur.

Another sound that informs us is the client's speech when we first meet. Consciously aware or not, we get an impression of the vigor of this person and how aggressive our treatment might be. Speech is part of a person's basic melody, the pitches and overtones, the rhythms, the tempo, the force of projection. I think of Novalis' notion, "Every sickness is a musical problem; the cure is a musical solution."

As teachers sometimes advise: "Listen to the flesh. What is it telling you about what it needs?" Here "listen" has a wider sense than just acoustic: the flesh itself is a modeling system, advising how the massage should proceed. For example, tight peroneal muscles on the side of a calf may be calling out, "Help us loosen up!"

Eventually, all these different perspectives feed into our sense of a body, as we first meet the clothed person, as we sense the warmth and vigor or a handshake, as we see a person walk, as we do an intake interview, as we see how she or he talks, lies on the table and, of course, what we sense by our hands as we work. When I study Qigong, I will learn still other ways of modeling and sensing a human body.

STUDENT CHATTER; MENTION OF "HOT MOMMA"

At lunch one day a student remarks, "You know, the average career of massage therapists is five years."

"I might have my student loan repaid by then," quips another, "but why just five years?"

"Some people blow out their thumbs. You gotta to be careful," adds another.

"Now come with the upper body," the Trinadadian man, Jocko, says to me one day. While I do a stroke on his back, he urges me to use the weight of my shoulders and chest to deepen the pressure of my hands. I try this and find that he's quite right. Using body weight is an important principle that helps protect hands over time. I've never forgotten this insight and his kindness.

I lie down for a woman from another class. As we chat, I learn that she is mother of a disabled daughter. She is learning massage so that she can better help her child.

Some of the guys drive off campus for lunch. Jocko has located a Caribbean restaurant south of The Center. Over jerk goat curry, someone mentions "Hot Momma."

"Who?" asks another.

"Don't tell me you haven't noticed the brunette in *February*."

I know exactly whom they mean. She is an older, Mediterranean-looking gal who has also caught my eye. No surprise here. A lot of school is boring, and men are—let's be frank—men are scanners and usually alert to the presence of attractive women. I'll call her Elena. If I were 20 years younger, I might have developed a schoolboy crush on her, as harmless as it would be hopeless. Although she's in the class ahead of us, we see her in the hallways between classes. And, yes, our class has lain down for *February*.

The men agree she's "awesome," but I recall an older expression: *the belle of the ball.*

AT AN ACADEMIC CONFERENCE, A QUESTION FROM A MAN PERHAPS JEALOUS BUT CERTAINLY HORNY

Years after massage school, I presented a paper on the various meanings of massage at an academic conference. A fellow academic, quiet and well behaved, accosted me afterwards and asked (more or less, I'm translating here): "*Isn't it a real turn-on to get your hands on naked, female flesh?*" At one level, I could understand this question, because, in other situations, my hands enjoyed cooperative female flesh, but I also remembered massage school and working with many sorts of people, male and female, young and old, attractive and not, sick and well.

I'm confident this man heard my sentences about the various kinds of intention and also relaxation versus stimulation, but in his psychology the notion of *naked, female flesh* still *cued his sexual appetite*, in a Skinnerian, stimulus-response fashion. (I privately understand this as "screwthink," although "think" would be an overstatement.) And it's not just men. A female friend told me that she discriminates between touches that are "comforting" versus those that are "triggering."

Now, I wish I had told him the following story that can serve as a parallel for understanding desire and intent.

BAKERS DON'T SNACK ON THEIR WARES

When I was a kid, my Dad took me to a bakery run by Tex Wagner, a large man always dressed in white. He was friendly and invited my Dad and me to his shop right behind the retail space. He showed us the ovens, the sacks of flour, and whatever project he was working on at the time. He demonstrated how to make delicate roses, squeezing frosting from a tube with his big hands. As the years went by, I often dropped by there without my Dad. Tex was a bow-hunter who shot deer. I was a fledgling archer, shooting only at a target range, but he seriously discussed archery with me.

One day he was working on a large sheet cake destined to represent the United States—all 48 in those days. The U.S. borders were traced in it, perhaps drawn by a knife.

"We need to cut out the Gulf of Mexico," he declared, as if I were a full participant. He sliced away a large, curving piece on the side closest to him. That must be South, I thought. I liked maps and usually made them for school projects about a country. He carried the chunk to an immense container nearby, pulled a large spatula from it, and slathered it with icing. He handed it to me.

Munching blissfully away, I asked him about how he learned to be a baker. He said that he went to school. I asked him, however did he keep from eating everything he was working on? He said, "Oh, that lasts just a couple of days." This shocked me at the time, because all I knew was the gluttonous joy I felt in eating the frosted Gulf of Mexico. Slowly I understood that there was a different state of consciousness for a working baker creating goods day after day. He had changed his lenses for perceiving baked goods.

Massage therapists (and healthcare workers in general) are like the baker when they look at naked or near naked bodies: they've changed their lenses. Professional massage is about relaxation, not arousal both for the client and the therapist. In giving a massage, there's plenty to think about in the job at hand…approaches, strokes, what is felt beneath the skin, a neck muscle that feels like a pencil or a sore place the client didn't mention earlier. Equally important is the *intent* of touch, to provide care in a calm, supportive way. Is the change complete and absolute? No. I still perceive by sight and by touch attractive features on women I massage, a slim ankle, a tapering waist, soft fat on thighs, but I note these and move on. I perceive features on men that I admire as well, affirming the solid texture of their muscles, the breadth of their shoulders, their big, heavy bones, and then I move on. What do these people need? What techniques should I use? How can I best serve them?

Baker or massage therapist, understanding *intention* is important. In Quantum physics, it's part of mind that is a basis of reality. We'll explore this in Sections IV, VII, and VIII.

INTENTIONS: MORE POWERFUL THAN YOU MIGHT GUESS

Years later, I gave a massage to a cancer patient. During our intake interview she complained about how she received her diagnosis. "They said it was melanoma and they needed to cut it out, just like I was a car that needed an oil change. Melanoma! A serious cancer! And not a word of comfort or any kind of care for my emotions!"

What were the intentions of that office, I wondered. Strictly technical? They probably believed they were doing a good job—but patients need more than technical care.

In the old days the phrase "bedside manner" referred to a doctor's care beyond technical expertise. Many docs still have this, most nurses do, although not Nurse Ratched in Ken Kesey's novel *The Cuckoo's Nest* (1962), or Milos Forman's fine movie version (1975).

Intent exercise

At a continuing education course, the instructor wanted us to experience different kinds of intention. "Is this possible?" I wondered. The exercise went like this: we broke up into pairs, then person A massaged person B's right arm, then the left arm. For one of the arms, Person A had the intention to care and heal, but for the other arm no intention at all, although using the same mechanical strokes. Person B was not to know which arm got which intention. To my surprise, I could tell the difference right away; so could my partner, and so could everyone else. I've led this exercise with rehab doctors; they, despite their scientific outlooks, said they also could tell the difference.

Is the intention in a person's touch caring or technical only? Sensual or sexual or hostile?

Talking to muscles

A Yoga instructor often says, "If an area is tight, breath into it." The actual air from our lungs doesn't go there, of course, but our minds can send messages of relaxation to the tissues involved.

Some years ago Joanne, my massage therapist said, "Well, Howard, this may sound weird, but I think we need to talk to these muscles. Are you willing to try?" "Sure," I said. She was working on my tight neck and shoulder muscles. This did sound odd, but what could I lose? In mind, I asked them to relax, and they slowly did just that. I was both glad and amazed. My tight muscles, we could say, had already been receiving subconscious messages to be tight, to carry the world like poor old Atlas—discussed in a few pages. By sending conscious messages, these muscles had new guidance.

Sports visualization

Visualization is common in sports psychology, mental rehearsals of golfers before they strike the ball or divers before they dive. In doing so, they ask their muscles and nervous systems to work in specific ways.

Mind moves the Qi

Such intentions are important for the Qigong world, and visualizations of warmth, color, pressure, flow, etc. work well. (See Section IV below.)

REAL CLIENTS AT LAST, SOME WITH BODY BOOGERS

By careful steps, we have been trained in stroke after stroke, building up our repertoire.

Several months in, we start to do a full sequence, guided by a video made by one of our instructors. Our eyes shift from the screen to our partner as we work. I go over the sequence mentally while driving home and at other odd moments. Sometimes I move my hands in the air, building muscle memory. Our class has been gaining skills and confidence, although not every student: three have dropped out.

The idea of clinic is exciting to us. We'll give an hour massage to a complete stranger, putting all of our strokes into a full treatment. At the same time, however, we are growing tired of all the directions and exacting requirements for every stroke. One day we learn that we don't have to follow the video version exactly but can "freelance," as long as we treat all of the required body parts. We like this because we feel less shackled and because we are gaining awareness of different needs in different bodies. (I have had massages from persons trained with a set, non-varying routine and find these mechanical and less satisfying than a massage from a pro who knows how to respond to my body.) As the wise saying has it, *Let the flesh talk to you and advise you what to do.*

The first clinic day finally arrives. Driving in, I'm excited about a prospective client. Who will I get? What will he or she need? Soon I am disappointed because I will be, instead, one of two students checking in clients and finding their folders. No hands-on for me today! Never mind, two other students will do these chores for the clinic next week and I'll be a therapist then.

As I ask names and check off the list, I learn that the clients are mostly older people, glad to have a half-price massage with a minimum of fuss. They will all be massaged in the same room, with one exception: there is also a "private room," where, before long, we will see a client one-on-one, just like in real, professional practice. Despite my displeasure at missing my first massage to a real client, I am glad to learn in other ways. Because we are the first to greet clients, we can *help set their mood.* Because we need to pull all their folders (with charts inside), it's important that they are in perfect order. In fact, a handful are misfiled. At the end of the session, therapists must write a progress note about the massage—another aspect of our training. When they return the folders, we file them carefully.

A week later, I will have my first outside client. I take a folder from the helpers and scan over the patient information and therapists' notes…an older woman…healed hip injury…comes regularly. I go to a hallway where 17 people sit, more women than men. When I call her name she smiles and stands up. I smile and we shake hands. We go into the practice room and I lead her to a table. Music has already started. Therapists and clients fill the room. My client and I do a brief intake interview and we make a plan. I get her on the table face down and start to work. The massage goes well. I'm grateful that I remember what to do. Now and then I sneak looks at my colleagues to see where they are in the sequence; they are all earnestly giving massages. Halfway through there's the noise of tapotement to the backs of our clients. The room feels wonderful.

At another clinic I have an older man. His left thumb has an unusually large muscle mass. I ask about it. He says, "Yes…I was a dentist and that was my mirror hand."

Another clinic and I have an older woman. I keep wiping my hand on the sheet…and wondering why. It appears that her body is shedding small bits of skin shaped like orzo pasta. Later I ask an instructor about this. He laughs, "Oh yeah, your first case of body boogers. Some older people do that."

BUT NOT CANCER PATIENTS

I've heard that interns from our school have gone out to give massage to cancer patients. I'm glad for those patients, but the idea scares me. First, I'm not ready technically or emotionally. Second, my Dad died a long painful death from brain cancer and grueling treatment of surgery, chemotherapy, and radiation, none of which cured him. I'm relieved when I hear that such internships no longer exist.

Third, as a cancer patient myself (discussed in the next section), I am afraid of seeing people worse off than I, persons representing a downward path that might soon be mine as well.

Four years later, I'll be ready…and will provide massage for cancer patients.

POOR OLD ATLAS, ALIVE TODAY: CAN WE IMAGINE HIM HAPPY?

In Rant 1 above, we looked at Yankee malaise, an antithesis of massage, and the cultural values involved. Now we can consider stresses on people, many because of those values, but also because of universal human needs. Humans from all times and places have felt stress because we act by instincts, yes, but also because our large brains perceive, assess, calculate, even obsess. Besides such internal stresses, there are always stresses, social and cultural, even physical, such as noise. There's no avoiding all stresses. How may we limit and manage them?

Atlas alive today

When I ask massage clients—whether sick or well—what they need, nine times out of ten they will say, "Oh, it's my neck and shoulders!" "Neck and shoulders! "Shoulders and neck!" They are like the figure of Atlas who holds up the earth or the entire heavens, depending the version of the myth. Atlas was one of the Titans who rebelled against the Olympians but lost, hence his eternal punishment. We can imagine two symbolic stresses for him: the external, material weight of the world and his internal awareness of failure, punishment, and being stuck forever.

It's easy to start a list of external stresses on modern people: commuting, sitting at desks, crowds, noise, pollution, and pressures of job, money, family—and all of these intensified by demands of time. Pope Francis refers to "rapidification" in his encyclical *Laudato Sí* (2015). In their classic *Type A Behavior and Your Heart* (1974), cardiologists Meyer Friedman and Ray H. Rosenman, wrote extensively about "time urgency" as a pervasive stress in modern life. They identify many internal drivers, several overlapping with the values of Yankee malaise discussed above. We feel our unrealized ambitions and dreams, demands of perfectionism, feelings of duty and responsibility, and our self-assessments of powerlessness and general unworthiness.

All these add up! They can make us feel isolated and alone. We tense our neck and shoulders, intending to protect our neck and upper chest—our hearts!—from real or imagined threats. Our private worlds symbolically weigh on our head and shoulders.

More positively, the story and image of Atlas assure us that we humans have felt burdened with stress for a long, long time. To be human is to carry burdens. How can we creatively deal with them?

Corroborations from language, biology, climatology, geology.

Dictionaries tell us that the word "stress" comes from an Indo-European root *streig*, meaning to "stroke" or "press." This root is the source of other similar words, "strike," "strict" "strain," and "strait." We can be in "dire straits" or feel that life sometimes resembles a "strait jacket." Indo-European words are used today from India through Europe to the North and South Americas as well as anywhere Spanish, French, or English is used. I assume that other language families have their own words for stress because there can be no human life without basic stresses: atmospheric pressure is always on us, also the pull of gravity. We must have food and rest. We have to get along with other people....

Still further, all life forms, from mammals down to single-celled marine, feel stresses, and some of these are very dangerous. The earth has always had storms, blizzards, floods, earthquakes, volcanic eruptions, even meteorite impacts, and now Climate Change threatens (see Section V).

Tuning the head and neck

The word "relax" includes "re-" for "again" (or as an intensifier, as in "research") and "-lax" means "loose," of course, or "untightened." How loose? Dead people—after rigor mortis—are much too loose.

Muscles and other tissues should be in a healthy range of "tonus" or tone, enough to balance the head on the cervical spine, allow for movement, and avoid discomfort. We may think of neck, shoulder, and other muscles as needing to be *in tune*, like the vertical strings on a cello. Tuning can be done by working with those materials (stretching, massage, etc.) but also by the states of our minds: how do we view ourselves in relation to the world? In tune? Stressed?

Imagining the happiness of Sisyphus and Atlas

Albert Camus's essay "The Myth of Sisyphus" (1942) concludes, "One must imagine Sisyphus happy." Like Atlas, Sisyphus was punished for eternity with, in his case, the burden of pushing a rock up a hill again and again. Camus urged us to imagine him happy, but I think we can go even further. Let's imagine that Sisyphus gets massages between trips up and down his hill, that he has hobbies, and that he takes vacations.

Similarly, let's say Atlas is not only happy, but he has a varied lifestyle and a different job from holding up the earth so that he enjoys the labor and finds creativity in it. When not working he can relax. In general, Atlas recognizes stresses upon him and he finds ways to manage stress.

The classic myths of Sisyphus and Atlas emphasize tragic elements of human existence—pain, punishment, separation, and lack of hope for the future. In contrast, our thought experiment imagines them free of punishment and happy; they are reaffiliated to the comic world of the well.

A THREESOME INCLUDES "TEMPERATE MOMMA"

Today *July* lies down for *February*. What luck! *February* is well ahead of us and already doing full-body massages. I'm very ready for an hour-long massage…by anyone in the previous class.

"Hot Momma" comes to my table.

Excellent, I think, as we make chitchat.

Why does she pick me? There are 17 other tables for her to choose from. Are we somehow linked?

Before long, another woman also comes to my table. Evidently there is one more therapist than there are people lying down today, so the two women work on me simultaneously. For a brief time I try to keep track of which hands are Momma's, but of course I cannot…and should not. She is now Temperate, the right temperature now and forever, and her proper name is Elena.

Lucky me. The two women give me a wonderful massage, with double the usual strokes.

HOW DOES MASSAGE WORK? MATTER, INTENT, AND MORE

Americans like to know how things work…Yankee know-how…rational control… illusions of "being on top of" any given topic. Here's what I've learned so far.

Massage works in multiple ways, and these work synergetically. I'll discuss four basic areas. We may call these the two-dimensional overview. After these, I'll discuss three other areas that we might consider in other dimensions, three at least. None of these seven are wrong; they all provide insight into massage. More important is that they all work together, and that it is wise to use as many as possible…and to allow for the possibility that there may well be others.

Massage is not a uniform field; there are variations and specialties. My class was studying therapeutic massage and not, for example, sports massage. The variations can be related to the most of the areas discussed below. Therapists often specialize in areas they find most congruent with their backgrounds, interests, and gifts. As their careers evolve, they often add new techniques and sometimes change their approaches and even clientele. My descriptions are a basic overview; there are technical and detailed descriptions elsewhere, for example Deane Juhan's superb *Job's Body: A Handbook for Bodywork* (1987).

On the first day of my school, a student asked whether there would be any attention to energy as part of massage. The teacher said that energy would not be presented in any depth. Today, I understand that there is always an energy dimension, and I'll discuss that later.

Figure 1 shows the four basic areas.

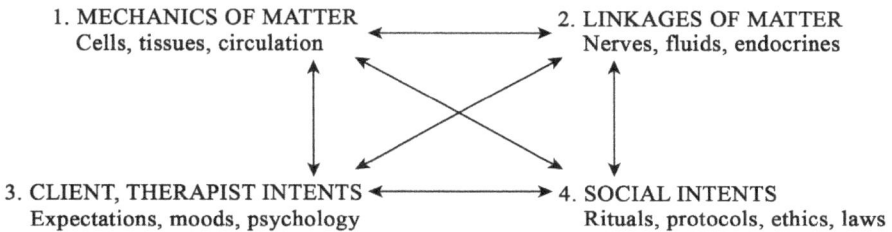

1. MECHANICS OF MATTER   2. LINKAGES OF MATTER
   Cells, tissues, circulation   Nerves, fluids, endocrines

3. CLIENT, THERAPIST INTENTS ←→ 4. SOCIAL INTENTS
   Expectations, moods, psychology   Rituals, protocols, ethics, laws

Figure 1: Basic Resources of Massage

Let's start with the basic *stuff* of a body.

1. Mechanics of matter; "Decongest that muscle!"

Our training stressed the basic materials of our bodies, the clients' and ours. Much instruction was about cells, tissues, and muscles. We learned basic anatomy of bones and muscles, especially the larger and ones near the body's surface, 50 or so, almost all in pairs. Many strokes compress muscles between the client's bones and the therapist's hands. Indeed we were taught that our "tools" (as if we were mechanics) were our fingers, knuckles, braced thumb, side of hand, soft fist, forearm, etc. Carefully used, they focus and modulate pressure.

Our teachers spoke of "decongesting" a muscle, a sausage swollen within its wrapping of fascia. They spoke of "hypertonic" muscles, meaning too much tonus or tension that usually shortens the muscle, even at rest. Our job was to relax and lengthen the muscle so that joints or body position would work properly, for example, lengthening neck muscles to free up the shoulders and head.

Typically effleurage strokes start out light and superficial and go deeper as an area is "warmed up." *Warm up* means not simply temperature (which may not change much) but also a fascinating change from gel (semi-solid) to sol (solution), a function of collagen fibers and liquid ground substance. The technical name for this is "thixotropy," and it is rewarding to feel muscles loosen up when massaged.

Types of massage emphasizing these dynamics include relaxation massage, sports massage, connective tissue work, myofascial work, and cross-fiber friction techniques. Thomas W. Myers' book *Anatomy Trains* (2001) is a good example of this approach.

Our training emphasized this mechanical, material realm. We learned that the body was made up of systems: circulation, lungs, digestion, nerves, musculoskeletal, etc., much like a building, but there's a difference between a *house* and a *home*. Besides their structural bodies, living people have consciousness.

All massage varieties work on the materials of the body, although there's a range of techniques from physical strokes to merely laying on hands. Licensed massage therapists have a similar core of training. Beyond that, personal interest and/or continuing education courses may add skills in imagery, tapping, aromatherapy, kinesiology, taping, hot stones, prenatal massage, or chair massage. Further training for hospital and cancer patients adds still other skills.

2. Linkages of matter; "Lay 'em down and knock 'em out!"

Muscles don't exist by themselves, of course; there are physical linkages of nerves and the circulatory system, and all these link to our minds, both unconscious and conscious aspects.

Here's a quick overview of a typical scientific explanation. Much of the power of massage is neurological: sensory nerves are stimulated so that the parasympathetic branch of the autonomic nervous system activates and slows down our breathing, heart rate, and biological activity in general. (This is the domain of "rest and digest," as opposed to the sympathetic branch's "fright, flight, or fight.") Muscles relax and can be stretched; fascia—which invests just about everything—becomes more pliable. (Fascia is the transparent tissue on raw chicken flesh.) Circulation of blood and lymph increases, bringing more oxygen to tissues and cells and carrying away wastes. The brain releases into the bloodstream "happy chemicals" such as oxytocin, serotonin, dopamine, and endorphins (opiate-like hormones) that make us feel good all over. Oxytocin (from Greek for "quick birth") has been called the bonding, "cuddle" or "hug" hormone. It is a mammalian hormone that acts as a neurotransmitter in the brain. In humans, it is released during hugging, pleasurable touching, and orgasm in both sexes. It may be associated with trust between people. While massage increases such happy chemicals, it also decreases cortisol, a stress hormone. These mechanisms have been carefully studied by researchers, for example Tiffany Field, Ph.D., Touch Research Institute, School of Medicine, University of Miami. Her books on massage include *Touch Therapy* (2000) and *Touch* (2003). Her *Massage Therapy Research* (2006), synthesizes scores of research studies about massage in chapters ranging from "Pain Reduction" to "Increasing Attentiveness" and presents some 160 abstracts of studies about massage applications to more than 40 subjects, from "Aggression" to "Yoga." Her *Complementary and Alternative Therapies Research* (2009) has chapters on "Massage Therapy, Acupressure, and Reflexology," "Acupuncture," and "Tai Chi and Qi Gong." She finds that massage can help premature babies, aggressive adolescents, and tense adults; some mechanisms include improving immune functions, reducing pain, and much more.

The slogan "lay 'em down and knock 'em out" comes not from teachers but from students. It reflects a distaste for all-too-chatty clients who talk about their latest operation, current aches and pains, the traffic on the way to the clinic, a bad meal last night, or their cat. Students are anxious to get to work, to get in all our required strokes and body areas, to manage time just right, and so on. Later, as professionals, we would have more latitude to listen and proportion our time depending on a client's needs.

Student or pro, one of the rewards for a massage provider is seeing a patient relax, get mellow, or "veg out." Some people become so relaxed they have trouble writing checks to pay for the session; I have permanently written the amount in my checkbook so I won't have to ask my therapist every single time.

Massage varieties in this area include therapeutic, relaxation, and Swedish massage.

More specifically, neuromuscular approaches include St. John work, craniosacral work, SOMA neuromuscular integration, NMT (Neuromuscular Therapy), PNF (Proprioceptive Neuromuscular Facilitation), and Reflexology. Approaches to circulation include techniques for moving lymph from arms and legs back into the venous return of blood to the heart. The Vodder technique is a well-known form of lymph drainage. Many of these were mentioned at my school.

3. Client, therapist intents

We've previously discussed how intent and emotional state can effect perception of touch. I recall a massage I gave to a woman. It appeared that she enjoyed it throughout, but at the end, she said the pivotal moment was at the very beginning when I first laid my hand between her shoulder blades. "I felt such relief at that moment!" she exclaimed. Clearly, that was a point in her psychology of expectation, trust in the moment, and desire—all priming her to interpret positively my first touch. In that sense any massage begins before the massage, and many cues can be important: the physical space, the first person to greet a client, as well as sounds, smells, and so on.

We've discussed therapists' intents of care and kindness. One of our teachers said, "Don't become bored while you give a massage. That will affect the quality of it." Discussion of our state of mind was, however, rare. Notions of intent were implicitly demonstrated, not explicitly taught: our teachers illustrated by their behavior kind and healing intentions. Our main stroke instructor Bonnie often ate lunch with students and made small talk or answered questions in a friendly way. We liked and respected her.

In sum, intent was a hidden or tacit part of the curriculum.

Intent is important for all kinds of massage.

4. Social intents

Beyond personal intent, there are social intents—traditional values commonly influencing massage. These include small rituals of greetings, the interview, how clients disrobe, use of drapes, opening and closing touches, and more. Some of these evolve with a massage therapist's experience, receiving care from other therapists, or noticing offices that are lacking in these areas. Such rituals shape the shared expectations of client and therapist and form an implicit contract of values and behaviors that gives comfort to both. Music, soft lighting, even incense, are other examples. Again, these were not so much discussed as illustrated in my training.

We were, however, specifically taught ethical and legal provisions for massage; we learned about requirements for licensure and ways we could lose our license. These came from strongly held cultural values including the separation of massage therapy from prostitution.

Social values that inform intent are important to all kinds of massage.

These four areas are basic to the nature of massage and they reinforce each other. While they are all true enough, we need to look at other areas. We might imagine that our flat, 2-D chart of Figure 1 is turned sideways—now on edge—and with arrows going through it in a third dimension. We can picture arrows for three more resources: (5) behavioral aspects representing time before and after a massage, (6) deep mammalian history representing ignored aspects of humans, and (7) healing energy, an approach from ancient cultures as well as cutting-edge biologists and Quantum physicists.

Figure 1 turned sideways:

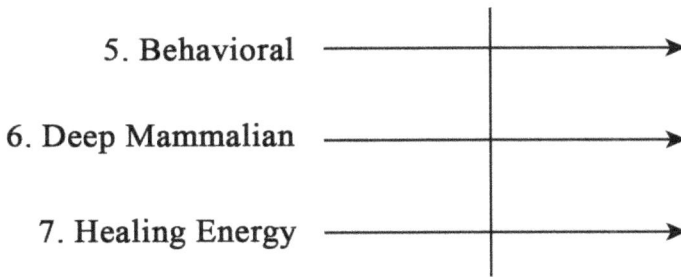

5. Behavioral ————————————→

6. Deep Mammalian ————————————→

7. Healing Energy ————————————→

Figure 2: Further Resources of Massage

5. Behavioral habits of patients, counseling for prevention

At a massage workshop many years ago, I saw a tag on a knapsack that read CIRQUE DU SOLEIL. Having seen those wonderful athletes perform, I spoke to the woman carrying the knapsack. Did she provide massage for the performers?

"Oh yes," she said.

"And what do you commonly see?"

"Ooooh," she said in her charming French accent, "I see lots of joints, especially shoulders and knees!"

Her athletic clientele would need help for particular stresses and injuries; they would also need counseling for protection, including stretches, strengthening, and rest.

Massage therapists may ask clients about work habits and risks, ergonomic questions about desks, chairs, and so on. A computer worker may have a stressful chair/desk/monitor arrangement. A musician may have a repetitive stress injury and need treatment by a specialist. Massage therapists may refer clients to physical therapists, occupational therapists, doctors, even mental health therapists.

In general, our largely sedentary society needs more ways to exercise, some as simple as more walking, using stairs, and stretching gently.

Further, our largely stressed society needs ways to relax, some as simple as counting deep breaths, 1 to 10 and repeat.

Probably all branches of massage can use this resource.

6. Deep mammalian history, cultures around the world, nature as context

Massage in many forms is millennia old and still, with many variations, widely practiced by people today. Parents hold babies, lovers cuddle, and just about everyone likes a back rub. When touching isn't threatening or ambiguous, it is pleasurable both in the sensations and in the feelings of trust and intimacy that accompany it. Massage allows people to take pleasure in their bodies. Recipients may feel some versions of: *I'm worth touching, my body isn't ugly, this touch does honor to my physical self, and my body is able to respond with pleasure and relaxation to a massage.* I believe massage is universal among humans and, it appears, many other mammals.

Cultures around the world have formal versions of massage, some of them with ancient roots: Shiatsu and acupressure from Japan, Tui Na and Qigong from China, Lomi Lomi from Hawaii, and Ayurvedic massage from India. These usually have a whole-person (even whole world) approach to wellness and understand that energy moves in the body and should be balanced. Many ancient cultures understood that *we are part of nature itself* and that we *routinely exchange energy with plants, all creatures, the earth, and stars.*

7. Healing energy in ancient and indigenous cultures and modern explanations from biology and physics

The approaches just mentioned from the East are based on theories of energy beyond our typical Newtonian worldview. Indeed, both ancient and indigenous societies of today typically understand energy as a basis of reality. An article by Tom Bender (2006) states that 65 indigenous cultures have been identified as using energy for healing purposes. Cell biologist James L. Oschman has written two technical books that explain how our cells send and receive energy within the body, within social settings, and with the earth (2000, 2003). A key to health is a dynamic balance of energies, for the ancients and for modern biology; too little hemoglobin, and you heart works too fast. Similarly, Quantum physics understands that there are energy fields around bodies, the earth, and throughout the universe. Modern approaches to energy flow and balance in massage include Qigong, Polarity therapy, Reiki, Therapeutic Touch, Touch for Healing, Touch for Health, Magnet therapy, Zero Balancing, and Reconnection Therapy. Typically these understand that *energy in nature (indeed the universe) is already balanced* and that we *need to rejoin*—or *reaffiliate* with—it. In sum, all kinds of massage—indeed all healthcare and all human interactions—have energy dimensions. We'll return to these topics in Sections VI and VII.

MASSAGE AS RESPITE, RESPECT, AND REBIRTH…EVEN RENEWAL AND REBIRTH

Yes, massage "works" in the all various ways just discussed, but there are still deeper resonances that contribute to the power of massage. Rectilinear figures like those above are oversimple because all the resources create a world of interrelating synergy on still deeper levels.

Respite, respect, renewal

A person scheduled for a massage looks forward to a *rest*, a *break*, a *mini-vacation*. Other similar words include a *breather, downtime*, a *pause, recess*, or *retreat*. Most of these expressions emphasize the departure from the normal, stressful world, and massage does, indeed, provide that.

The word "respite" carries these meanings but also further sense of *an improved future*. A respite from punishment can be a postponement or an outright stay of execution. Further, the Latin origin of "respite" gives us a related word "respect," literally a "looking over" and, presumably, seeing worthy qualities in a person. A massage therapist routinely looks for and observes the good in a client or patient. To give a massage is to have ways to acknowledge and respect the recipient.

In modern usage "respite care" describes relief for family members caring for a sick person at home: either the patient goes a facility for a short while, or respite-givers come to the home so that family members can have some free time for relief and renewal. After such respite, they are, at least to some extent, renewed and better able to give care in the future. When I was nursing my dying father, another family offered me a meal and a leisurely swim at their house. I was surprised to see how much stress I had been carrying, and, as I left there, I felt better able to go back home and help. It seems to me that massage is another form of respite care. Not only do we make an escape, we may return to the normal world better able to cope with it, and all its stresses.

Renewal and rebirth

There is a still deeper level; we might call this primal or mythic.

In *Touching*, Ashley Montagu writes, "In the womb the fetus has been constantly stimulated by the amniotic fluid and by the growing pressures of its own body against the walls of the uterus" (p. 38) and: "Within the womb the fetus is enclosed and intimately bounded by the supporting embracing wall of the uterus. This is a comforting and reassuring experience" (p. 232). These descriptions sound very much like massage, but Montagu doesn't use that word. Further, he writes that the baby "requires the continued support of his mother, to be held and rocked in her arms, and in close contact with her body" (p. 232). Even on a mattress, the newborn has a "great need for enfoldment, to be supported, rocked, and covered from all sides" (p. 233). I believe massage answers all these needs and reminds us of our first home, our mother's womb.

A massage, we might say, resembles a second gestation and a second babyhood, when we are secure, comforted and stimulated in just the right ways. The draping provides not only modesty but also the "enfoldment" that Montagu mentions, and that we desire consciously or unconsciously. We recall the pleasure of being "tucked in" as a child. The massage *table* is not, however, a soft bed; it is resilient enough to support massage strokes and to link us to our ultimate mother, the earth.

In a section entitled "On Being Stroked the Right Way," Montagu describes how the massaging uterus helps develop the fetus' various systems, respiratory, circulatory, digestive, and nervous. He concludes, "The short, intermittent stimulations of the skin over a prolonged period of time that are produced by the contractions of the uterus upon the body of the fetus thus appear to be perfectly designed to prepared it for postnatal functioning" (p. 51).

Similarly, massage also prepares us to return to the world not just relaxed, but also ready to re-engage.

MORE HUGS MEAN FEWER COLDS!

I never see the professional journal *Psychological Science*, but I learned about an article on hugs and colds from a brief mention in *Harper's*. Searching the Web, I found a dozen references to the study in the popular press, both electronic and print. Since then I've heard it mentioned in conversation, "Say, did you hear that hugs prevent colds?" Clearly people are interested because (1) we do not want colds but (2) we do want hugs—and don't get enough of them in this touch-aversive society.

According to the original article, four scientists designed an experiment with adults, analyzed the results, and concluded: "people who regularly receive hugs are more protected [from disease and severity of illness] than those who do not." (See Cohen, S. et al., "Does hugging provide stress-buffering social support? A study of susceptibility to upper respiratory infection and illness," 2015.) This study was a collaboration of two psychologists, a pediatrician specializing in infectious diseases, and a director of ENT laboratory research in an otolaryngology department; they represented three different universities. The experiment, described in detail, involved 404 healthy adults; their immune functions (and other variables) were assessed, and they were quarantined in a hotel for two weeks. During this time they were exposed to a cold virus and monitored for infection and illness signs. The article describes procedures, measures, disease outcomes, and lengthy data analysis.

Reviewing earlier studies, the authors found that nonsexual touch from trusted other people helped buffer stress from pain and had positive effects on hormones and brain function. Further, stress can harm psychological wellbeing and immune function. Would hugs help avoid colds or lessen their severity? The experiment provided the answer yes, because hugs provided (1) stress-lowering physical contact and (2) represented social support.

The final sentence of the article reads, "Moreover, that the buffering effect of hugs could explain much of the attenuating [lessening of illness] effect of social support suggests that hugging is a behavior that may be manipulated to provide the beneficial effects associated with support." In non-academic prose: *Because hugging clearly helps people's health by giving social support, we should find ways to provide more of it.* Wow! What if doctors, nurses, receptionists —many people—routinely hugged patients in their care? How much healthier and happier would we be?

This ingenious and important study does a fine job with the material bodies, and even the consciousness of both hugger and huggee, but it doesn't touch concepts of energy that we'll visit in Section IV below.

Smiles and kind words contribute to wellbeing as well: a touch on the arm, a mention of the weather, a joke—all these emphasize a shared human community that makes everyone feel better and be healthier, both the receiver of care and the caregiver. It's not just the huggee who benefits. Huggers benefit from the "helper's high" that come from altruistic acts, according to some 50 scientific studies funded through The Institute for Research on Unlimited Love (http://unlimitedloveinstitute.org/).

"YOUR HANDS ARE ICE-COLD," HE SAID, BUT AT LEAST I'M NOT SWEATING

Near the end of our training, *July* must pass the practical exam. We stand by our tables expectantly, while students from another class arrive. Who will be my client? It doesn't matter. Anyone will be fine.

A man comes to my table. We've chatted before. He's mid-thirties and looking for a second career following construction work of some kind. We do a brief intake. Our instructor starts the music. I get him comfortable on the table, face down. I think of *Lay 'em down and knock 'em out.*

I start a nice slow stroke on his back.

He says, "Your hands are ice-cold."

Damn, I think. *I wanted to have a good start!*

Sadly, however, I know that he's right, at least from the contrast with his newly unclothed skin. But, also, I'm nervous, and warm blood has probably diminished in my extremities.

I start off school sweating like a stuck hog, and end up with frigid mitts that no one should ever want on their body! How about that for an entire cycle of training?

"Oh, I'm sorry," I say, and rub my hands together to generate some heat. "Let's see if this feels any better."

The massage proceeds without any more complaints, and my instructor says that I have a passing grade.

STEEL DRUMS INSTEAD OF FOOD: GRADUATION DAY

There are multiple graduations yearly because there are always overlapping classes—the "straight-through" classes like mine and the "weekends only"— that finish at different times. Indeed my class has already seen the largest room converted for ceremonies two times. The graduating class sits front and center, flanked by two classes still in training. Our class has moved from one side position to the other and now triumphantly sits in the middle, entirely finished with the program and wearing good clothes, not shorts and T-shirts.

Our instructors have given careful instructions about guests, the time, parking, and so forth, and a very clear directive: "NO FOOD!" However festive food may be in some settings, it's never entirely consumed here and the remaining mess must be cleaned up by the staff.

"Never mind," says Jocko, "We're going to have steel drums." We look at him. "No problem...I know a guy. Gotta have it." He smiles and we hesitantly agree.

These graduations mean a lot for the younger students; some have no education beyond high school. I've attended perhaps 40 graduations, some for myself, some for relatives, and many for my college students. Nonetheless, I feel the excitement for this ritual, put on a rarely worn suit, and enjoy going to The Center with no bag, no linens, no texts, no lotion, and no possible tests.

There are balloons and other decorations. Sherry hands out diplomas and awards. I don't remember much else, but I sure remember the steel drums at the end. Jocko's friend hits them hard, and they make a wonderful, joyful music.

GIFTS MENTIONED IN MY GRADUATION TALK

My class has asked me to speak at graduation. This is what I say:

As I reflect on the experience of the last six months, the word that keeps coming up to me is the word 'gift.' And the gifts have been of many sorts. We have inherited the gifts of several streams of knowledge: anatomy of the human body and medicine (how the body works and doesn't work, things to look for in our clients), theories of touch and massage (Field, Summers, Chaitow, Barnes, and others), and, of course, the modified Swedish strokes for a relaxation massage as well as the therapeutic tests and treatments. These streams represent the research and practice of many, many professionals, and we are the lucky beneficiaries.

But we have not found these gifts under a cabbage leaf. Our teachers have given us the gifts of their clarity, their patience, their good humor, their kindness, and their enthusiasm for the disciplines and crafts that they teach us. They have created a safe haven for study, a special place where kindness and care permeate everything that happens. After struggling through nasty traffic in the morning, I always felt welcome when I arrived in this building and confident that everyone I'd meet would be glad to be here and that we all had interesting and important work to do.

I have felt from the first week that my classmates have been a lively crew of givers. We've helped each other study; we've lain down on the table for each other; we've given or lent each other towels, lotion, sanitizer, notes, sheets, whatever was needed. Classes senior to us have taken us under their wing. Classes starting since July have shown us their desire to learn.

As the first clinics started, we were all nervous. It seemed amazing that real people would pay money, show up on time, take off their clothes, and lie down on our tables for us to practice a craft that we knew only as beginners. They gave us the gift of their patient and willing bodies, and—miraculously enough —it became clear that our skills were gifts that we could give other people.

Finally, then, these wonderful bodies that we live in—whether we are lying down on a table or standing beside it to give massage—these bodies that we have studied for their magnificent architecture, their elaborate biochemical workings, their struggles with injuries, hypertonicity, lactic acid, trigger points, tender points, and so on—these bodies are gifts that we live in, live with every day. Six months of massage school is a wonderful time to consider how we can care for our own bodies, how we use our hands as therapists, and how we can help other people enjoy their bodies and care for them.

I have wondered why my time here has been so rewarding and joyful. One set of answers is the number of gifts we experienced during our studies at The Humanities Center.

During the talk, I see smiles in the audience, especially from the students still in training.

ESSAY 1: GETTING IN TOUCH AND IN TUNE: MATTER AND MIND, BUT NOT (YET) ENERGY

The visible sweating of beginning massage students soon faded away, but it took most of the six months for us to feel at ease with giving an hour-long massage to a complete stranger—*any person* who came to the school for the clinics. By then, we no longer worried about the seamy history of massage or sexual overtones, and we were much more confident when we put our hands on our clients. Perhaps we still worried that we might do something wrong and hurt someone, but the instructional design had worked well, gently leading us through the various strokes, exchanges with our classmates, and finally massage for outsiders. We slowly gained confidence in our techniques, our own routines, and how we could assess the flesh under our hands in order to do a good job.

I would love to reassemble my class many years later and hear everyone's reflections, but that's not possible. Instead I'll explore my thoughts about our training and my experience of massage in the intervening years.

Of the three proposed aspects of reality—matter, mind, and energy— massage training emphasized the matter of the body, both in its make-up of bones, muscles, nerves, and more, as well as in the physical tools of the therapists' fingers, hands, forearms, and body weight. It was right to learn the basics of flesh, truly a *sine qua non*. As for mind, we learned the social values and intents that defined licensure, proper behaviors, and ethics. Other values were transmitted indirectly, from the attitudes of the teachers, the school in general, and the growing collective (although unexpressed) wisdom of the class. These included intents of concern, care, and dealing with stress.

My massage school did not teach energetics and the director said so on the first day of class. It would be some years before I studied Qigong (Section IV) and brought that resource to massage.

1. Matter: material and technical aspects

Anatomy and Physiology were, of course, basic, foundational. We studied bones, joints, and muscles, and how they worked: range of motion, muscle antagonists, connective tissue, and cellular respiration. There were overviews of body systems (cardiovascular, respiratory, digestive, etc.) and information on landmarks on the body, anatomical planes, and how to palpate for orientation and assessment of tissue. While important and necessary, these approaches objectified the body as if it were a building, a dynamic collection of various *systems* (structural, electrical, HVAC, and so on)—but who lived and worked in the building? How did it fit into a neighborhood, a landscape? We gained an overview of bodily systems from explicit lessons, but we had to create our own sense of how they fit together in a person on our table. My experience now is that massage therapists enjoy seeing clients as a whole person: how does Mr. A. look and feel today? Is Mrs. B less stressed?

Is Person C trying any of the stretches I suggested?

As we came to the table for instruction on strokes, we learned that our hands, forearms, and elbows were called "tools," so we as therapists were also objectified. We brought the materials of our body into well defined—even mechanical—relationships with other material bodies through strokes that had requirements for beginnings, ends, speeds, and depths. It was easy to fall into this mindset. I found myself dragging my fingertips on various tables as I entered the treatment room in order to locate one the right height for me so that I wouldn't spend time and effort adjusting all four legs of another table.

Some critics have observed that medical training may push some of the humanity out of students, replacing it with technical and scientific information. My massage school didn't do that because it was less rigorous scientifically, students massaged each other regularly, and, later, we massaged clients in our clinics. When we had someone on our tables, we felt the joy of working with fellow human beings, not many trillion cells of matter.

Victoria Sweet criticizes modern medicine in general. "Modern medicine's framework for the body, then, is industrial, mechanical, and democratic: The body is a factory with workers; a machine with parts; a democratic republic of cells...." Sweet, a doctor and a historian, looked for a more synthetic and integrated model of humans and found it in Hildegard of Bingen, the twelfth-century Benedictine nun. Hildegard's framework was based on premodern medicine, the four humors of the ancient Greeks: earth, air, fire, and water. "It was a holistic system whose underlying metaphor was horticultural, and it corresponded perfectly to the rural and agricultural life of the premodern world. With the rise of industrial modernity, it disappeared quite naturally, if abruptly" (*God's Hotel: A Doctor, A Hospital, and a Pilgrimage to the Heart of Medicine*, 2002, pp.180-183.)

We will return to the synthetic power of premodern medicine when we look at Qigong that arose from ancient and agricultural China; like Hildegard's understanding, it also uses a system of elements that need to be in balance. Hildegard praised the color green (*viriditas*) as a symbol of nature's fruitfulness, life-force, and physical and spiritual health, and Qigong also emphasizes the power of the color green.

Montaigne favors body and mind "together" and in "mutual service." He continues, "Let the mind arouse and quicken the heaviness of the body, and the body check and make fast the lightness of mind" ("Of Experience," Frame, 455).

Humans are wonderfully made. Our material bodies are complex, with wonderful interconnections working day and night. They grow, if fortunate, from birth through adolescence to adulthood. They survive injury and disease. They experience pain and, fortunately, pleasure.

But they are not matter alone.

2. Mind, consciousness, awareness, intention, etc.

Philosophers and psychologists have written extensively on this area, well beyond my research and ability to summarize problems raised, let alone resolve.

In this discussion, consciousness and intention may be defined separately, but they also overlap, especially in thoughtful people.

Consciousness is the general mixture of our awareness, our values (personal and social), and assumptions about reality. These may be examined or not: some we explicitly know and can talk about, some we hold implicitly, and some are rooted in biological instincts and drives. During massage school, our consciousness about physical bodies deepened because of the study of muscles, the nervous system, and so on, and because we had our hands on many bodies, and because we felt other's hands on our own bodies. Consciousness emphasizes perception through various lenses, and awareness of self and non-self.

Intent is similar in having both explicit and implicit levels but differs in having clearer purposes and goals, in line with our desires, ambitions, notions of duty and the like. Our attitudes and behaviors are shaped by our intents, and these can evolve through life experience, education, social influence, and personal choice.

Clearly there is overlap between the two as they influence and are part of each other.

During our training, our consciousness and intent evolved as we learned about kindness in healing and professional caregiving in general. More specifically, I'll discuss three topics (and their opposites) that were part of our evolving consciousness.

Neither domination nor violence but gifts of relaxation

In 2016 and later cases of sexual harassment were in the news, and the Me Too movement arose to protest sexual misconduct, typically men against women. Our training in massage was far from this realm. From the beginning none of us had any intent to do violence to anyone. We were aware of unintended violence, however, and worried that we might accidentally hurt someone on our tables. Perhaps more common were some feelings of domination…being in charge…having control over another person. No one talked about this, but when we stood tall over a person lying passively before us, there was an asymmetry of power. When our hands were on a barely clothed person, when we gave directions, when we took care of draping for modesty, clearly we were in charge in ways that might be tempting for rescuers or power seekers. This challenge was never discussed, so each of us had to create our own understanding. As we lay down to receive massage, we understood the status of the client and came to respect that. As we were nurtured, so should we also nurture.

As technical aspects of massage came to us more easily, we could focus more on how they affected people on the table. We saw our clients relax, breathe more slowly, and say words like, "That felt really good!" As we gained confidence in our training and the art of massage itself, we understood better that we were giving gifts, gifts that had little to do with manipulation and power.

As human analogues of poor old Atlas came into our hands, we helped relieve their stressful loads.

Not sexual focus but appreciation of the human body

As we gave massage (and were massaged), we appreciated bodies in their masculinity, womanliness, and human-ness. Thus there is a deeper sense of the erotic than just the sexual aspect; instead we learned the joy of warm bodies that enjoy loving touch. As we massaged and were massaged, we appreciated bodies in their strengths and weaknesses, in their complexity, and in their abilities to relax. Although we never spoke if it, I think we sensed the wondrous universality of all human bodies.

Even the strange phrase "table slut" allowed us to acknowledge the acceptable pleasures of giving or receiving massage.

Not rescue, but celebrating our common humanity

As I got to know my colleagues, I felt that all of us were so-called "helping personalities," persons who enjoyed giving to others and felt rewards in so doing. The negative side of helping can be a self-aggrandizing role of rescuer, someone jumping in to help as an expression of power, but we were not jumping in. We worked according to a reasoned and time-tested tradition.

As we gave and received massage, we understood giving and receiving as mirror images. Therapist or client, we were all basically the same humans in our bodies and in our life experiences, including joy, pain, and tragedy. All of us—even the youngest, strongest, and healthiest in our class—had health issues at one time or another. One man had a short leg from an injury. One woman missed a class because of day surgery. All of our elderly clients for clinic had health problems. Obvious to me but to no one else was my recently diagnosed leukemia.

American Buddhist nun Pema Chödrön writes, "Compassion is not a relationship between the healer and the wounded. It's a relationship between equals. Only when we know our own darkness well can we be present with the darkness of others. Compassion becomes real when we recognize our shared humanity" (*The Places That Scare You: A Guide to Fearlessness in Difficult Times*, p. 50).

As we have been nurtured, so shall we also nurture. In my massage practice later, the greeting of clients and the intake interview are important for establishing our relationship. With loving intent, I shake their hand, look them in the eye, and make small talk, all to nourish our equality as fellow humans. Then we can move to the more asymmetrical relationship of client and therapist. Sometimes this shift hinges on two similar sentences. When I ask, "How are you feeling today?" they routinely answer, "Fine," a normal, social response. When I continue, "And how are you feeling in *your body*," they give me useful information for the massage.

When I have them on the table, my hands resting gently on their back, I often say, "Let's take a few deep breaths together." I use my breaths to help them pace their breathing slower. We breathe together as fellow human beings sharing the same air, the same protected time and space.

3. Rubbing the right way; massage for all; energy

"Stand up and turn to your left," choral directors will routinely say to singers to initiate massage throughout the group. "OK, now to the right." Whether a small church choir or a large community chorus, it is common for singers share massage with their neighbors at the beginning of a rehearsal. In terms of the body's matter, singers relax their backs and necks of each other so that they can sing better. In terms of intent, singers share caring for each other, their common humanity, and their commitment to the work of the group. A massaged choir sings better, commits to a shared purpose, and feels joy. Sometimes there is laughter, for example, if the director joins the massage lines, and the time given for it, therefore, runs *longer*.

The claim mentioned earlier, "If everyone had a massage every day, there would be no more wars" rings true for me. Massages change people's moods through relaxation and release of "happy chemicals." The consciousness and intent of massaged people change. Massaged people feel less stressed, angry, and lonely. Because massage—like other caring touches such as hugs and pats on the upper arm—communicates social support, people feel happier and healthier.

Americans (and citizens of other industrialized nations) are often stressed by their crowded, busy, and noisy lives. Economic pressures harm poor people who feel helpless and humiliated. Rich people feel they must protect what they have and/or get more. In societies with taboos against touching, people are touch-starved. I believe such stresses help explain tendencies to violence between people, within neighborhoods, and internationally. You don't stab, shoot, or bomb people for whom you feel kinship and equality. The English words "kin," "kind," and "humankind" are all related.

My intuition during massage training was that energies of some sorts were at work, but I didn't have any terms or concepts to discuss energy until Qigong came into my life.

A much more recent term, however, provides some insights.

A new perspective: reaffiliation

In the Introduction, We saw the notion of "affiliation" from the narrative medicine approach of the Columbia group (Charon et al., 2017); this term explains some of the benefits of massage. I expand the term to "reaffiliation" to emphasize connecting, once again, to something that was lost. Reaffiliation through massage is a source of healing: recipients reconnect to their own bodies, their sense of well-being, and the society of healthy persons. Massage therapists working with sick people gain also find reaffiliations, reconnecting to the gifts of appropriate touch, the cooperation and agreement of two people, and the pleasure human bodies may feel even when sick.

We saw earlier that the word "affiliation" traces back to the notion of adoption, adopting a commitment by one person to provide faithful care to another person in a caring, familial relationship. Similarly, "reaffiliation" may suggest a recovery of family members have became estranged, or recoveries of other sorts: broken relationships, an alienation from one's own body, or a disconnection with nature. Nature can be especially healing because to lose touch with nature is to be uprooted, no longer nourished by Mother Earth, even exiled from the world of living creatures. In reaffiliation, we acknowledge that we are all part of the same terrestrial family, a nuance of meaning that goes beyond the "integrative" of "Integrative Medicine." If we reaffiliate to a symbolic adoptive parent—be it nature, Earth, or the universe—we accept humility, protection, and the promise of growth. The implied adoptive parent has power over us, certainly, but it is a nourishing and caring power that promotes our maturation. For some people there are spiritual and/or religious dimensions to such sustenance and protection.

Caring touch reaffiliates both recipient and the giver, lowers stress, and heals body and mind for both parties.

Tuning matter, mind, and energy for harmony and coherence

Music became a leitmotif to massage: slush music accompanied massages we gave; our clients became percussion instruments when we performed tapotement on their backs, and steel drums celebrated our graduation. In discussing Atlas, we spoke of tuning his neck muscles like the strings on a cello. This can be done mechanically by loosening his hypertonic muscles but also mentally by making him happy.

The words "tonic," "tone," and "tune" are all related. Music provides many ways to create order and pleasure through many variations of tone color, rhythm, and theories of harmony. Sound waves are orderly energy waves of compression and rarefaction; we feel these specifically in our ears and mind but also throughout our bodies.

Massage, hugs, and caring touch in general help tune our minds and bodies. They are the tactile equivalents of lullabies that also comfort and solace babies. Ashley Montagu discusses lullabies in three places of his chapter "Tender, Loving Care." While praising the many powers of music in his Preface, Oliver Sacks focuses on the mental aberrations of neurological patients in his *Musicophilia: Tales of Music and the Brain* (2007). He also acknowledges that music activates widespread neural networks, as shown on PET scans of healthy persons (p. 78). To maintain my license, I took a weekend workshop in tuning forks for massage. There was an elaborate theory about which pitches stimulated which areas of the body when the vibrating forks were directly applied to a client.

Music is a good metaphor for harmony within the body and also between therapist and client as they perform a duet together.

Sick or well, healing or dying, human bodies are wonderful.

Women seem to know better how to care for themselves, and they seek out massage more than men. They're usually easier to work on, smaller in size, slimmer of limb, and less muscular, but I like to work on men as well, so big boned and meaty solid. Furthermore, I'm proud of men who show up for a massage, because many men would never do such a thing because they can certainly take care of themselves—thank you very much—or maybe even because they're reluctant to be massaged by another man. I especially like massaging another therapist or anyone who is aware of the multilayered process. Some of these people breathe with my strokes and send energy to areas I touch.

I've touched a lot of bodies, in the thousands by now. They range widely by size, age, and condition. I've worked on people so skinny I can feel their bones all too well. I've worked on people so fat I can't feel their muscles specifically. People receiving chemo often have no hair, but some men are so hairy I need a lot of crème to keep my hands moving. I've worked on healthy weightlifters with enormous muscles and such low body fat that they provide a walking anatomy review. A one-legged woman had a well-muscled upper back because of long use of her Canadian crutches. A man who was a stonemason had hands, arms, and shoulders almost as dense as the rocks he habitually moved. A woman who had both breasts removed asked me to work her pectorals above and below the horizontal surgical scars on her chest that looked like openings for pockets. One young man tented my sheet with his erection. I've felt flat feet, very high arches, webbed toes, even conjoined toes (two that never separated and the same on each foot). I've seen scars from accidents and operations. I've had crippled people, frail people, weak people. Also athletes with wonderful bodies, toned and fit, but also often with injuries.

Most people, however, are ordinary with minor variations in how their muscles lie, how their bodies are shaped, and what their energy is on the day I see them. All of them—usual or unusual—have in common two things: (1) the ability to relax when massaged and (2) gratitude for being carefully touched.

In summary
Because massage can help bodies reaffiliate, tune, and relax and therefore heal, massage should always be available in hospitals as a resource for patients and caregivers who desire it.

SOURCES 1

Arsan, Emmanuelle. *Emmanuelle*, trans. Lowell Bair. New York: Grove Press, 1971.

Bender, Tom. "The Physics of Qi." tombender.org/energeticsarticles/qi_physics.pdf. accessed March 9, 2016. This is a "brief summary" of his DVD "The Physics of Qi (2006).

Biel, Andrew. *Trail Guide to the Body, 2nd ed.* Boulder: Books of Discovery, 1997.

Braun, Mary Beth, and Stephanie J. Simonson. *Introduction to Massage Therapy, 3rd ed.* Philadelphia: Lippincott Williams & Wilkins, 2013.

Camus, Albert. "The Myth of Sisyphus" (1942). Available in various editions and on the Web.

Chödrön, Pema. *The Places That Scare You: A Guide to Fearlessness in Difficult Times.* Bostonand London: Shambhala Publications, 2001.

Cohen, S., Janicki-Deverts, D., Turner, R. B., & Doyle, W. J. (2015). "Does hugging provide stress-buffering social support? A study of susceptibility to upper respiratory infection and illness." *Psychological Science*, 26(2), 135-147.

Downing, George. *The Massage Book*. New York: Random House/The Bookworks, 1972.

Field, Tiffany. *Complementary and Alternative Therapies Research.* Washington D.C.: American Psychological Association, 2009.

_____. *Massage Therapy Research*. Edinburgh: Churchill Livingstone, 2006.

_____. *Touch*. Cambridge, Mass.: MIT Press, 2003.

_____. *Touch Therapy*. Edinburgh: Churchill Livingstone: 2000.

Pope Francis. *Laudato Sí: On Care for our Common Home*. Huntington, Indiana: Our Sunday Visitor, Inc., 2015.

Friedman, Meyer, and Ray H. Rosenman, *Type A Behavior and Your Heart*. New York: Fawcett Crest, 1974.

Juhan, Deane. *Job's Body: A Handbook for Bodywork*. Barrytown NY: Station Hill Press, 1987.

Kesey, Ken. *One Flew Over The Cuckoo's Nest*. New York: Signet: 1962.

Linden, David J. *Touch: The Science of Hand, Heart, and Mind*. New York: Viking, 2015.

Metzger, Deena. *The Woman Who Slept with Men to Take the War out of Them & Tree*, Culver City, Calif.: Peace Press, 1981.

Montagu, Ashley. *Touching: The Human Significance of the Skin*, 2nd ed. New York: Harper and Row, 1958.

Myers, Thomas. *Anatomy Trains: Myofascial Meridians for Manual and Movement Therapists*. London: Churchill Livingston, 2001.

Oschman, James L. *Energy Medicine: The Scientific Basis*. Edinburgh: Churchill Livingstone, 2000.

_____. *Energy Medicine in Therapeutics and Human Performance*. Edinburgh: Butterworth Heinemann, 2003. For a description of the two titles, see my annotations in the Literature, Arts, and Medicine Database from NYU at http://endeavor.med.nyu.edu/lit-med/lit-med-db/topview.html.

Sacks, Oliver. *Musicophilia: Tales of Music and the Brain*. New York: Alfred A. Knopf, 2007.

Shem, Samuel. *The House of God*. New York: Richard Marek Publishers, 1978.

Sweet, Victoria. *God's Hotel: A Doctor, A Hospital, and a Pilgrimage to the Heart of Medicine*. New York: Riverhead Books, 2012.

II "IT AIN'T NO PICNIC. "
THE MORTAL AND MENTAL WORLDS OF
CANCER PATIENTS

WHAT DID I LEAVE OUT OF MY GRADUATION TALK?

The last few weeks of massage school overlapped with another adventure in my life, the beginning of chemotherapy for cancer. I had been earlier diagnosed with Chronic Lymphocytic Leukemia (CLL), a fairly nonaggressive cancer of the bone marrow, and my oncologist had said, "Let's watch and wait."

And that is what we did for five years.

In 2001, some of my lymph glands began to swell up, and my oncologist said it was time for chemotherapy. All this was frightening, of course, but when you are a patient you do what's needed…one day at a time. I didn't mention any of this to my teachers or fellow students at massage school, but my energy, once I started chemo, was in steep decline. One day at our afternoon break, I lay on a massage table pretty well exhausted. Another student said, "Oh, are you tired today? Let me give you a back rub." I gratefully accepted and felt her energy infusing me…enough to finish the day and drive home.

"WE NEED TO GET YOU FIGURED OUT," THE DOCTOR SAID

Here's the background.

In the summer of 1996, I had a series of infections: sore throats, coughs, skin problems—quite unlike me. My family physician sent me for allergy testing, which showed nothing. Blood work showed very low white cells.

Why would that be? I went to some infectious disease specialists; they looked at one possibility…then another…then another. I was tested for AIDS, a frightening disease at that time because it was unmanageable, let alone curable, and AIDS suggested homosexuality, sexual profligacy, or use of street drugs—none applying to me.

I've forgotten some of the doctor visits. What I do remember is a puzzled specialist saying, "We need to get you figured out."

Hell yes, I thought, although I was annoyed to be considered a puzzle… an intellectual exercise…*an interesting case.*

Actually, I was more than annoyed—I was frightened. Because my immune system was not working, I was vulnerable to, it seemed, almost anything. Uncertainty loomed in my life as never before. As usual, the Florida summer was hot and humid, and I traveled uncomfortably to doctors' offices for tests…and a continuous lack of news and diagnosis.

I had left the secure road of health and normalcy. I had entered some strange, uncharted territories that were chaotic, absurd, and scary. Was it possible I could die? Was there a heavy, sharp sword of Damocles hanging above me suspended by that single horsehair…poised …ready to skewer me? What kind of peril surrounded me? How much? After all, I told myself, I was a serious long distance runner, a veteran of 5Ks, 10Ks, two marathons, even a triathlon. I was a nonsmoker, moderate drinker, and at good weight—all guarantees (surely!) against disease. My self-image was that I was a family man, a professor at a good college, a reliable person, a good citizen, etc. Being mysteriously ill challenged my being: I was distressed at levels beyond the cellular. The medical professionals wanted to figure out my disease, but the specialists paid no attention to the turmoil I was in. *Figuring me out* was like working with a petri dish…or a lab rat; their focus was on the *materials* of my body, and not my *mind* or *energy*.

I was sent to an oncology clinic. "I'm here to rule out cancer," I announced to the nurse. She smiled, knowing that anything was possible. Looking down a hallway I saw a pleasant room with people receiving drugs through IVs. So, this was the infusion room. I felt superior to those patients because *surely I was not going to be one of them!*

Dr. Andy Peterson greeted me in a friendly and hearty way. He asked about my work.

"I teach at Eckerd College."

"Oh really, what field?"

"Literature."

"Great. I had some undergraduate English courses. Do you like Ambrose Bierce?"

"Well, sure…in his oddball way."

"Oddball…that's it."

A wise physician, Andy knew how to treat me both technically and socially. He let me feel that I was a fellow human being and not just one more puzzle to be solved.

He said, "We need to do a bone marrow biopsy to see what's going on with your marrow. It should be producing more white cells, but it just isn't getting the job done."

"OK," I said, but I had no idea what I was agreeing to.

FIGURING BY SKEWERING

Andy explained the procedure to me and we went into a small room. I climbed on a padded table and, as instructed, lay on my side and facing the wall. Andy asked me to draw up my knees to provide good exposure to my lower back. The nurse assisting offered to hold my hand, and I accepted. It was, indeed, comforting. I sensed her care-ful intent and it calmed me.

Andy injected lidocaine to the puncture area…and inserted some instrument, certainly smaller than the sword of Damocles…and with sterile technique. I couldn't see what was happening and—unusual for me—didn't want to see. I felt a *crunch* when he punched through the top of the pelvic bone (posterior superior iliac crest, or PSIS, as we called it in massage school).

"Great!" he said. "We are in and can get a good sample."

He's happy, and I'm skewered.

He got his sample…and I was soon bandaged and sitting up on the table. He showed me a little pink wiggly strand of tissue in a specimen jar. "Very nice sample," he enthused. I took his word for it, since it was the only one I had ever seen or was likely ever to see.

What would the pathologist learn from this?

In a few days, I knew.

My phone rang.

"Hello?"

"Howard, this is Dr. Peterson, and I have some results for you."

"Yes...OK...."

"Well, there's good news and bad news. The bad news is that you have cancer. The good news, though, is that we have an excellent treatment for it."

I was unstrung. I fell on the floor.

"What?" I managed, perhaps trying to buy time.

"You have a kind of leukemia." He sounded cheery, as cheery as if I were entirely well.

"You're sure?"

"Yes. We asked them to run it twice, and their diagnosis matches all of your history and symptoms. We're quite confident."

Confident! He's confident...and I'm on the floor.

"Unh huh."

"It's a very slow growing cancer, Chronic Lymphocytic Leukemia, a cancer of the blood and bone marrow. We call it 'indolent,' as opposed to aggressive, and the treatment is very effective."

"How soon?"

"No rush at all, and you may never even need treatment. We see this disease commonly. The treatment for now is simple: we watch and wait."

"Watch and wait...doesn't sound too hard."

"Well, it's not. And you may never need treatment."

"That would be good."

"I know it's a shock," Andy said, "but we have a good plan for dealing with this."

"Yeah, it is a shock. No kidding. A shock."

We said goodbye.

I hung up the phone and slowly got to my feet.

I couldn't think of how to move or why.

The word "cancer" echoed in my head like gunshots. I also heard (or directed myself to hear) the phrases "excellent treatment" and "a good plan." From those I tried to create a context, perhaps a melody that would include the gunshots.

I NOW LIVE IN THE WORLD OF CANCER, EVEN IN MONTANA

My childhood globe symbolized clarity, control, perhaps even purpose. The earth was smoothly round, divided into land, ocean, and nations…etc.

The world of cancer is nothing like that. My leukemia has no shape or size. It is not even a specific solid tumor that could be excised by surgery or fried by radiation. It is pervasive in my blood stream traveling to every cell in my body. It's in the marrow of my bones, my pelvis, my vertebrae, even my skull and brain. The world of cancer is now an amorphous phantom in my consciousness, sometimes lurking, sometimes front and center. Some days I feel well, even normal. Some nights I start to worry…and worry a lot. My mind is in turmoil because my cancer does not *make sense*. It has disrupted my life. Emotions well up: *just how much trouble am I in?*

My leukemia is called "chronic." In medicine "chronic" means over a period of time, as opposed to "acute," or immediately dangerous. What messages about *time* am I getting here? Is chronic leukemia *good* because I'm not in immediate danger? Or is it *bad* because I'm *stuck* with it for a long time, maybe forever? And just how much time does my threatened life have?

My leukemia is also called "indolent." In my normal Western world, indolence is frowned upon; it means lazy, unproductive, idle, even self-absorbed. Is it good to have an indolent (lazy) disease? Or is it bad because the disease is just waiting in ambush before becoming acute?

Indolent…or *lazy*…or *lethargic*…or *slothful*. Like most Americans, I've received a zillion messages to avoid such qualities. Rather, I should be *up and at 'em,* because *the early bird catches the worm*: witness Benjamin Franklin, Horatio Alger, and other *hard-chargers*. My high school coach's highest praise was "That boy's got DESIRE!" We hear praise for the "fire in the belly" of "high achievers" and of urban, Type A personalities in general. I had absorbed such values and, aware or not, lived by them. Suddenly the reins have pulled back on this hard-charging horse. It's time to reconsider priorities. If my time on earth has limits, what, at base, is most important to me?

No big deal? Or heart will get me?

When I see my family doctor (who already has the news), he seems unperturbed. A wise and gentle man (indeed formerly in a religious order), he says, "CLL is no big deal. Lots of old people get that, but you're not old. Besides, heart'll probably get you."

What?

He gives a smile.

In a death-aversive culture, it's hard to imagine that *each one of us will die*, and the U.S. statistics show that the main causes, by far, are heart disease and cancer. So maybe I'm normal, with my "early" cancer, just a bit ahead of some statistic curve, but invited, so to speak, to do the work of considering my mortality.

The first part of my treatment is called "watchful waiting." I had trouble with this at first. What American likes to *wait* at a restaurant, for a promotion, in any line...or *anywhere*?

Anyway, on to Montana

In the spring of 1998, I was invited to be a visiting distinguished professor in the Institute of Medicine and Humanities, a joint project of St. Patrick Hospital and the University of Montana in Missoula. I wanted to accept but I worried about my diagnosis. Andy said, "No problem, you can get the same treatment there if it is needed." My chemotherapy would be a commonly used formula developed by the University of Texas MD Anderson Cancer Center.

My wife and I went to Missoula for the fall semester of 1998. We had adventures in Glacier National Park, Yellowstone, and in our local Bitterroot mountains. We basked in the Lolo Hot springs like Lewis and Clark did as they came through Lolo Pass roughly two centuries earlier. While I wasn't glad to be a cancer patient, I became fond of the word "indolent" as it applied to my disease and also because I could enjoy the Rocky Mountains, new friends, and my work there. I taught one course at the university, gave a set of public lectures, and did research for my next book on hearts.

BAD NEWS FROM THE CT MACHINE

Back in Florida in the late fall of 2000, I could see my lymph nodes were swollen in my neck...also in my armpits and groin. In fact, *anyone* could see that I was sick and especially Andy. He sent me for a CT of the neck.

I lay on my back in a sliding tray that took me through the large, vertical donut. There were various noises as the machine worked, creating "transaxial" images, that is to say, as if a very sharp guillotine cut across my neck slice after slice. This was on December 6th when I would have preferred to be shopping for Christmas.

Before long I have the FINDINGS, several lines of pitiless, medical prose:

> There is extensive lymph adenopathy. There are multiple lymph nodes in both posterior cervical regions, internal jugular regions, and supraclavicular regions where are abnormal by their number and size. The largest lymph nodes are seen in the left jugular digastric region. Left internal jugular chain and bilateral submandibular regions which measure up to 3 cms in size.

> A 1.8 cm round mass is noted in the superficial portion of the left parotid gland consistent with intraparotid lymphadenopathy.

No shit, Sherlock! Hell yes…any idiot can see I've got bulges under my jaw and down my neck. I look like a freak. Clearly sick! Children should look at me and run in terror.

Then follows, in bolded, italic caps:

IMPRESSION:
EXTENSIVE CERVICAL LYMPHADENOPATHY, CONSISTENT WITH LYMPHOMA

The phrase "consistent with" is standard radiology reporting: the physical evidence, as shown by ultrasound, X-ray, MRI, or CT—might fit with one or more diseases, which the referring doctor—Andy—would surely know… or suspect. Patient history, physical exam, labs (including the bone marrow biopsy), imaging, and any other tests should ordinarily yield a solid diagnosis —which is what was happening for me.

Am I glad it's decisive news because I no longer need to get "figured out"? Or am I sad and angry because I definitely have cancer that was once *indolent* but now is *actively causing trouble that anyone can see?*

I make the mistake of asking Andy whether there was a "stage" for it.

"Oh, we'd have to call it a IV because it's in your blood stream, but don't worry. We'll take care of it."

Don't worry? Stage IV! The worst! Bad! Very bad!

Nonetheless, it's time for action, and in a way I'm glad—especially with the swollen, visible lymph glands. Dr. Peterson explains that Lymphoma and CLL are on a diagnostic border to each other; the treatment is the same for either one, but—guess what?—there's no rush.

"*What?*" I think…and say to him. Shouldn't we leap right on this ghastliness?

He suggests I should enjoy Christmas first.

What?

"The treatment is chemotherapy, and we can wait until after Christmas." I'm quiet.

"The timing is not crucial at all. In the new year we'll treat you."

The new year. I like that: something new and possibly good!

"We treat this disease in our clinic here all the time and get good results."

"Good results."

"Good results!" He affirms. "Call us in the new year and we'll schedule some appointments."

That's good enough for me…I've bought some time…for the moment anyway.

"OK," I say. "I'll call you."

A little voice in me says, *the hell you will…I'm going to Jamaica or maybe even Nepal.*

CROSSING THE RUBICON TO CHEMOTHERAPY

Remember Julius Caesar? He crossed the Rubicon near Ravenna, coming back south toward Rome…a rebel against the current government. This action led to a four-year war, which he won. "Rubicon" means the "big red" river. I've got my very own red river of blood, and it's full of cancer.

Well, let's clean it up. Caesar said (or is said to have said), "*Alea jacta est*," or "The die is cast." I muse about the English word, "die," the singular for "dice," and also a dreadful pun on "die" as in "end of life." The Latin *alea* reminds us of "aleatory" or "chancy." Who or what threw the dice on me?

Well, all right. I will cross a bridge over troubled waters to the promising shore of treatment. What land might there be on the other side? Complete cure? Regained health? Limited healing...debility? Death? How far down will treatment take me? I've seen people worn out by chemo...bald, eyebrows gone, weight lost, haggard. Early chemo was based on mustard gas. *Kill the cancer and hope you don't kill the patient!* Terrible nausea, vomiting....

No more watchful waiting: I am now certifiably sick and entering treatment. I recall my earlier view of people in the infusion room and how I felt superior to them. Now I'm going to be *one of them*—a suitable lesson for my earlier pride.

INFUSION: GETTING STARTED

My boon companion George offers to go with me to my first chemo treatment, and I gratefully accept. Fellow professors, we have lunch once a week during the academic year and enjoy gossip, silly remarks, and a lot of laughter. I'm sure none of these will happen today.

By some quirk of scheduling today, I must go to another clinic across town, where I have never been before, so I am especially glad for his company. A nurse I've never seen before leads us to a small room and we all sit down. She asks if I have a port.

"A what?"

"A port in your chest for the chemo."

"No, I don't. Am I getting one?"

"Not today. You need a surgeon to put that in."

What the hell is that?

It's all in a day's work for her, but I'm *on the moon* right now, or maybe a distant galaxy. I'm very glad George is here, an anchor.

Nurse continues, "We need to give you some pre-meds. They help prevent nausea."

I spot a redundancy in her medi-speak, because pre-meds would be some *meds* before some *other* meds...in the same bizarre locution of airports, where there's "pre-boarding" of people who are clearly "boarding." Nevertheless, I take the pills then (and with every future visit) because they prevent nausea—except for one day, which I'll describe later. In earlier times terrible nausea routinely accompanied chemotherapy.

I have a new schedule in my life: three days of chemotherapy in each new month, Monday, Tuesday, and Wednesday. I soon learn that the rest of that first week is shot, because I am tired and useless. As the month moves on, however, I progressively gain energy and interest in life, although I know another knock-down is scheduled.

Each time I go to the clinic, I check in, have my blood drawn, and receive the pre-meds. I take my place on a Barcalounger and a nurse accesses a vein in my hand for my IV. Before long, this routine seems routine. The entire visit takes four to five hours. Like most cancer patients, I learn that I can't rush things and must give myself over to the treatment, however long it takes. My sense of time changes. I stop wearing a watch when I go to the clinic. A Type-A person, I was previously "efficient," doing many things quickly. Now I sit on my ass a lot and learn benefits of slowness and reflection.

I have time to ponder, to woolgather. What about this strange word "infusion"? For an infusion something is "steeped" or "melted" into a solution, for example tea leaves giving their essence to a liquid. The chemo drugs (three of them) are dripped into me with the intent of melting the cancerous cells into oblivion. This is the meltdown Andy and I hope for, a process that has happened for many patients before me.

I think of the final text of the medieval Latin poem "Te Deum" I have sung in various setting with various choruses: *In te, Domine, speravi: non confundar in aeternum*, usually translated "in thee, Lord, have I trusted, let me never be confounded." "Confundar" in Latin and "confounded" in English have a sense of melting, as in a *foundry* for steel or a *fondue* of cheese. If a person is confounded, he or she loses structure…melting away like the Wicked Witch in *The Wizard of Oz*. Am I melting away like her, or is it just the cancer that melts?

CONFUSION: INDIGNITIES OF CHEMOTHERAPY

A nurse mentions that side effects can be "cumulative." I pay no attention because I see no difference in the first few rounds. Before long I learn that she is correct, as each set wears me down more than the last.

Andy hasn't mentioned a port, and I don't either. I receive chemo in a large vein on the back of my hand, sometimes my right hand, sometimes my left. When the normal saline starts, this room-temperature liquid makes my hand feel cold. Next, my forearm. After all, I'm running some 98 degrees (F) roughly 26 degrees warmer than my medicine. As more and more liquid flows into me, my whole body's warmth is compromised. Further, I'm sitting still, doing nothing while cold fluids go into my blood, my kidneys, my bladder. Like other patients, I become chilled; I learn to ask for a blanket, even two blankets. Like other patients, I have to get up to urinate, pulling my IV stand with me. One bathroom door is narrow, quite tricky to get the five wheels over the threshold going in, going out.

I am not used to being a patient…not used to being sick. I'm used to being active, healthy, productive. Now my job is to lay low, be indolent, lazy even, so the drugs can help my body get well.

In some ways, I am now a battlefield…the petri dish…even the lab rat, and that is (more or less) OK with me, given the task before Andy and me.

I am on a retreat of some sort, a hibernation.

My IV displays my blood

One day my blood flows back into the IV line. It appears as thin, dark-red curved stripe in the air, a *catenary*. Indeed, I'm *chained* to the meds bag hung over my head.

I call a nurse.

"I'll meet you in the bathroom," she says with a smile.

I meet her there and she fixes the dilemma. I don't remember the physical specifics, but I do remember her kindness.

Although initially alarmed to see my blood going in the wrong direction, I was later glad to see that *I did, indeed, have blood*, and that the infusion of chemotherapy was entering the Rubicon of my circulatory system to heal both blood and bone marrow.

The blood was dark because it was venous blood, not the bright red of oxygenated blood in an artery. When we give blood, it's venous. Because there's little pressure in the vein, we sometimes need to pump our fist to keep blood moving. Because of my blood cancer history, I will no longer be accepted to give blood.

Insomnia

Some nights during treatment I cannot go to sleep—unusual for me, and inappropriate for my fatigue because I truly need rest! Instead, I'm wired, jittery. Worse yet, there's nothing good on TV at 1:00 or 2:00 a.m.

The next time I'm in the clinic I ask a nurse, "Say, am I getting anything that disturbs sleep?"

She chuckles, "Oh for sure, your anti-nausea medicine includes a steroid, but it doesn't bother all patients…only some lucky few."

"Well, I'm very lucky, I guess. Can I do anything about that?"

She suggests an over-the-counter drug. I get it, take it, and it does the trick. Next visit I ask a nurse why I wasn't warned about this and she says, "There is such variation between patients, we never know who will react which way. Besides if we gave everyone the full list, it would be depressing." I later learn that caregivers know that suggestive patients can experience side-effects just because they were mentioned to them, so they don't offer them. Over four centuries ago, Montaigne commented on how patient expectations could change perception. (See "Montaigne on Placebos, Nocebos, and More" a few pages below.)

Welcome to Chemo Brain

Another day the anti-nausea meds hit me like three martinis on an empty stomach; I am stoned.

Some days I am able study for my national exam for my massage license, but not other days because I'm in a fog. I learn the phrase for this: "chemo brain." For me, this is sporadic and unpredictable. I learn that other patients have it more often, even extending beyond treatment. One day, my buddy George brings a poem he's excited about. When he reads it, I can't grasp it at all.

Tired on my ass

Tired, tired, tired.

DEFUSION: STRATEGIES AND MOORINGS

My emotional turmoil and my rational efforts to deal with my disease are both telling me that my mind is, willy-nilly, involved.

What can I decide to do for myself?

I come up with the following strategies to *defuse* (melt away, if possible) the pressure, the stress.

Here's what I carry to the clinic each time: a CD player with headphones and CDs, get-well cards from friends, also books, magazines, a small blanket, and a little tube of peppermint oil. One of the nurses makes fun of this: "Oh, Howard's coming here camping again," and I reply, "Damn right. Whatever it takes."

One of my strategies is to engage my senses in pleasant ways, thus the peppermint oil and the music. Another is to connect with symbols of relationships, thus the cards from friends. I have two dozen of them. They all say the same things: *get well, we are thinking of you, we send our love,* but do I mind repetition? *Hell, no.* I read them over and over.

All these are cases of *something small means big when you are in hard times.*

I think of Tim O'Brien's *The Things They Carried,* the superb book about soldiers in the Viet Nam war. They carried things that connected them back to "the World" they left behind. That war was one I might have gone to; in the late 1960s I was the right age and classified as 1-A (AVAILABLE FOR MILITARY SERVICE!) during a very nervous time of my life. I thought, "Well, if I have to go, I'll write a book about it." I went to graduate school instead. O'Brien did go…and wrote a masterpiece. The things I carry to the clinic are, similarly, my tactile and visible tokens or hope—guardians and links back to the world of the healthy.

What no one could see, however, are prayers and mantras I've created. I repeat these while in chemo, while going to sleep at home, while jogging or walking in the morning:

ANTIBODIES ANSWER "YES!"
CLEAN UP ALL THIS CANCER MESS

LOVELY DARK AND NARROW
IS MY HEALING MARROW

HARMONIOUS TONES
HEAL HOWARD'S BONES

I chant these in my mind or aloud, a music of continuity and orderly rhythm. Perhaps they will unify mind and body, bringing intent to heal my disease. Whether I'm at peace or distressed at the moment, these couplets calm my mind and give some sense of control, whether or not they actually work on a cellular level—although later I think they may have had good effect because of improving coherence of my energy, mind, and the *materials* of my body.

My basic strategy is to throw everything possible at the cancer: Western medical technology, of course, but also all the kindness of my doctor, nurses, and friends, also the things I carry, my chants, and, of course, prayer. I pray for the medical team, all family and friends supporting me, and for a good outcome of treatment. These make up my homemade version of "integrative medicine," whether my doctor knows it or not.

I attend a support group, diverse by age, sex, education, et cetera, but unified by cancer.

"What kind of cancer do you have?" the woman next to me asks.

I'm surprised at her question but answer, "I have a leukemia? And you?" I am surprised also that I ask this.

"Breast cancer," she says, touching her breast.

Well, we're all honest here…and without our social masks.

One man says with a smile, "You know, this cancer ain't no picnic."

"You've got that right," another answers, "but we are *still* here, aren't we?"

"Still here" is a phrase I hear often…and more and more with my disease…and later as I age.

"You know," our group leader says, "all of you are *cancer survivors right now,* from the very day you get that diagnosis."

"No kidding!" a man says.

"No kidding," the leader says. "I think it helps to consider yourselves as *survivors*, not as *cancer patients* and certainly not as victims."

We all smile and nod in agreement.

Years later I read Stewart Justman's article about Montaigne.

MONTAIGNE ON PLACEBOS, NOCEBOS, AND MORE

Prof. Stewart Justman has studied Montaigne's essays. He finds that

Montaigne described contextual aspects of medicine of his day, some positive and some negative ("Insights of a 16th-Century Skeptic," 2015.) The positives include: the influence of belief for doctor and patient, the recruitment of the patient's mind in healing, and the attending rituals. The negatives include misleading notions of society and medicine. Justman finds that all of these are applicable to patients and doctors today.

The "placebo effect" is the healing or alleviation of symptoms that comes about from the patient's belief, not from a particular medical intervention. A negative result is called a "nocebo": the patient experiences harm or pain apart from the medical intervention. (Not showing patients the list of side-effects of chemo illustrates this concept.) Ideally, a patient supplies a placebo effect to any medical intervention, thus enhancing its positive effect. Also, ideally, any health or medical care should promote benefits of the placebo effect.

Among misleading notions Montaigne saw was the separation of mind and body. Instead, he saw the two as interactive, as we saw earlier. He wrote that we should not "divorce a structure made up of such close and brotherly correspondences." Rather, "let us bind it together again by mutual services (Frame, p. 455). Justman argues that Montaigne was prescient: division of mind and body—intensified by Descartes in the following century—was a factor in Montaigne's time and is now paralleled by the modern reduction of patients to their bodies and their diseases. Justman sees a further link in the creation of the modern randomized clinical trials (RCTs) that now dominate medical research. At the end of his essay, he cites Montaigne's emphasis on health, not disease, and his criticism that the medicine of his day was limited and easily "shaken…by change" (Justman, p. 504). Of course medicine is always changing, and I will argue that the emergence of "Health Humanities" in American and Europe may improve and enrich medical care (see Section IX).

DEALING WITH OUTSIDERS

Should I say I have cancer to anyone outside the clinic? Should I confess that I have failed to maintain my good health? A sports injury would be more honorable and fixable! Cancer has a dreadful history, including mistakes people made by exposures to Bad Things, like tobacco, asbestos, and radiation. Many people today are afraid that *any bump* on their body *might be this ghastly and fatal disease*. Although Andy says I'm not contagious, sometimes I feel that I'm a walking symbol of doom, disease, and death.

I decide to be open about it and, anyway, word travels fast within various communities. Suddenly I start to hear from people who are cancer survivors that I never knew had been sick. They are especially supportive. Indeed, most families have a cancer patient at one time or another and, nowadays with improved diagnosis and treatment, there are many, many survivors.

I live in two worlds: the normal world that both fears and ignores cancer and also the world of cancer that turns out to be quite large. On the edge between the two, a liminal state, I often forget that I am sick because I am functionally more-or-less well.

I learn of the Leukemia and Lymphoma Society and contact it; I get support there that means a lot to me. I'm matched with another CLL survivor, called a "cancer buddy." She is a nurse well across the country. For a while we email several times a week. It seems that we trust each other in a way that no two other people can.

I do some searches on the Internet but soon learn that information there can be scary, even terrifying. Some information is out of date and some deals with averages that are misleading for any given patient. The Internet world of cancer is so varied that all of it seems suspicious to me.

I am frightened but somehow holding on.

Years later, I encounter Stephen Jay Gould's essay playfully entitled "The Median Isn't the Message." Upon his diagnosis, he goes to the library and reads a study that suggests he would die in eight months from his abdominal cancer. This frightens him, of course, but he is versed in statistics, so he can re-envision the corresponding graph of mortality. He imagines himself in the "long tail" on the healing side of the bell-shaped curve. I love this; patients can imagine healing! Gould lived 20 years beyond his initial diagnosis. I heard him speak at my college. He died in 2002 from a different cancer. Not versed in statistics, I decide to get scientific news only from Andy; he knows me well and can select the news that will be best for me.

What I take from Gould's essay is twofold: patients with strong markers for health do better in general and, in his words near the end of his essay, "The swords of battle [against death] are numerous, and none more effective than humor." Humor! Excellent. Jokes, smiles, any signs of kindness! (See my book, *Clowns and Joker Can Heal Us: Comedy and Medicine*, 2011.)

The chemo nurses are routinely polite and cheerful. They are a new society for me and I like them. In fact, they and the patients are all in the same clinical boat. I honor them for dealing with all sorts of patients—some clearly very sick—day after day.

At the clinic I suddenly recognize a friend of my daughter accompanying her husband who—much too young!—has testicular cancer. We exchange pleasantries that carry more freight than usual: *cancer patients support each other in times that are bizarre and absurd.*

I'm tired all the time; my energy is low. I end 20 years of jogging and running.

With treatment (and my own placebo efforts?), the swollen nodes on my neck shrink back to normal size.

WORSE THAN INDIGNITIES, LASIX AND A HOSPITAL STAY

Side effects come in unpredictable ways. Sometimes I have diarrhea…sometimes I'm constipated. One week I have mouth sores. I seem to be working my way through the list of side effects the nurses wouldn't show me.

One day, with no warning, I start to vomit, bringing up sour slop briefly and then shaking with dry heaves. I feel helpless, as if a demon had taken over my body. My wife suggests I pinch my thumb web and the nausea stops. I don't know if the pressure worked, our joined intent worked, or the nausea had somehow run its course. (Later I learn that this point is Large Intestine 4, usually understood to stop headaches. Pericardium 6 three fingers up from the inside of the wrist is usually described as the point to stop nausea, as in the commercial product Sea-Band.)

One day my nurse says I need a diuretic.

"What?" I say. I was actually proud that I was not losing weight, but today's weight gain means that there is too much water retention in my body. Untreated, this could be dangerous for my lungs and heart. She says she must give me Lasix. I say that racehorses sometimes get this drug, and we share a little chuckle…as if I were one. At the end of my usual treatment, Lasix goes into my IV set-up, neat as you please. After the IV is removed, I dawdle around the office, ready to urinate and be *normal*. This is a mistake. I should have run to my car and raced home, because when I do drive home, I have a terrible urge to urinate a dozen blocks away from my house. The urge is overwhelming; I'm afraid I'll wet my pants, the upholstered seat, perhaps more. I lurch the car into the driveway and run for the bathroom where torrents from my body give relief.

And then there is fatigue…bone-tired exhaustion.

And a general feeling of absurdity, my body and the world *out of tune*.

And occasional fear and anxiety.

Was there a nadir, hitting bottom? I don't think so…more like a yo-yo up and down, but it is clear to me that a cancer patient has two illnesses, the cellular disease and the accompanying mental turmoil.

My hospital stay

One day I feel hot.

I ignore this…it's temporary, surely! Later in the day afternoon I am hotter. After supper I am hotter still. I take my temperature. The thermometer reads 101 degrees. A little later it says 102. *Damn*. My instructions for any fever over 101 are to call the hospital to arrange admission. I don't want to admit my weakness, but with low white cells, an infection could be dangerous. I call the hospital, and the admission officer agrees to stay late to process me. My wife and I drive to the medical center where I was a part-time pastoral-care volunteer for a dozen years.

The woman admitting me is very kind. Although "non-medical," she is the first person to extend care to me and her kindness means a lot. From my perspective, she is also a healer.

When I shake her hand, it feels very cool.

I'm put in a room on the top floor. I have a roommate by the window. He is inert.

I'm fitted with an IV that will deliver antibiotics day and night.

From time to time the hospital helicopter lands on the roof directly above me with strong vibrations and huge noise. I get little sleep the first night because of this racket as well as the night cleaning folks who seem to use the hall as a yelling chamber and bowling alley. Nurses take vital signs in the middle of the night. I'm awakened early in the morning for a blood draw.

I ask my wife to bring a radio; the second night I tune it between stations so that white noise fills the room, and I sleep better. Strangely, I don't complain about the hallway noise even with a large sign on the wall, staring me right in the face. It has the phone numbers for me to call with any "concerns." I should do that, but I do not. I learn that *it's hard to be assertive when you are sick.*

My roommate doesn't do anything, it seems. Am I glad there's someone sicker than I am? The third day, he is gone. To his home? To the ICU? To the hospital morgue? How many people have died in this room? In my bed?

Hospital patients have a lot of time to think, although some of the thinking is irrational. Just how sick am I? Can medicine save me? Will I get out of here soon? Alive?

On the other hand, I'm so tired, I can barely think.

I hibernate—in Florida!—sleeping the time away.

Over three days IV meds clear up my infection. I go home grateful to be better and to sleep in my own bed.

RANT 2: I'M ALL FOR PRIMARY RESARCH, BUT THIS SUCKS BIG TIME; ALSO: AM I GOING TO DIE?

In Rant 1, we looked at social and personal values that drive loneliness, stress, etc.

Now we look at the emotions attending a serious illness.

Regret: My life and the lives of my family are disrupted.
Shame: I'm not as capable and in control as I assumed.
Self-pity: Poor me, a victim.
Anger: Why me? This is not fair! My life is ruined!

I feel the loss of my habits for ordinary living and a loss of my sense of self. Were all these a sham, a delusion? Had I been living a life of tawdry and pathetic ambition and unconscious self-congratulation?

These emotions have, however, a positive side: I can assess my life and current losses, even while I am humbled. I was humbled when undiagnosed. With every visit to the clinic, I am humbled. I am humbled by my work schedule wrecked. I am humbled each time chemo takes my energy and confuses my mind. I am humbled knowing that I am a cancer patient— sick!—every time I think of it, week after week, month after month.

These are *quality of life* issues, and yet…I still have a life.

So to all these, I learn to say: *Tough! Deal with it! You're getting good medical care, you have good support, and time is on your side. Also: you are still here, alive and kicking even at half speed.*

Also: *I learn to ask for help and receive it with gratitude, not shame.*

The next set is harder.

Worry: What about tasks I am not able to do now?

Dread: Will the nasty side effects get worse, even continue indefinitely? Will I feel pain? Will my treatment work?

Fear: Will I lose ability, function, my long-term health, my life?

Terror…especially at night

These emotions are harder to resolve; they draw on primal instincts for survival.

These are *quality of death* issues, and death is out of fashion in this culture. We view death as a failure, as in the phrase "he didn't make it," when, clearly, *he should have.*

When I move only at half speed, am I half dead?

If my pain and suffering get worse, with no relief possible, would I consider killing myself?

I remember my father, just in his mid 50s and dying of brain cancer; he was bald, wizened, and very fragile following the trifecta of chemo, surgery, and radiation. While the matter of his body was in decline, his generous mind set the intention to be a thankful and gracious patient. I had flown home to be with him and to assist my mother during his last summer. At supper one evening, I drank a bottle of wine and went into his bedroom. I lay down in bed next to him.

"Dad, do you know you are going to die?"

"Yes," he said. We held each other for a while.

This was in 1970, before hospice was available in Florida and before speaking about death became easier.

Although my own cancer wasn't the worst kind, the illness caused me to consider death. When we are under threat, we tend to consider the worse outcomes.

While we rationally consider that we are mortal, we nonetheless have strong instincts for survival. We all hope to live a long, long time and in very good health; this is our "default position."

At a lesser level, much of my life was disrupted with the treatment schedule, the loss of energy, and a lightened teaching load. These were, I might say, small deaths.

DEALING WITH "INSIDERS" OF THE MIND

When I was about 14, I visited a friend in a hospital. A college student, he had driven his car off the road, rolling it over. He was banged up but not grievously hurt. I looked at his swollen face and the dried blood in his bandages. He said, "You know, you have a lot of time to think in a hospital bed." Made sense, I figured, but I thought no further and went back to my normal teen-age life.

With my long hours in the infusion clinic, my hospital stay, and the times I must rest because I'm out of energy, I am forced to retreat, to pull back, and reconsider. Men don't have menstrual periods or similar, programmed times to take it a little easy; men *charge right on ahead* with manly duties, responsibilities, and many postures of power and control, some of them effective and useful.

So what about this welter of thoughts, emotions, mysteries, these strange "insiders"?

Has my modern life been so busy that I haven't taken time to find my foundations and to learn from them?

Mind, consciousness, intent, emotion, awareness…the terms overlap and all flow into each other. Now I have time to be aware on my inner life and make some choices about thoughts and my emotions. I could dwell on the unfairness of it all, I could wallow in despair or cultivate anger—and there is some of all that. I have time to look at this turmoil, the instincts for survival, and possible futures after chemo.

I resolve to practice gratitude for my life, the world that goes on, my caregivers, my family and friends, and the world of medicine that steadfastly seeks treatments for disease.

My "grace capsule" and other matters of faith

A religious person, I imagine a "grace capsule" that protects me on my long space flight. It is made up of my wife on one side, Jesus on the other, and the Holy Spirit out in front. I take this *vehicle* to every chemo session, doctor's visit, and times when I might entertain negativity. I imagine it will take me a long way, even into the world beyond this one. A friend has said she'll be my friend "forever," even beyond earthly life, and this too comforts me.

My religious faith holds that there is *a surrounding presence of love,* no matter what the specific trials. I affirm this context and try to recall it, especially at my unhappiest moments.

Like language, faith is another lens for seeing the world—even if used sporadically.

As an undergraduate, I found the world of reason and logic was exciting and insightful. For a while I considered myself agnostic or even atheistic. Suddenly *I was too smart for religion!* During those years, however, I sang in the choir of Rockefeller Chapel at the University of Chicago on Sundays and in concerts of major works by Bach, Brahms, Beethoven, Fauré, etc. We singers were paid, and most were older and paid little attention to the service below us and far away—we sang from a loft in the back of the chapel. The music stirred my soul, and I slowly became interested in the sermons. I heard some of the luminaries of the day: Karl Barth, Paul Tillich, and Reinhold Niebuhr. Faith and intelligence could cooperate, it seemed, and I grew interested in the entire liturgy once again, hymns, texts, prayers, chants, anthems—the whole lot.

I have students and friends who state that they are "spiritual," but not religious. I respect this but wonder whether their spirituality might be deepened and extended by religious practice that can provide corporate worship, forms of confession, esthetic variety, and caring between congregants. I have found all these very valuable.

Harold G. Koenig, M.D., has been the director of the Center for Spirituality, Theology and Health at Duke University Medicine Center for many years. In his *The Healing Power of Faith: How Belief and Prayer Can Help You Triumph over Disease* (1999), he writes that hundreds of studies at Duke and elsewhere show that religious faith can have many benefits. These include: lower blood pressure, less stress, less hospitalization, less depression, fewer illnesses and quicker healing more well-being and life satisfaction, stronger immune systems, and longer lives (pp. 24-25). He writes that Magic Johnson, the renowned basketball player who was HIV positive became "intensely religious" after his diagnosis but led an active and productive life. In Koenig's words, "'I'm not cured,' Magic Johnson recently commented, "'But I am healed'" (p. 230).

A DIGRESSION ON CLASSICAL PHILOSOPHY AND KINDNESS

Pondering how my medical care provided both technical excellence and the loving care of people, I recalled a professor from my undergraduate years. Similar to the baker in "BAKERS DON'T SNACK ON THEIR WARES," he also illustrates the notion of *intent*, an important aspect of *mind*.

This memory involves abstract rationality in philosophy courses with Richard McKeon, a champion of Aristotelianism at the University of Chicago. He taught a rigorous, neoclassical system of philosophy, a crystal palace of fine distinctions with wonderful terms like "ontic" and "entitative." In a small seminar room, a handful of undergraduates looked at dual-language texts of Aristotle and other classical philosophers. From time to time McKeon pointed out subtleties in Greek that none of us could read. We worked through the English version on a facing page sentence by sentence, each student taking turns interpreting the words, and then McKeon would comment.

One day after a student gave his interpretation, McKeon whipped off his glasses, leaned forward and exclaimed, "Mr. Rabinow, I have been reading Aristotle for over 40 years, and your idea never remotely crossed my mind! Would you care to try again?" Paul Rabinow did so…and later become a distinguished professor of anthropology at University of California and author of several books.

McKeon had a reputation of being tough. In his wide-ranging and wonderful novel *Zen and the Art of Motorcycle Maintenance: An Inquiry into Values* (1974), Robert Pirsig presented a caricature of McKeon as a domineering professor, indeed "a holy terror." In my class, I felt none of that. I think we understood his gestures and speech as a form of *theater up close*, closet drama involving all of us in a play of minds. McKeon, a superb intellect, could be clownish, even a little silly, all for the aim of nurturing our (and his) learning.

A week after the course ended, I passed McKeon on the sidewalk of the Quadrangles and nodded politely. He nodded back. I continued on…but heard, "Mr. Carter!" I turned around.

"Mr. Carter, I thought you might like your final back."

I was surprised. These were usually not returned.

We went to his office—wonderfully messy—where he handed me my blue book. It had a B + on it. (I had already received a plain B from the Registrar). That B +, in his own hand, was worth a lot to me—an excellent grade in a difficult course—but his kindness in stopping to take a lowly undergraduate back to his office was worth even more.

I believe that *kindness* and *humor* are the cardinal virtues of humans. They can permeate even the most abstract, remote, or arcane pursuits of the mind, whether classical philosophy or medical science or any realm of healthcare.

THE POWERS OF INTEGRATIVE MEDICINE

There is a larger world of treatment beyond the medically technical. Some approaches are called "alternative" because they avoid standard medicine. Another term, however, is "complementary," when medicine and other approaches somehow cooperate. There are informal ways of complementing, such as basic human kindness just mentioned or a doctor's "bedside manner," but also formal programs, such as pastoral care for patients and massage in hospitals. Often these are parallel tracks, however, without attention to how they relate to strictly medical care.

Best of all, then, is "integrative care" a combination of normal healthcare and other supportive approaches. We can see this evolution of concepts (from alternative to complementary and integrative) in the National Institutes of Health's website, including the specific change of the old name of the National Center for Complementary and Alternative Medicine (NCCAM) to the new name, the National Center for Complementary and Integrative Health (NCCIH) in 2015.

In the chapter above "Defusion: Strategies and Moorings," I described my homegrown efforts (cards, music, peppermint, etc.) to keep up my spirits. My aim was to complement the medical treatment I received. I also learned that kindness of my doctor, my nurses, the receptionist, and the admitting officer at the hospital buoyed my spirits—all these seemed integrative and supportive to me as a patient.

Valet parking, fish, and more

My oncology clinic did much to make my visits pleasant and supportive.

The very first kindness was valet parking. I'd drive up to the main entrance, give my keys to a young man, and step right into the air-conditioned building. Sure, I could have parked in the lot across the street myself, but to be treated like a big shot gave me a boost. I'd take the elevator to the top floor and check in with a receptionist who was always friendly. The waiting room was like a hotel lobby, spacious, with nice furniture and—amazing!—up-to-date magazines. Not that there was ever time to read them: a cheerful nurse called my name almost immediately. She'd be at the door, holding my chart and smiling.

The clinic always had blankets available, and I often used them.

There were also drinks and snacks.

There was a large aquarium with lazy, slow, fish—not just symbols of nature, but alive and active creatures of nature swimming gracefully in this remote and alien space.

All these features buoyed my spirits. We might say, "They boosted my placebo energy."

The dark-haired surviving woman, a clown

One day an energetic volunteer brought cookies and drinks to patients in their Barcaloungers. She was a small woman with dark hair. Then she strode into the middle of the room and announced, "I lost my hair three times!" All of us knew what this code meant: she had chemo for cancer at three times and survived to be healthy and active. She became an enabling image for me, and I'll probably never forget her. I wrote about her as a clown figure knocked down but risen again to survive and prosper and help other cancer patients. (See *Clowns and Jokers Can Heal Us: Comedy and Medicine*, pp. 14-15).

The barbershop quartet

Another day a barbershop quartet with red and white striped vests and straw hats moved through the clinic, singing lively songs. (I was especially alert because I sang barbershop one summer long ago.) Their harmonious and upbeat music energized the space. As I drove home, I thought about the four men who had taken time out of their lives to come to a non-musical space and to sing in all the treatment rooms. Besides the energy of their music, there was an aura of kindness then and each time I thought of them later, even many years later.

Dinner with the docs

Yet another day, there was a sign out at the clinic, an invitation for dinner at the house of one of the doctors.

"No kidding?" I asked a nurse.

"No kidding," she replied. "He does this once in a while. You just have to sign up."

I signed up.

Nancy and I went to his house, where there were other patients and—if I recall right—all of the physicians in that practice. The meal was buffet style...extra tables set up. We sat at a table with another doctor (other than Andy), and we all made cheerful small talk, *as if all we were all entirely well and social equals.*

Driving home, Nancy asked, "How often does this sort of thing happen in America?"

"Not often enough," was all I could think to say.

Nourishment, nursing, nurses, mothering

Whether sick or well, all of us need all of these.

In summary

All these features integrated into my experience as a patient, my emotions and my resolve to keep my spirits up. *Sometimes small means big.*

MORE COMPLEMENTARY MEDICINE; I GET MASSAGED

When I had cancer, I received massage and it meant a lot to me. As I wrote above, massage is safe, caring touch within a supportive human relationship. It means your body is still worth something, even if you are sick and weak and you feel unattractive, perhaps worthless. Massage means you can still feel pleasure. It means you can relax and allow body and mind to do their own healing. For a brief while, you feel normal and healthy.

For several years, my massage therapist was Diana Grove. She treated my runner's legs and my nervous psyche. My wife and I decided that we'd get massaged about every three weeks as a way of lowering stress in our busy lives. For one thing, when we went off campus it was like being on vacation. Besides, it felt good.

Over time, both of us became friends with Diana; we'd ask about her professional life, and she'd ask about ours. She'd start me face up on the table, going straight for my scalenes, small, recalcitrant—one might almost say *nasty*—muscles on the front of the neck. These are tight on most people. Having decongested those, she'd move on to more pleasant touches. When she came around the side of the table to do my left hand, she'd say, "Well, how's the burn book going, Howard?" referring to a project that took forever to finish and see print. Actually, besides my wife, she was the only one who ever asked about it, another example of *sometimes small makes big*.

After my cancer diagnosis, I phoned her and asked if I was any risk to her. Uninformed about cancer transmission as many people are, I had heard the notion that it was communicable through touch.

"No risk at all," she said. "There used to be a belief that it was communicable but that's not true."

Diana massaged me through my years as a cancer patient. I felt that she nurtured my body as a whole, even when a part of it was sick. She helped keep my spirits up, both because she touched me as she always had done before but also because, in the privacy of the massage room, I could talk about my illness with her and my feelings of anger, despondency, despair, or hope. Her massage helped me integrate body and mind, illness and wellness.

One day, Diana said, "I'm going to do a little work above your body."

What? I thought.

Two dilemmas sprang to my mind: (1) I would lose the tactile sensations that I liked, and this would be a loss, and (2) what sort of work could that possibly be and how would I benefit? I didn't ask about either, and she didn't volunteer any explanation.

In a half a dozen years, I'd be providing massage for cancer patients. Some years beyond that, I would be trained in the Qigong, which uses work both on and off the body.

ILLNESS AS AN OCCASION FOR RESPITE, RESPECT, REBIRTH, ALSO THE COMIC

In "Massage as Respite, Respect, and Rebirth" above, we spoke of primal meanings of massage against a general background of stress in modern life. Massage, of course, is a chosen activity, something we want to do, and we allot time, travel, and money for that purpose. Serious illness or a trauma, on the other hand, comes to us against our wishes and separates us from both health and normal life. It also brings deeper and more pervasive stresses, which we may label "tragic."

Tragedy…and comedy

In *Clowns and Jokers Can Heal Us: Comedy and Medicine* (2011), I described tragedy as separation from what we want, especially inclusion in a desired society. This definition comes from literary critic Northrup Frye, who saw comedy as the opposite: a reintegration of a person into a desired society. (For my purposes, Frye's "reintegration," Charon's "affiliation," and my "reaffiliation" are all much the same, but I especially like the latter two because of the family resonances.) Shakespeare's *The Taming of the Shrew* presents the initial tragic separation of Petrucchio and Kate followed by their shared love and marriage, a comedy for them, their family, their society, and all viewers and readers of the play (Carter, pp. 14-20). Tragic drama, on the other hand, shows the deaths of Antigone, King Lear, and Willie Loman. Modern solitary confinement is a horrible deprivation from human society, with the loss of contact and touch from fellow humans, even the sight of them.

A sick person exists in a halfway state between tragedy and comedy, a liminal (threshold-like) state (pp. 21-23). The world of the sick is tragic by separation from health and the clear signs of pain, debility, disfigurement, and the possibility of death. The polar opposite, the world of the well is comic by integration into the world of the health with the clear signs of joy, activity, vitality, life, and, joy. In Shakespearean criticism, this is called "the green world," an ideal world that is healthy, fruitful, fecund, and filled with love (pp. 2, 3, etc.); we may recall Hildegard's vision of green, *viriditas*.

The power of comedy helps explain why *small means large*, when a well person brings a meal to the home of a sick person, sends a card, makes a phone call, or holds a hand and says, "I hope you are feeling better." For those moments a material object symbolizes a loving intent from the well person that is a gift for the sick person. When the sick person feels separated from the world of the well, such comic actions show acceptance and suggest reaffiliation—if only for the moment—back into the world of the well. This explains the power of get-well cards arranged on a table or taped on a wall so that the sick person can see them and draw sustenance from their symbolism any time they look at them.

Massage as comic

Massage takes patients into the world of the well for a brief time, reaffiliating them into normal society. After the respite of a massage, they return to the world of the sick but better equipped to deal with it.

My massage was as an outpatient: I went to Diana's office. I would have loved to have massage in the hospital when I was an inpatient. I'll discuss hospital massage in Section III.

HOW DOES CANCER WORK? ALL TOO EFFICIENTLY

Cancer isn't one disease. As the American Cancer Society website puts it, "Cancer is the general name for a group of more than 100 diseases." Any human cells—skin, bone, blood, organs—are vulnerable to cancer, whether solid tumors or blood cancers, such as my leukemia. Cancers have been known since ancient times. The word "cancer" has Latin and Greek roots meaning "crab," symbolically a nasty creature that bites into flesh and makes a mess.

One day, Andy says, à propos of nothing, "you know, your cancer is caused by a DNA error." Just like that! He might be saying, "The sun came up this morning" or some other mundane news. For me, however, it is BIG NEWS with meanings that reduce my stress. First, it appears that he (and all oncologists and many scientists) actually *understand* the basic causes of cancer. His remark suggests that medical science had "gotten cancer's number" and, therefore, has some (much?) control over it. My understanding is, of course, limited, and colored by my emotions wishing for medicine to be very much in control.

Second, I feel relief that my disease was not my fault, my failure, a result of bad habits, even a lack of character. It is *an absurdity* that came upon me—as with many other people—in a roll of genetic dice. When Andy says "your cancer," he means my leukemia, which is different from some cancers that do, in fact, have behavioral and environmental causes. While questions about many modern pollutants are still not yet answered, I am relieved to feel that I am not to blame for having somehow caused my disease.

Third, I am glad that I did not somehow inherit cancer. My risk came neither from my parents (who both died from cancers) nor from my maternal grandfather who died of it in 1939. He had a colon cancer that very likely could be well treated today. There are some other inheritable risks, such as the genes BRCA-1 and BRCA-2 that can mutate and cause breast cancer, but these do not involve me.

How does cancer work?

The short answer is "very efficiently," even "all too efficiently."

Cancer physician and researcher Siddhartha Mukherjee has written a magnificent book *The Emperor of All Maladies: A Biography of Cancer* (2010). This large (almost 500 pages) book details the history of the disease and the long history of scientists and doctors seeking to understand the disease and find treatments. Mukherjee describes, for example, the chain of investigations and discoveries that led to the identification of the "proto-oncogene" by J. Michael Bishop and Harold Varmus. They were awarded the Nobel Prize for this in 1989.

Mukherjee lists six "rules" for cancer from an article "The Hallmarks of Cancer" by Robert Weinberg and Douglas Hanahan (2000). I'll give a simplified version here. 1. Cancer cells grow by themselves because of their oncogenes. 2. They defend themselves against tumor suppressor genes. 3. They don't die like normal cells. 4. Indeed, they become "immortal." 5. They create their own blood supply. 6. They grow into neighboring tissues and travel (by blood stream or the lymphatic system) to colonize other organs— the well-known (and dangerous) metastases or so-called "mets" (p. 391).

In short, it would be hard to imagine a better-designed disease to damage living creatures and to resist medical treatment. And yes, other mammals, birds, fish, and reptiles can also have cancer.

When the Human Genome Project sequenced the normal human genome in 2003, later research could, by 2009, sequence genes for various tumors, including brain, lung, melanoma, ovarian cancer, pancreatic cancer, and some forms of leukemia (p. 451). Some cancers have 50 or more mutations, but acute lymphoblastic leukemia has "only five or ten genetic altera-tions" (p. 452). One breast cancer has 127 mutations. Mukherjee writes, "Every patient's cancer is unique because every cancer genome is unique" (p. 452). While such genetic mutations appear to be endless, the biological pathways that turn them on or off are some 11 to 15 "core pathways" that may be easier to understand and influence with drugs (pp. 451–459).

Small wonder cancer research is difficult and stretches over decades and, assuredly, into the future

With aging, our risk of cancer goes up. In "some nations" it is "one in three persons," even "one in two." (p. 459). "Cancer, we have discovered, is stitched into our genome," Mukherjee concludes; *it is part of who we are*, an unavoidable risk for all of our many, many cells that continuously reproduce and sometimes make errors, some of which persist and cause cancer. This description makes sense and gives me comfort. I like the power of words to control cancer on a symbolic level and a narrative that provides a clear and compelling story.

But does cancer deserve a "biography," as Mukherjee styles it, as if cancer were a person with intent, mind, or personality? I don't think so. I understand cancer (with some exceptions) as an illness caused by random, genetic errors, a set of dice, not a person with mind or intent.

Andy's remark was comforting at the time. Now, after reading Mukher-jee's book, I understand that cancer, while a collection of abnormal cells, is, as a disease, quite common and, even—we might say—*normally* occurring in humans and other creatures. Cancer patients are not absurd outliers from normality but part of regularly occurring events.

SENT HOME WITHOUT TREATMENT!

It's a Monday, the first of three days of another treatment. This will be fourth set of the six prescribed. I'm on a countdown! Only two more!

I arrive expectantly at the clinic with all my comforting gear.

After my blood draw, however, Andy says my counts are too low for treatment and that I should go home and rest.

Damn.

I go home unhappily. I am sad because my program for getting well has been disrupted and, somehow, my pride is hurt. After all, I've shown up in a responsible way, but I am—so to speak—sent to bed without my supper.

Furthermore, I am distraught that lab values declare, implacably, that I am weak and vulnerable.

Never mind: it's best to be patient when you are a patient. My job is to rest and relax.

A week later I come back; my counts have risen enough for me to have treatment.

REMISSION: THAT LOVELY WORD

In May, I finish my fifth set of treatments.

On the third day of that sequence, Andy has some news for me.

"Great news, Howard. The lab values show you are in remission."

I can hardly believe my ears. "Wow, please say that again," I say.

"You are in remission."

"Oh, that's great!"

I am elated. Tired, but elated.

I ask, "Does that mean we can skip the sixth go-round of treatments?"

He says, "I think we'd better go ahead, just to be sure there aren't any stragglers."

"OK," I accede.

"Remission" suggests "sending back," as in remitting payment to pay for a bill. It's hard to image just *to where* cancer would be remitted, a realm as abstract as where it came from in the first place. Never mind. Remission means *I do not have cancer,* which is a gigantic relief from stress. To be in remission means *cancer-free*, a state I've wanted for five years. Five years! I understand this news as comic acceptance back into the world of the well.

According to current usage, however, remission does *not* mean *cure*, which is would suggest a permanent state of health from that particular disease—not that anyone is ever in perfect health. Further, some remitted cancers can reoccur sometime in the future, and reoccurrences are often very stressful. After all, this is America, and when something is fixed it should stay fixed!

Never mind…no quibbling on this happy day.

"In remission" today means celebration, and I drive home on Cloud 9. My wife and I drink champagne.

Something else remission does not mean: it does not mean that I spring back from chemotherapy hale and hearty.

SICK…STILL OR AGAIN? AFTERMATH OF CHEMO

In the early days of chemotherapy, the hope was that patients would survive both the disease and the treatment. My particular case hearkened back to those days. While lab work showed continuing remission from cancer, it showed also that my bone marrow was *blasted*, barely creating the red cells and white cells I needed.

The lack of red cells was clear to me every day. I felt as though I was living at 10,000 feet, where the oxygen was thin. Any effort taxed me. If I walked up a flight of stairs, I had to stand a minute and gasp for breath, sometimes spitting out phlegm and saliva. And me—a former runner of 10Ks and marathons! In short, I was anemic, lacking sufficient red cells to carry oxygen to all my tissues.

As for low white cells, that meant that my immune system was weak and that I was vulnerable to infections. I had a bad cough that continued for weeks. And shingles.

Shingles!

Perhaps the worst sequel to treatment was a case of the shingles, common enough in people with weak immune systems, and remarkably unpleasant. I presented a string of painful blisters to my family doctor who recognized it right away. He put me on an antiviral that did nothing to heal the sores and a painkiller that did nothing to relieve the hurt.

Shingles is a painful rash, usually in a line on one side of the body. Mine was curled around the left side of my head. (The dictionary tells us that "shingles" comes from the Latin *cingulum*, meaning "belt" or "girdle."). Driving the car, I could not have my window open because wind touching my hair would cause sharp pain. I called the office and got another drug that made me woozy, but it did dull the pain. My wife and I went ahead with a B&B reservation down on the coast; I had to wrap my head in a towel to protect the pillowcase from my absurdly seeping blisters.

Eventually the affliction healed.

(N.B.: there is a shingles vaccine that I heartily recommend and now have; persons over 60 have a one in three chance of getting this bizarre disease, caused by chickenpox virus that has reactivated in the body.)

Once again in my academic life, I had an opportunity that I feared could not happen because of my cancer. This time it was to be a visiting professor at the University of North Carolina at Chapel Hill. I discussed this with Andy, who said I could go there, and also with my chairperson at Chapel Hill, who found me an oncologist, Dr. Eugene Orringer, for continued care at the university.

We moved to North Carolina and settled in. I functioned at half speed, maybe less. Patients respond differently to treatments, and I was a very slow responder. My bone marrow was very slow to recover. Nancy was predictably worried. She fed me iron-rich dishes in the hopes that my red cells would increase: when someone in a family has cancer, there's a sense in which the whole family has the disease.

Recovery

While CLL isn't the worst kind of cancer, my treatment knocked me down more that the average patient.

About eight months after treatment, lab results showed my red and white cells improving. In another two months they were close to normal.

My marrow had woken up and gone back to work.

I had energy once again.

Doc Orringer later said, "You know, we were worried about you there for a while."

ESSAY 2: ADVENTURES: SOME WILLED SOME NOT

Joseph Campbell described a hero's adventure in response to a call. We think of Luke Skywalker hearing the call to leave home, choosing to go, and questing about the universe. Other adventures do not come like that, however: instead, they are thrust upon us. Some are pleasant, like a phone call from an old friend; then we are happy. Others are unpleasant, like a car accident with injuries or like a bad diagnosis; then we are cast down, with many emotions and uncertainties. Life, of course, is a mixture of pleasant and unpleasant, the heroic, the unheroic, and just making it through any given day.

As for massage school, I chose to attend that, made plans, and carried them out on a timeline. I became a licensed caregiver, a therapist helping others. Perhaps I was heroic in Campbell's terms or the general norms of my society. Whether a doctor, a nurse, or a massage therapist, a professional caregiver has tools, procedures, and acceptance by society. From the point of view of that role, the world makes sense. The caregiver can work from a place of safety, surety, and reward. Caregivers have their own illnesses and accidents, to be sure, but their roles in life allow them to give comfort to others. In terms of comedy, caregivers happily belong to the society they desire to be in, the world of the well. Yes, well people have stresses, and there are many ways to be, on balance, well.

Cancer, by strong contrast, came upon me mysteriously, by unclear steps, and certainly against my wishes. The cause, Andy said, was a microscopic DNA error, a roll of invisible dice that broke the norms of cellular reproduction. In no way a hero, I was suddenly a patient needing help both from medical specialists and from anyone else who could give me solace, comfort, or support. I became, worried, tired, and weak. I was a diminished person, a new member of the world of the sick. In this world, the stresses are large, pervasive, and continuous. People who are seriously sick often worry…sometimes worry a lot. Must I stop work? Can I pay for all my treatment? Will I die? If I survive, will I have deficits? Sick people may have little energy and feel useless. They are not likely to be working or *doing* in the usual American sense. They are no longer in control, but rather in a place of absurdity, even chaos. They spend time in doctors' offices, clinics, even hospitals. They often feel pain. Sometimes there is no promise that things will improve. Indeed, things may get worse, with debility, disfigurement, even death.

From a head cold to a terminal illness, the world of the sick has many variations, all with stresses in leaving the world of the well and joining the world of the sick: the body suffers and also the mind.

In the first essay on massage school, we started with *matter*. In this essay we start with *mind*. Certainly the matter of my body was also affected by cancer and the treatment, but as I look back on my experience, I think of the large impact on my mind and the mind's resources in responding to threats. While our death-aversive society shies away from thinking about death, my illness urged me to consider it.

1. MIND

In our thought experiment about Atlas, we considered the material impact—the weight pressing down on his neck and shoulders—but we also allowed his mind to be happy.

Let's now understand him not frozen in time, but with a past and a possible future. Nor is he always a man and only a man; let's imagine (à la Virginia Woolf) that he has a sister Atlanta, also forced to hold up some symbolic world. Her mind, like his, is active, wondering how she got sick and whether her behavior was a cause. She wonders *where she stands* with her illness, and what various futures are possible, including death. The globe she supports symbolizes both the illness of her body and all her mental activity and distress.

People in the world of the well experience stresses with jobs, family, relationships, money, bills, etc., etc. Compared to a bad diagnosis, however, these are routine, even background noise. If (or when!) any of us enters the world of the seriously sick, the stress is well beyond normal background stress, and our minds must suddenly deal with *a clear and present danger*.

We think of our bodies as functioning well for a long, long time. Because of this attitude, we are disappointed (or sad! or shocked!) when something declines or outright fails.

Our minds, however, are built for survival. It seems to me that they work on at least four levels.

Instincts

Many years ago I was running near a nature park in Florida. Suddenly my body leaped in the air and stopped still on the sidewalk before my rational mind interpreted that there was a large snake crossing the sidewalk ahead of me. I backed away quickly…then gingerly approached and saw that it was a diamondback rattlesnake, much thicker and more ornate than, say, a black snake. Some deep instinct had kicked in and kept me from danger. This was part of "fight or flight," proposed by Walter Cannon (1915), and later revised to "Fight, flight, or freeze." I think I experienced both the freeze and flight with the jump into the air well before my conscious mind could perceive and analyze what I saw.

Here's the physiological mechanism: the hypothalamus activates two systems, one nervous, the other endocrine. For the autonomic nervous system, it activates the sympathetic branch. (We saw its partner, the parasympathetic branch, play a role in relaxation from massage.) The sympathetic branch then charges up our sensory and motor nerves, putting us on high alert, ready for action. The hypothalamus also activates the adrenal-cortical system, shooting endocrines into the blood stream to stimulate muscles and glands throughout the body, ready for immediate and forceful action. Anyone who has experienced an adrenaline dump when frightened knows about this. It could be a near miss in traffic, a physical threat to the body, or terrible news—even a bad diagnosis. When I fell on the floor at my diagnosis, that may have been a freeze response, a version of "playing dead."

We have instincts for survival in general. News of cancer that might be fatal touches our primal instincts for survival. A visit to the infusion clinic may include the sight of some very sick people; at some level, instincts tell us we don't want to be like them: *we want to live!*

As time went on, cancer was for me an ongoing stress for my mind, an Atlas/Atlanta-like load. There was no physical *flight* available to avoid the dilemma—even though I briefly thought of fleeing to Jamaica or Nepal. Nor was I frozen. I resolved to deal with my disease as best I could and with all the help I could find.

Our minds are hard-wired by instincts to avoid threats—real or imagined—to our lives.

Emotions

Emotions are also ways we assess and evaluate the world; some are familiar and pleasurable, but others are disorderly and disturbing. We may believe we are basically and predominately rational, but emotions are central to each one of us.

In Rant 2, I reported on regret, shame, self-pity, anger, worry fear, and terror. I think these emotions came from instincts for survival but also the very bad reputation of cancer in general and specifically from my family's history of cancer.

It was healthy for me to recognize these feelings and to talk about them in the support group. It was comforting to learn that other patients had similar feelings. I felt that I was not alone, but part of different communities, friends, family, even my fellow patients as we all received chemo—colleagues in the world of the sick. When the man spoke in an understated way about cancer, "It ain't no picnic," we all shared the same understanding and were grateful that he had spoken of it, even in his laconic and ungrammatical way.

My disturbing emotions came and went randomly, but they were slowly (if partially) replaced by hope, patience, and trust in my caregivers. A well-known scheme for such change is Elisabeth Kübler-Ross's five stages of grief: denial, anger, bargaining, depression, and acceptance (*On Death and Dying,* 1969). Emotions, especially when dealt with, can evolve. They don't always change in a neat linear progression, nor do they entirely disappear, but they can change to become less dominant.

Underlying much of my turmoil was the notion of *abandonment*. I felt I was abandoning my normal life, and I feared that other people might abandon me. The first was, perforce, true. Normal routines were upset, displaced. The second didn't happen, thanks to family and friends. Abandonment has to do with separation and losses we count as tragic. People who accepted me in my illness were, by contrast—and in my terms—comic.

Mortal emotions

Like *cancer*, the word *mortal* has a very bad reputation. It dwells in the ghastly company of such synonyms (from Roget) as: *dire, grievous, grim, malignant, terrible, bitter, monstrous, murderous, pestilential, poisonous, fatal, grave, terminal,* and *lethal*. What a dismal crowd!

At least in the West, social (and therefore personal) values take for granted that being mortal is *bad*. Our society has an unhealthy attitude about death and often ignores it—until it suddenly intervenes. We define death negatively: it is *not* part of life; it is a *failure*; it should avoided *at all costs*. Instead of saying, "She died," we say, "She didn't make it," which, in a very "making" society suggests a failure of effort or will or character on her part. Perhaps attributing the "failure" relieves us of guilt? Mourners sometimes say, "I wish I could have done more… seen the signs earlier… given more care." Would such have saved her? Maybe so, probably no.

Sooner or later all people die.

"Do everything!"

When our family cat Goldie was very sick, I took her to an all-night emergency vet. The man there said the cat was pretty far down, but there were some things he could try. He also said they could run into considerable expense. "*Do everything!*" I exclaimed, without hesitation or calculation of cost. Clearly my emotions had spoken, not my common sense, and I cried out *to defeat death at any cost.* Goldie died immediately, saving effort and expense, although there was of course sadness and grief to follow.

The notion to defeat death at any cost pervades modern medicine. In his helpful book *Being Mortal: Medicine and What Matters in the End* (2009), surgeon-writer Atul Gawande makes a strong case against medicine's orientation to "defeat death" and similar values in modern society. He calls "the most aggressive treatment available" a "default setting" for clinicians, patients, and families (p. 220), because of traditions and values as well as legal vulnerability for caregivers. That's the norm, the unreasoned "Do everything" that I urged for my cat. He calls the assumption that medicine can defeat death a *delusion*, a *fantasy*, and the cause of many tragic deaths during futile treatment in, for example, an ICU, instead of, say, at home, surrounded by family.

He provides reality checks. "Decline remains our fate; death will some-day come" (p. 44). For medicine, "Death is the enemy. But the enemy has superior forces" (p. 187).

Instead of current norms about death, Gawande argues that we should listen to dying people and help them enjoy their lives up to the very end. We should ask them what are their reasons to live. These, we might say, are their *vital* emotions; they include absence of pain, happiness, time with loved ones, and dying at home. One man wanted to be able to eat ice cream and watch football on TV (p. 183).

Vital emotions

We may consider—and celebrate—that humans are both *mortal* and *vital.* To be vital—even while sick—is to be alive, living, dynamic, vigorous, alert, breathing, and here on this earth—*still here.* These are assets, in contrast to the deficits of our illness.

Our emotions reflect both our mortality and vitality.

When death does come, it can be an expected and honored event. The patient has, indeed, made it to the end, not only of life, but also of suffering and pain. So, we could say that, in dying, the patient did "make it"—achiev-ing a normal, even healthy, ending to life.

Intent and emotions

While sick, my intent was to understand my emotions, work with them, and evolve them toward protecting me whether I lived or died. I also resolved to accept the good intentions of other people, medical and non-medical. My doctor and all the nurses seemed to bring "good mind" to me as they supported my physical and mental health. My family and friends urged me to get well soon. During some of my hardest times, my friend Mary Ann said, "I'll be your friend forever." Unlike measurements of my blood cells, these supports had no numbers but were powerful and helpful as they influenced my mind.

Rational and/or reflective thought that faces death realistically and practically

Our linear minds seek strategies to deal with adversity. Some are not helpful: procrastination, reasons for denial, or extrapolations that catastrophize, thus creating frightening stories of the worst possible outcomes. We feel a lump in our bodies and wonder *could this be cancer*. TV advertisements for drugs assail us (and especially older people) about dangerous diseases we *might* have.

Our sense of autonomy and control of our lives makes denial easy. According to Gawande, many physicians fear not doing enough even with patients who are clearly dying.

Other strategies, however, can calm us and shape our plans, our intents, even our understanding of our lives.

Gawande describes hospice, palliative care, and end-of-life care in general that support the best possible lives all the way to the end, whether at home, in a hospital, or in a well-run nursing home. He devotes an entire chapter to "hard conversations," which I'll discuss below.

Being seriously ill gives a person time to think about many things we often ignore: pain, limits of life, and death. Such thoughts may lead to an adventure of sorts in considering *what might I do with my life, limited as it surely will be?* It is a time to step back from life…to ponder, to weigh, to assess… to get to know yourself more deeply as self, not as an actor in a series of roles. I think of Thoreau's line, "Love your life, poor as it is" (*Walden*, Chapter 18, Conclusion).

I heard other cancer patients say, "You know, lots of little things don't bother me any more."

And: "In a funny kind of way, I'm glad I got sick, because now I am more grateful and have a new sense of direction"

And: "Looking into the asshole of death makes a lot of other things more clear."

And: "We're still here!"

Preparations for death including "hard conversations"

Gawande emphasizes "reasons to live," even for very sick patients (p. 125) because he wants the best possible life for them, even as death nears. This may mean stopping treatments, especially those with little promise for extending or improving life.

In Chapter 7, "Hard Conversations," He lists five questions palliative care specialists recommend for frank conversations about end of life (pp. 182-183). Family and sick persons take time and careful attention to answer these together:

- What do the sick persons understand their prognoses to be?
- What concerns, issues, or fears do they have?
- What trade-offs are they willing to make or not make, including stopping treatments?
- If health worsens, how do they want to spend their remaining time?
- Who should make decisions when they cannot?

Gawande illustrates these with two patients. In one case, he is the treating physician, and he finds guidance in medical care because the patient's "aims…were clear." In the other case, the patient is his own father, Dr. Ram Gawande. In both cases, the patient and the family find much relief from the process. (pp. 211-213, 234-242).

Hard conversations put the patient, the family, and caregivers all on the same page, minimizing uncertainty, unexpressed emotions, and conflicts about care. They are hard because they break through denial and taboo to face harsh realities. In Tolstoy's *The Death of Ivan Ilyich* (1886), this routine denial of death is called "the lie." Hard conversations take time and effort, but the rewards can be many, including relief from delusions and fantasies of holding off death. In these efforts, both emotion and rational thought are combined, and all participants come to a similar vision that can avoid emergency treatment by paramedics, an unneeded ambulance trip, and distressing scenes in hospitals, including futile care in an ICU. Too often relatives are unwilling to let a person die, because there is unfinished business, they didn't get go say goodbye and/or they believe that *surely there is more that medicine can do*…any new drug!…an experimental study!…something at a larger hospital!

In her touching and painful book *Dying: A Memoir* (2017), Cory Taylor has harsh words for Australian physicians who won't use the word "death," Australian law that makes euthanasia difficult, a psychologist who doesn't help her, and medicine in general that sees death as a failure.

There is now in common usage a form called Physician Orders for Life Sustaining Treatment (POLST), readily found on the Web and available in some 45 states. These orders can avoid needless treatment by paramedics at any scene and by doctors at any hospital.

Hard conversations allow for sharing, dealing with emotions, and making plans together. Participants then share the same intents, the same mind, and form a small comic community. They have expressed their love and agreed upon the values and goals. They have looked death clearly in the eye and taken charge by making choices.

As an illness heads toward death, it is a time to heal relationships and make plans for death: estate planning, gifts to others, and so on. There can be plans for hospice. When my mother was sick, I had little experience with very ill people, and I put off calling hospice. When I did call, a worker came, saw my mother, and immediately said, "Oh yes, she's ready." Hospice care was expert, supportive, and loving to my mother and family. After my mother's death, my wife and I spoke of the hospice workers and volunteers as *angels*.

There can also be plans for a funeral, the disposition of the body, and the like. When my father-in-law was ill with Parkinson's, we were uneasy (and silent) about asking if he had any wishes for his funeral. When we did ask, he had specific ideas that he was glad to share, but he, too, had been reluctant to mention them.

We typically plan for retirement from work. We should equally plan for retirement from life.

With such care and consideration, death is less feared as a failure and more accepted as a natural path and perhaps even as a new adventure, especially for people with spiritual lives.

Gawande speaks of courage as part of this process. I believe love is equally important.

Spiritual worlds

In the last 65 pages of *Being Mortal*, Gawande movingly describe the illness and death of his father, Ram. In contrast to his descriptions of deaths made difficult—tragic, even ghastly—by excessive medical care, Ram's death is presented as well managed as possible. The main reason for this outcome was a hard conversation. Ram, with family agreeing, made clear his exact wishes for treatment and the main goals. He also said that he wanted his ashes taken to the Ganges and spread according to Hindu ritual.

The family makes the journey and does the rituals, but Atul, while impressed with the ceremony and the tradition, is not religious. Indeed his book as a whole is secular, with little mention of patient's spiritual lives and no mention of hospital chaplains or ministers, rabbis, imams, deacons, or other clergy who might support a family and the dying person.

Gawande emphasizes "reasons to live" (p. 125) but not "reasons to die," which might include the end of suffering—as well as, for the religious, attaining life beyond this one. Many obituaries in North Carolina speak of the dead person going to heaven, to the arms of Jesus or meeting the Maker. Some older gravestones depict a hand and finger pointing skyward with the phrase below "GONE HOME."

Gawande uses the word "soul" as something to be sustained in life by medical professionals (p. 128), not as quality sustained by spiritual or religious experience, nor as a quality existing beyond earthly space and time. One of his dying patients tells him, however, that, "She was at peace with God" (p. 242).

Although we live in a mixed state of *vitality* and *mortality*, there's the possibility, for some, that our souls (or minds) live—and will continue to live—in a state of *immortality*. Poet and priest John Donne concludes "Holy Sonnet 10" with the words, "Death, thou shalt die."

For persons of faith, soul includes many or all aspects of mind, and death can be a passage to a new world of rest and reward, reunion with the other dead, and seeing God, however perceived and named.

Universal mind?

While Gawande stresses dealing with the realistic here-and-now circumstances of mortality, there may be afterlife circumstances as well: many religions consider that there are.

When my Dad was in his final illness, he told us that he had a vision of his dead relatives in a tunnel leading to the "other side." He said they wore white robes and motioned for him to come across. This was in the summer of 1970, well before Raymond Moody's book *Life after Life* (1975) that introduced many readers to the concept of near death experiences (NDEs). Since then researchers has studied NDEs around the world. Some scientists provide a materialistic causation: neurological oddities produce a hallucination that has no relation to reality. By contrast, persons who believe in an afterlife (or at least the possibility) may see NDEs as an accurate vision of another world.

If the elements of the earth and our bodies are the same as all elements in the universe, and if the energies of the earth and our bodies are the same as all the energies in the universe, why could not our minds by related to a vast universal mind, often known as God, the Creator, or many other names?

I recall Pascal's famous wager (*le pari*): if you're not sure about an afterlife, it is prudent (and costs nothing) to imagine that it exists. When we die, we'll find out the truth either experiencing the next world or learning nothing at all because our minds have ceased to exist.

Still further, we can consider this: *what if that which we believe is exactly what we get?* You imagine a heaven, then it will be yours. Imagine reincarnation, you will get that. Imagine nothing, you will get nothing.

We are clearly both mortal and vital…but, possibly, even depending on our minds…also immortal?

It is good common sense to understand and accept that our bodies slowly, eventually, and inevitably break down. Such wisdom is an example of *mind over matter*.

2. MATTER

Contrasting the worlds of the well and the sick gives some clarity, but there is, we noted, considerable middle ground between the two. Being *entirely well* would be an illusion, both pleasant and useful, but our material bodies are never entirely well. They are continuously busy with repairs. We have bumps and bruises, scrapes to our skin, sore muscles or achy joints. Also bowel trouble, caries in our teeth, sleep disturbance, and/or dandruff.

Nobody is perfectly healthy or symmetrical, but, despite ups and downs, our bodies, like the minds just discussed, are built for survival: they are remarkably durable. Today many people live well beyond our basic biological role of reproducing ourselves in children because of the resources of these bodies, as well as good nutrition, good public health (sanitation, inoculations), and medical treatments. In so-called "developed" countries, average ages have risen sharply in the last century.

Indeed, we live in a dynamic balance of harm and repair every day especially on the cellular level. The 2015 Nobel Prize for Chemistry was given to three scientists who specialized in the repair of DNA molecules, Tomas Lindahl, in England, Aziz Sancar, and Paul Modrich. The latter two work in my corner of the world, one at UNC-Chapel Hill, the other at Duke University. My wife and I watched them receive their awards in Stockholm on North Carolina Public TV.

Nor can we be perfectly sick; what would that mean? Not even death is "perfect sickness," because, in various ways, we have survived—we might say—whatever ended our lives, and the materials of our body, one way or another, will live on in other forms.

Matter, aging, and mortality: like cars, humans wear out

Although various accidents may occur at any age and various limitations can exist from birth, many of us live (more or less) in the world of the well for a long time. In middle age, we may assume (or hope) that our good fortune will last for many more decades, but this is not often possible. The plain truth is that the matter of our bodies becomes more and more compromised. Skin, vision, and hearing can have changes; hair turns white…falls out.

Gawande lists several body parts or systems that decline over a long life: teeth, hands, muscles, heart, brain, bowels, lungs, etc. All of our senses become dulled…also our immune system. He writes frankly of "the unfixables in our life, the decline we will unavoidably face" (p. 46).

Mortals live with constant risks and, *if they are lucky*, grow older year by year.

When I went to my 50th high school reunion, one third of the class was known to be dead. When I went to my 50th college reunion, one quarter of the class was known to be dead. At each reunion, other alums were said to be unable to travel, and some who came looked unwell. My childhood friend Phillip died not long after we shook hands and spoke of the old days.

At the college reunion, I spoke with a man who had become a doctor, an endocrinologist. I asked him how business was. "Oh, it's great," he said. "Because people get older and then have diabetes, I have plenty of work. In fact, anyone who lives long enough will get it, because the pancreas just wears out."

"Getting old isn't for sissies."

I hear this slogan now and then, almost always from older people. In an understated way, it refers to the stresses of the inevitable losses of aging, as well as to a claim that older people are, in fact, not only mature but grateful for and even proud of their survival.

The Greek myth of Atlas froze him for eternity. Modern products often bear his name to suggest their durability: housewares, snowshoes, electronics, carpets, even roof shingles—all nonetheless doomed to fail sooner or later. In our earlier discussion we imagined Atlas happy and productive. Just above, we imagined him as Atlanta. Now, in unfreezing her, we understand that she, like all of us, moves through time, aging and slowly wearing out. Over her lifespan, the materials of her body suffer illness and injuries, but she survives and prospers. Finally, however, she falls apart, and the world she has carried crushes the materials of her body. The stories we tell—of her, of him, of others—live on in our minds, and we can tell and retell these in endless variations, thus keeping them alive.

Actively dying...healthily dying

Now and then in the hospital I hear that a patient is "actively dying." On first hearing, this phrase shocked me: no patient should be dying, I felt, and certainly not *actively*. What a paradox! Our culture typically considers death a failure of vitality and a singularly passive event, as in "giving up."

I asked Dr. Laura Hanson about the term; she is a Professor of Geriatric Medicine at UNC and leads the Palliative Care Program for UNC Hospitals. She said, "Actively dying is the standard phrase we use to describe the observed process of organ failure that occurs in the final hours to days of life. It is active in the sense of a process that is common and irreversible also that it progresses over time. There are commonly changes in skin, temperature, vital signs, alertness, and breathing patterns."

So…the body knows what to do and has various sequences for shutting down, sequences that not even medicine can stop. From this perspective, *death is normal and natural*: the materials of our body finally wear out and call a halt to life. Medical people like Laura see death often and understand it better than almost all modern people for whom death is a rare event, both disturbing and tragic. When I worked a hospital massage therapist, a nurse told me that a patient was actively dying with the implication that this was expected, normal, and to be honored.

When patients are actively dying, they are headed toward death despite any medical interventions. Laura recommends comfort measures only for the actively dying, but sometimes family members, nonetheless, choose treatments to try and prolong life, and she follows this wish, no matter how futile the treatment.

In fact, all normal cells of our body have individual and overlapping lifespans with a beginning and an end. The fancy term for this ending is "apoptosis," with Greek roots meaning a "falling away" (the word "ptomaine" is related). In cancer, this normal, programmed death of is disrupted, and cells are "immortalized" and cause trouble by using blood, taking up space in the body, and sending out metastases—when they should be dying and dissolving away like all normal cells.

As for the normalcy of dying, Gawande uses the phrase the "dying role" in another positive sense: "People want to share memories, pass on wisdoms and keepsakes, settle relationships, establish their legacies, make peace with God, and ensure that those who are left behind will be okay" (p. 249).

It would be stretch for this culture, but can we imagine the concept of *healthy dying* not only for cells but for our entire bodies?

It is natural for us to fear death; many instincts want us to live, to keep going, even choosing treatment that is futile and extends dying.

It is even more natural to understand that death comes to us all and that there are advantages in accepting and working with it.

Physician Bernard Lown made similar observations in 1999, and he went a bit further in concepts and in rhetoric. He referred to "the pornography of death" that kept dying patients needless and expensively alive because of the reasons Gawande discusses and also patients' lack of knowledge of their rights (*The Lost Art of Healing: Practicing Compassion in Medicine*, 1999, p.

270).

Not actively killing patients

We can actively nourish the dying mind. We can support the dying body, but Gawande and many others (myself included) do not believe in killing patients, even when close to death.

As for suicide, my instincts are against it, although I can imagine why some people might choose it.

Can we become dual citizens of two worlds, the well and the sick?

The word "patient" sounds like someone who is patient, as in philosophic or understanding, but "patient" has Latin roots meaning to suffer, to undergo, to bear a burden.

All of us, whether caregiver, patient, or well person, would do well to consider themselves as *dual citizens* of two worlds, that of the well and that of the sick. Even when we are well, we have much potential for becoming sick, either slowly or all too quickly. Denial of death helps no one.

The two worlds are interconnected, interwoven, and dependent on each other. In *Gulliver's Travels* (1726), Jonathan Swift describes the immortal but still alive Struldbrugs, who live forever but unhappily, sick in body and cranky of mind. It would be better for them to die. The commonly used phrase, "Well, s/he's in a better place now," suggests some acceptance of death, however grudging.

For believers in an afterlife, there's yet a third world, one to be yearned for and sought.

Heaven, Valhalla, Elysian fields, Deva Loka, Tian, Nirvana, reincarnation… humans have imagined a wide variety of homes beyond ordinary life.

3. ENERGY: ANOTHER APPROACH THAT PROVIDES UNIVERSAL CITIZENSHIP

At the time of my illness and recovery, my understanding of energy was very limited. I had a layman's sense of energy in medical uses: the gamma rays of my CT scan, chemical energy of drugs acting on cancer cells, even the mechanical energy of Andy skewering my pelvis. These were applied energies, selected by the world of medicine as important tools to discover and treat cancer.

For me personally, there were other areas about energy. The first was my physical energy, which was much diminished; this was all too clear during and after treatment. My layman's understanding was that my body was reacting to the drugs during treatment and, afterwards, my bone marrow was anemic, producing fewer red cells that should bring oxygen to my cells. The second area was chemo brain, fortunately only a handful of episodes.

More generally, I felt the good energy of my caretakers, family, and friends, even though I had no way of defining or quantifying that. I also had a sense of which people greeting me didn't want to shake my hand or wanted to say something quick and leave me; they were in the energy realm of the hippie's phrase "bad vibrations, man." I also sensed "good vibrations" from family and friends. And from many others: my doctor Andy, the nurses and techs, the hospital receptionist, even the man who parked my car—all these were polite and caring. My cell buddy from the Lymphoma and Leukemia Society shared good energy with me via email, even though I never saw her. I recall specific events that buoyed me up: the woman who lost her hair three times, the barbershop quartet, and the dinner with the docs. All of these provided symbols of care, coherence, entertainment, or even comedy, in the sense of re-integrating patients into the world of the well.

Further, my home-grown strategies supported my personal energy, affirming my spirits up through mantras, prayer, meditation, and the things I carried to treatment. If my physical energy was low or if I couldn't sleep, these were times when it was easy to be sad or to worry. Prayer or a mantra could put my mind in another direction, boosting good vibrations and keeping bad vibrations at bay. The material things I carried to treatment had, I felt, symbolic value and gave me positive actions to do, as simple as playing a CD or putting some peppermint oil on my upper lip. They may well have had placebo power.

Later I began to think more about energy, especially when I was a student of Qigong and when I learned more about scientific work on energy (see Sections IV, VII, and VIII).

IN SUMMARY

Being sick is a time to reflect on illness, life, and death and to find strategies to deal with all of them. Integrative medicine, from the patient's point of view, is a collection of ways to deal with body, mind, energy, and, for some, spirit. While alive, we will always exist in the liminal space between sickness and health, citizens of the overlapping worlds of the sick and of the well.

With luck, we can shape an understanding that makes peace with both our precarious but rich vitality and also our certain mortality.

As a high school student I was startled when I read William Cullen Bryant's poem "Thanotopsis" and learned that it might be possible to accept death. All I knew from my culture were values of fear, avoidance, and taboo in general. These are the closing lines:

> So live, that when thy summons comes to join
> The innumerable caravan, which moves
> To that mysterious realm, where each shall take
> His chamber in the silent halls of death,
> Thou go not, like the quarry-slave at night,
> Scourged to his dungeon, but, sustained and soothed
> By an unfaltering trust, approach thy grave,
> Like one who wraps the drapery of his couch
> About him, and lies down to pleasant dreams.

SOURCES 2

Campbell, Joseph. *The Hero with a Thousand Faces*. Princeton: Princeton Univ. Press, 1968.

Cannon, Walter. *The Wisdom of the Body*, 1932. Various editions available.

Carter, III, Albert Howard. *Clowns and Jokers Can Heal Us*. San Francisco: Univ. of California Medical Humanities Press, 2011.

Frame, Donald. See Montaigne below.

Gawande, Atul. *Being Mortal: Medicine and What Matters in the End*. New York: Metropolitan Books/Henry Holt and Company, 2014.

Gould, Stephen Jay. "The Median Isn't the Message." Readily available on the Web. See https://people.umass.edu/biep540w/pdf/Stephen%20

Jay%20Gould.pdf, accessed March 1, 2016.

Justman, Stewart, "Montaigne on Medicine: Insights of a 16th-Century Skeptic." *Perspectives in Biology and Medicine* 58.4 (Autumn 2015): 493-506.

Koenig, Harold G. *The Healing Power of Faith: How Belief and Prayer Can Help You Triumph over Disease.* New York: Simon & Schuster/Touchstone: 1999.

Kübler-Ross, Elisabeth. *On Death and Dying.* New York: Macmillian, 1969.

Lown, Bernard. *The Lost Art of Healing: Practicing Compassion in Medicine.* New York, Ballantine: 1999.

Montaigne, Michel Eyquem, de. *Essays and Selected Writings; a Bilingual Edition.* Trans. and ed. Donald M. Frame. New York: St. Martin's Press, 1963.

Moody, Raymond. *Life after Life: The Investigation of A Phenomenon-survival of Bodily Death, 2nd. ed.* San Francisco: HarperSanFrancisco, 2001.

Mukherjee, Siddhartha. *The Emperor of All Maladies: A Biography of Cancer.* New York: Scribner, 2010.

O'Brien, Tim. *The Things They Carried.* New York: Houghton Mifflin, 1990.

Pirsig, Robert. *Zen and the Art of Motorcycle Maintenance: An Inquiry into Values.* New York: William Morrow, 1974.

Swift, Jonathan. *Gulliver's Travels* (1726). Many editions available.

Taylor, Cory. *Dying: A Memoir.* Portland: Tin House books, 2017.

Thoreau, *Walden* (1854). Many editions available.

III "EVEN BETTER THAN SLICED BREAD." THE WORLDS OF MASSAGE FOR SICK PEOPLE

VOLUNTARY LIFE CHANGES

As my energy slowly returned from my disease and treatment, my wife and I made choices about our lives: it was time for some new adventures. In 2003, we finished our teaching careers at Eckerd College, and we moved the Chapel Hill, North Carolina, where we had stayed on academic leaves twice before. I was appointed adjunct professor of Social Medicine, School of Medicine, at the University of North Carolina-Chapel Hill.

What might I—now unpredictably both a cancer survivor and a massage therapist—do with my life? A connection of the two seemed unlikely, perhaps implausible.

Although I knew how valuable massage was in general, when I had cancer I learned specifically how comforting and affirming massage was for a sick person. My massage therapist Diana Grove had provided caring support that helped my body and mind—matter and consciousness. It seemed to me that care like hers would have much improved my hospital stay. Why not massage in hospitals? I learned that there was an emerging specialty of hospital massage and decided to pursue that training. Nothing would happen without more training, that much was sure. But would I ever find work in this unusual field?

I found a program at the Oregon Health & Science University in Portland. My wife and I had friends there; we used to share meals together when we lived on the same street in Florida. I asked if I could stay with them while I went back to school. They agreed and before long I was on the West Coast embarking on my new adventure.

HOSPITAL MASSAGE, A VOLCANO, AND BRIDGES

The Oyster Bar in downtown Portland, Oregon, has a wooden interior much like an old sailing ship and nautical memorabilia everywhere; it's a fitting location for eight students who are about to embark on a five-day workshop: massage for cancer patients. Gayle MacDonald has been offering this workshop for several years; she's written two books on the subject, *Medicine Hands: Massage Therapy for People with Cancer* (1999) and *Massage for the Hospital Patient and Medically Frail Client* (2014). She'll be our leader—captain, one might say—on this voyage, which, at the moment, feels to me very uncharted. Gayle has found that this Saturday evening gathering is a good way for a class to meet and become acquainted. Her associates Liz Davidson and Kelly Burke also join us. As we gather, we swap basic information: we are from Hawaii, Florida, Kansas, Utah, California, the Greater Portland area, and North Carolina. We students are seven women and one man—a common proportion in massage gatherings. Trained and licensed therapists, we are eager to increase our knowledge and skills. We wonder about the next day, when we will gather at the Oregon Health & Science University, or "O H S U," as the locals call it, on Marquam Hill, locally known as "Pill Hill."

Sunday is a slow day at OHSU, a good time for us to try to get our bearings. Because it's a bright sunny day, we can see across the Willamette valley to Mt. Hood, the spectacular snow-covered mountain some 45 miles to the east. It's an active volcano just over 11,000 feet high and covered with snow, ice, and glacier. The OHSU campus, on a much smaller hill, is a labyrinth with buildings sprawled everywhere, even in three dimensions: I enter from the street to a lobby—presumably a ground floor—but learn that it is, in fact, a ninth floor of a building that descends into a deep ravine.

Our group gathers in the lobby and traverses a long hallway that is actually a bridge to the Doernbecher Children's Hospital. On our left is another bridge, a longer blue affair with jaunty suspension rods; it goes all the way across the ravine to the Veteran's Hospital. I hope that bridges, necessary on this crazy campus, may be a good metaphor for this week: we've departed from the solid base of home and suspended ourselves in time and space, all the while yearning for some kind of solid ground on the other side.

We make our way circuitously down halls and up an elevator to an elegant conference room. Eight large tables form an enormous rectangular table, with one end bathed in blue light from a projector. Inveterate problem-solvers, we therapists look for a remote to extinguish the light but find none. Ignoring signs DO NOT MOVE THESE TABLES, we pull off four tables from the other end to make our own, non-blue space. Gayle spreads a multicolored cloth in the center, an altar she has mentioned in her emails. She has asked each of us to bring a symbol representing why we have come to this workshop.

As we go around the table describing our symbols, at least half of us mention people we know who have cancer or who have died from the disease. One woman's husband died within the past year. In my case, my father died of brain cancer many years ago and I myself am four years out from chemotherapy for my leukemia. Several of the women cry as they talk about "this work," their phrase for massage for sick people. The expression seems a bit odd to me because massage is not a 9-to-5 job. Nor do we work to gain wealth, fame, or power. Nor, when I provide it, does it seem like work to me, but an honor, a sacred ritual. I'm moved by the tears and the passing of tissues, but I don't cry. I'm not immune to crying, but not here, today.

About half the symbols are hearts, emblematic of the caring, love, comfort that touch can bring. I have brought two small pieces of sea glass from the North Carolina coast, one blue, one clear; these are small cylinders frosted by tumbling in the surf and sand. For me they symbolize beauty in persons who have been damaged by injury or disease but live on with their own unique riches.

After a break, Gayle hands out intake forms so that we can ask patients and their nurses the right questions; the answers will guide how and where we touch. She also shows us various medical paraphernalia that patients may have, lines and ports that we must be careful not to touch. Then we review three basic "alerts," warnings about risks to a patient that correspond to the three main cell lines: clot formation (red cells clumping up internally), infection (low white cells), and bruising (low platelets). Patients undergoing bone marrow treatment are especially at risk for the last two.

After lunch in the cafeteria, it's time for hands-on. Back in our room, I survey the blue light still shining obstinately at one end. Somehow it symbolizes for me the worst of bureaucracy and medical care, one-size-fits all theories that view patients as test tubes to receive drugs, more drugs, and still more drugs for all the previous side effects. I step onto a chair then onto the tables; my feet make strange, hollow noises as I stride over to the projector. I find a switch and shut it off. The whole room feels better.

We set up three massage tables and another commandeered table to simulate hospital beds. We pair off, one as therapist the other as patient, who has, Gayle says, low platelets, and edema in one leg. We are to give comfort-oriented, light-pressure massage. My partner is Iris, the nurse from Hawaii. The four women "patients" leave the room to put on their dinky hospital gowns. I stay by my table, take off my shirt, and put on a yellow paper gown and bright purple gloves. I've never given massage wearing gloves and think it must feel strange, but Iris says the warmth of my hands comes right through to her back. In general, we learn that hospital massage is light, slow, and given only to safe places of a patient's body. These are ideas we may need to explain to patients and hospital personnel, especially if they have received only vigorous massage or seen karate-chop versions in movies.

We review a process for working with a patient who is immunosuppressed, including washing our hands or otherwise cleansing them; some units have sterilizing lotion in a bottle mounted on the wall outside patients' rooms. We are to knock, identify ourselves, and offer a massage. If the patient is willing, we find that patient's nurse for information and permission, then decide what approach makes sense. We return to the room, putting on gown and, if needed, gloves or mask. We talk with the patient and create a plan. We survey the space and any medical equipment, especially IV lines. We adjust the bed as needed to a good working height. (Massage tables are lower, allowing use of upper body weight; here, leverage is not important.) We position the patient with pillows, towels, even the overbed table. Because we don't wear our germy watches, we need to locate the clock. We breathe deeply, center, and start "the work." How can I remember all this? By contrast, massages in a massage room are easy, so much being the same every time. Besides, even though I have a dozen years' hospital experience as an ER pastoral care volunteer, I'll probably be nervous.

Monday and it's raining. I don't mind because we'll be inside all day, but I do miss being able to glimpse Mt. Hood with its dramatic range of temperatures, glacier over hot lava. Somehow it's become my beacon of surety: I trust that it's there amidst the clouds. Our class meets in the lobby, now very busy with weekday traffic. Medical people in a variety of costumes—scrubs, lab coats, jackets and ties—flow purposefully by us. Family members, some with flowers, drift along. Patients move at various speeds, some on crutches, in wheelchairs, or in large beds pushed by transport staff. Ethnicities abound, both in caregivers and in patients: Hispanic, Asian, Anglo, African-American, Native-American. Injury and disease may come to any of us... and will come to all of us.

We meet with the volunteer coordinator Ivy Nelson for paperwork. She's a small, energetic woman who immediately makes us feel at home. A good listener, she asks each of us where we're from and what we're hoping for during the week. This is all very welcoming, in contrast to the considerable paperwork in the previous weeks, so much that we students have complained to each other via email: we have dealt with our measles/mumps/rubella status, two TB tests, a fire-safety quiz, two online trainings in confidentiality and respect, self-reports of any misdemeanors or felonies (should I report a traffic ticket?), a background check, and so on. Evidently the OHSU folks are very careful about who might be working with their patients, especially any that are immunosuppressed. Ivy checks off these requirements and gives us more paperwork for badges at the Security office, which is some hundred yards away through, of course, the rain. She gives us gray shirts that read VOLUNTEER SERVICE, and we walk or jog through the drizzle for our photos and badges. As we wait in line, Kim unexpectedly observes, "You know, cows are out-standing in their fields." We chuckle, then guffaw. A little loony humor often helps dealing with bureaucracy.

Officially garbed and labeled, we spend time after lunch practicing intake forms. We role-play in pairs: a therapist and a patient who has various limitations, for which massage techniques must be adjusted. We also practice charting, the brief, written reports of our visits to patients. This is an exacting craft because our entries become part of the legal medical record.

Then we go on a tour of the adult oncology units, including the Bone Marrow Transplant unit or BMT. Sounds like the subway, I think, but the glass doors and walls look like a very static ICU. As with burn units, such places are often in lesser-used parts of a building to keep germs to a minimum. Patients here have had their bone marrow destroyed by radiation and/or chemo so that they can receive transplants of healthy marrow harvested from themselves earlier, from a related donor, or a "MUD," a matched, unrelated donor. Somehow the transplanted marrow cells know how to migrate into the bones of the recipient and set up shop making new red cells, white cells, and platelets. It's truly magical. But until the graft gets going, such patients have no defenses against infection. If we therapists bring them germs, such patients could become very sick or even die. This news gets our attention and motivates us to use the transmission-based precautions we've been taught.

That afternoon I have my first patient. She's an adult oncology patient in the other, less strict unit. After speaking with her and her nurse, I massage her feet. She's in her bed, and I do a poor job arranging the space so that my back is bent and starts to ache. There's so much to remember, it's easy to forget something.

We return to our classroom for a short debriefing. We're all pleased that we performed a massage and made someone happy. Gayle reminds us to wash our shirts tonight; the hospital is aware that the bacterium *Clostridium difficile* is currently a threat. This is a superbug that is hard to kill and can make patients very sick. I see also VRE warnings posted on the units: *vancomycin-resistant enterococci* are other very bad bugs.

Tuesday. It's still raining, and I look in vain for Mt. Hood. Today we to start our rotations, a commonly used technique in medical schools. Med students typically have two years of scientific study called the "pre-clinical years" before they see patients. Because an entire class would flood any particular part of a hospital, groups of students see different services at different times. Similarly, our class breaks up to go different ways. I'm assigned to the BMT isolation rooms, and Liz has already found a patient I can work with. I wash my hands and consult the patient board, which tells me that he's neutropenic: low white cells; I must be careful. I slide back the glass door to his small room and ask his permission, then go find his nurse to learn what precautions I must take. I also check his chart for age and diagnosis.

I return and carefully glove. Entering the room I see the blue Skybridge that parallels the white bridge we took to the Children's Hospital.

"Nice view," I say.

"Yes," he says, "and before that was built, patients from the VA had to come over here by ambulance." Evidently he's been here before.

"Wow," I say. How strange: an ambulance ride between "next-door" hospitals, necessary because they are separated by the deep ravine.

"And it's 660 feet long," he informs me. I ask him what he'd like and he says a back rub. He lies down with all his clothes on. OK—we need to be flexible in this work. I raise his bed and rub his back. He likes to talk, so we discuss his past work. He's a scientist, used to measuring things. Great, I think. Because I'm a bit at sea with all the new experiences and information, I like looking out the window at the dramatic bridge and knowing its exact length.

After lunch, I meet up with the others. "How did it go?" we ask each other. "Mine was good, but later I realized I forgot to lower the bed," I report. "Yeah, everyone does that," a colleague soothes. It's an easy mistake because my own table at home doesn't go up and down electronically, but here this error can be dangerous for patients getting out of bed later. Although the control is within their reach, they might not notice the different height.

That afternoon Kim and I go to the Bone Marrow Outpatient Unit. Our teacher goes over the intake form we'll use there and turns us loose. Now we have to offer massage and coax people to accept our service—chores previously done by our teachers. I'll call one recipient Maria, currently receiving IV medication; she speaks Spanish with, I assume, her mother but accented English with me. Since she's wearing red tights, I don't glove or use lotion as I sit on a stool in front of her. I give her a good foot massage.

Kim and I go next to the ICU waiting room and join some of our classmates. Gayle recruits some massage candidates from the people scattered through this large but pleasant room. They all have friends and relatives who are very sick, some possibly dying. We stand behind chairs and rub shoulders and necks. I massage a woman named Milagro, which, I recall, means miracle. Perhaps she's praying for a miracle for the patient beyond the wooden door.

That night I wash my shirt and the two pairs of pants I've worn in patient rooms. I also massage the husband and wife I'm staying with. I've borrowed a massage table and am happy to do the hour-long routine I'm familiar with. My experience at OHSU, however, urges me to try more holds and placements—energy work, not just the Swedish strokes I learned in massage school.

Wednesday. It's raining again. We have a short class, a check-in. Gayle reads us an inspiring passage from Rachel Naomi Remen's *My Grandfather's Blessings* (2001), and then we walk to Radiation Oncology. We enter the simulation room where a large CT scanner allows the patient to lie down for imaging. I recall my CT scan five years before. Our host Julie Rettinghouse, the chief radiation therapist, explains the ways masks, tattoos, and other strategies allow for exact placement of high energy X-rays into cancerous tumors. In the next room we see the actual linear accelerator, large and uncompromising. Julie shows us custom-made chunks of lead that shield healthy tissue; these are labeled with patients' names. She demonstrates the awkward lift she and her colleagues must do to place the heavy shield in a tray high above the patient. (Later, one of us will massage her hardworking shoulders.) Looking at all this equipment, the masks and shields, and the dosimeter badges the workers must wear, I have a new insight into the serious business of cancer treatment: X-rays, chemical poisons, surgery—these are the main weapons in the Western arsenal. How can massage—in all its gentleness—complement such treatments in order to comfort patients and contribute to their healing?

My rotation today is the chemotherapy infusion room for outpatients. I feel right at home here, because of my experience as a patient. Our teacher today is Kelly, who shows us how to lay lotion bottle, gloves, and a pill cup in a towel and roll it up. When we sit down on a stool in front of a patient, we set this roll to the side while we chat and do intake. Having dropped some of these items already, rendering them *dirty* by hospital standards and hence unusable, I'm glad to learn this trick of the trade. My first patient is a shy woman who agrees to my offer of massage but without enthusiasm. I forge ahead even while she keeps her small feet close together. I do the best I can but her reticence rubs off on me and use my thumbs mostly, not my whole hands. After I've finished, Gayle (who has been observing) demonstrates this lack of contact on my hands. Always encouraging, she mentions that my patient did warm up to the touch and seemed to be enjoying it by the end. My second patient is a fully cooperative and outgoing man. He and I share many interests, and I lose track of time as we chat and I rub his feet with full-hand contact.

At lunchtime, I recall a café for hospital folks on the third floor. I take the elevator down and have a healthy salad. Wishing to stretch my legs, I return to the ninth floor and find the entrance to the Skybridge. It has an attractive interior, with a linoleum floor, walls of glass, a lowered ceiling with lights, vents, and air conditioning. The whole shebang is, in fact, one very long room with wonderful views over a misty Portland. Is it really 660 feet long? That would be an eighth of a mile exactly. I stretch my normal stride to roughly a yard and walk across and back. My first patient, the scientist, was exactly right. There are other folks purposefully striding with me or toward me, but I (visiting Southerner) am the only one peering out into the rain. When I reach the far end, the entrance to the VA hospital, I turn around and see some of the same faces. Other folks use this bridge for a dry walk on their lunch hour.

I've mentioned to my teachers that I have yet to do a real back rub, skin on skin, for a patient in a bed. Ever resourceful, they find a young woman for me. I do all the formalities and give her a back massage. We're doing great, her energy and mine, and her breathing slows. Suddenly the phone rings: it's her husband. She answers in a very relaxed voice. Having called on cell phone, he arrives almost immediately. We shake hands and I take my leave. Again, *flexible* is the watchword. I'm happy and proud until I realize that I forgot, once again, to lower the bed. Perhaps her husband will cover my tracks. How am I going to learn to do this right?

In the hall Gayle gives me another lesson on strokes; she's remembering looking in on the man who kept his shirt on the day before.

"Even under those conditions you can get more going," she says rubbing my forearm with both hands. It feels good.

"You don't mind the feedback?" she asks.

"Hey, I welcome it. That's why I'm here," I say.

With another 45 minutes until debriefing time, I head over to the ICU waiting room and freelance. Three women accept massage. Upon seeing my gray volunteer's shirt, the first throws down her magazine and moves her back away from her chair: she's already been massaged today by one of my colleagues and is ready for more. I stand behind the chairs and rub shoulders and necks. Another woman urges her husband to have a back rub, but he declines and gives increased, noisy attention to his newspaper. The third woman has both her shoulders jutting forward. Ah, tight pectoralis muscles, major and minor, I assess, eager to make some repairs, but that's not my work here today. Much of my training, including a sports massage workshop, has been about manipulating muscles to loosen and lengthen them. This whole realm of comfort massage and the implied energetics is new to me. I think of Liz's words: *we can do a lot for a patient just by putting a still, caring hand on them.*

At the end of the day I am tired. After supper I wash my shirt and go to bed early.

Thursday, and the sun is out! Gayle reads another selection from Remen, and we scatter to our assignments. From comments from my classmates, even how they look, I see that we are a group with increasing confidence about this work, this hospital, and these patients.

Gayle takes Kim and me to a new area, the apheresis service in the Mark O. Hatfield Clinical Research Center. We enter a large room where a man covered by a blanket sits in a chair; two lines connect him to the apheresis machine. That's the basic idea: blood is taken out of this patient, treated one way or another, then returned to him. A cheerful nurse greets us and we get to work. Kim, wearing purple gloves, rubs the man's feet. I massage the woman who gave him a ride to OHSU today. She lies down on a bed, and I work on her through her clothes. While I massage, I look at another apheresis machine nearby. A permanent chart shows a complicated series of arrows for blood flow through the machine, including, as needed, a centrifuge or a UVA irradiation light, which can damage the DNA of white cells so they won't replicate; nonetheless, the cells are still capable of killing germs and are returned to the patient's body. Amazing, I think. But if high-tech medical technology is miraculous, so is massage, which is about as low-tech as you can get.

When we finish, we retrace our steps. While waiting for the elevator I look out a window and see the two bridges. The white one to the Children's hospital is ingeniously fastened to the main OHSU hospital by two triangular buttresses, one above, one below. What ending will I find to the bridge of this week of training?

I eat a salad and walk the blue Skybridge twice, giving me a half a mile. The sun is out, and I can find, through the tops of some Douglas Firs, Mt. Hood resplendent in noonday light.

I'm confident about the day and all I've done. Part of me wants to take it easy by going to the familiar ICU waiting room, tapering off, so to speak, but my teachers wisely steer me to more challenging patients in the Bone Marrow Transplant Unit. The first is from Egypt, and I've been warned that he's "crotchety." He's an elderly man who has never had a massage. Soon I'm rubbing his back while he lies on his side. With a great sigh, he flops on to his stomach, better to receive my strokes. A young woman knocks and enters; she asks to take his vital signs. He roars, "My God! Can't I have ten minutes of peace?" She retreats hastily, and under my hands he relaxes once again. And then goes to sleep.

When I meet my second patient, he talks a mile a minute. His nurse says, "Howard's here to give you a massage. You should take this time to relax and take it all in." Luckily, he does so. He goes to sleep too. When I mention this to Liz, she tells me that the nurse is also a massage therapist, one of our tribe.

One of my fellow students has observe, "I tell my patients that I'll take it as a compliment if they go to sleep or at least drool."

At our last class session, Gayle suggests we go out into the hospital for 30 minutes to reflect on the week and then come back and report. I go the surgery waiting room, which has a good view of Mt. Hood, and make my notes. I have a great victory to report: today I lowered both of the beds I raised. I also celebrate that I'm taking home new knowledge (about oncology, medicine in general, modifying massage strokes, positioning of patients, infection control, hospital sociology), a new sense of inquiry (what questions to ask about a patient, how to keep ears open about medicine, healing, health in general), confidence in doing this work, inspiration to do the work, and a new sense of what love is, love for the sick and love for and among care-givers. My colleagues have been kind to each other all week. *How's it going? What did you see this morning? Isn't this work powerful?*

And then there's a sense of wisdom from our friends the cows out standing in their fields. Mama cows nuzzle their calves. Some cattle sleep leaning on each other. Yes, simple bovines, doing what bovines do. Long before there were hospitals, doctors, even shamans and ancient healers, there were mammals, including humans, who comforted and healed each other by touch, by stroke, by gentle pressure…and there still are. Today's humans still touch babies, pets, and children this way, but as a culture we have largely forgotten that people, especially in medical settings, can benefit greatly from caring touch. Nurses were trained, as late as the 1950s to give back massages, but this too is typically a lost art.

Now, to redress these losses, there is a new movement in hospital massage. I want to be part of it. Family members can give loving touch, but not professional massage. During my week's workshop I have seen the immediate benefits to patients and family members. It seems to me that massage should routinely be part of complementary and integrative medicine. I want to provide massage to patients in hospitals and cancer patients in support centers. That's the other side of the bridge I'm crossing.

A CALLING?

Sometimes people ask me how I got involved with massage, especially if they know that I was a college professor for many years. Some say, "Was this a second career?" Perhaps I should answer, *no*, because it's not a career in the usual American sense that I'll work full-time, hoping to advance ever higher in rank and salary. Perhaps I should say yes, because it's a path that's important to me, and one I've invested time and money in, one that is now a focus for my life.

In briefest terms, I like massage because it provides direct, sensory pleasure to the recipient and it makes both of us happy. Is that a calling? It's not a celestial voice in a dramatic moment asking me or even requiring me to serve. If calling it is, it's something more rooted in the tissues and genes of humans, mammals, and our related precursors who have known forever that comforting touch is a good thing.

What is clearer to me is an on-going stream of rewards, especially spoken thanks. When a person I've massaged thanks me earnestly, I know I'm on the right track. That is calling enough.

PROVIDING CANCER MASSAGE

While receiving chemo in the infusion room, I studied—when not limited by chemo brain—pages of typical exam questions for the national boards. It seemed odd that massage mimicked the licensing traditions of physicians and nurses, but the medical model is powerful and legislators have chosen to protect society through strict regulation, especially given the history of "massage parlors."

One hot day, I drove to a low-rent office, where I typed multiple-choice answers into a computer, while on either side of me an accountant and a real-estate assessor took exams in their fields. I remembered taking my emergency medical technician exam many years ago under strict security. Fifty or so of us were scattered through a large room. Only one at a time could use the restroom.

The massage questions were on familiar topics, except for a handful about meridians, about which I knew nothing. In another ten years, I'd know about these.

I passed my exam but didn't practice right away, given that I was still teaching. After a year of getting settled in North Carolina, I applied for the state license and learned about the Cornucopia House Cancer Support Center. I had an interview, provided a massage for my prospective boss, and was "hired," perhaps not the right word, because I initially worked as a volunteer. I liked the name, Latin for "horn of plenty," since we offered many supports to patients free of charge. One day a participant (as we called them) said, "You know, I visit a lot of clinics and hospitals, and I always feel some kind of fear. The nice thing about Cornucopia is that there's no fear because there can't be any pain or bad news."

Ten years later I still provided massages on a Wednesday afternoon. I never knew whom I'd get, a man or a woman, a young person or an old one, someone I've treated before or someone new. I enjoyed them all. The new ones provided a challenge in starting a relationship. The repeat clients allowed me to ask about particular concerns. *So, how's that knee doing now? Are you still receiving chemo? How did that MRI come out for you?* These may sound like intrusive questions, but since I've said that I too am a cancer survivor, we had that in common and patients appreciated interest and concern from a fellow survivor.

And, yes, there were patients who slowly went down…and died. One man I massaged regularly always reported his weight, a symbol for him that he was either getting better or at least not getting worse; eventually he died and all the Cornucopia staff mourned the loss.

NURSES USED TO PROVIDE HOSPITAL MASSAGE. HOW COULD WE BE SO FOOLISH AS TO LOSE THAT?

The book *Putting Patients First: Designing and Practicing Patient-Centered Care* (Frampton, et al., 2003) provides a brief history of hospital massage. American doctors formerly used massage as a medical tool, but—as medicine became more complex in the early 1900s—they delegated it to nurses or assistants. By the mid 1950s, it was disappearing from nurses' duties, replaced by further demands for charting as well as more patients per nurse: "Oddly, it evolved from accepted medical practice into an 'alternative therapy'" (p. 106).

Massage for sick patients has been making a comeback in the last 30 years, inside and outside of the hospital. There is a National Association of Nurse Massage Therapists, founded in 1992, for nurses providing massage within healthcare institutions and in private practice—often with more flexibility. More specifically, the Society for Oncology Massage, founded in 2007, lists some 325 massage American therapists on its website.

The Hospital-Based Massage Network, founded in 1995, lists over 100 programs on its website. According to Laura Koch, LMT, the founder of the HBMN, some 400 hospitals now use massage on way or another. She wrote me in an email: "According to the *2010 Complementary and Alternative Medicine Survey of Hospitals,* created by the American Hospital Association Health Forum and the Samueli Institute, massage therapy was offered at 64% of outpatient CAM programs and 44% of inpatient programs. Approximately half of all hospitals reporting CAM programs offer massage therapy. This indicates possibly 451 hospital massage programs nation-wide." With some 5700 hospitals total, however, that means only 8 per cent or so of U.S. hospitals offer massage. Further, some of these programs are probably minimal, by number of available massage therapists or access by patients. Also, none can be called permanent; regrettably, some are discontinued because of changes in leadership, economics, administration, or availability of therapists.

In contrast, Memorial Sloan Kettering Cancer Center, New York City, has an Integrative Medicine Service that offers some 20 therapies, including Mindfulness and Nutrition, but also many touch therapies, including Reiki, Reflexology, Medical Qigong, Manual Lymphatic Drainage, and different kinds of massage. I believe that other cancer centers and hospitals—indeed all—should provide similar resources.

For a multi-service hospital, let's consider the Mid-Columbia Medical Center, The Dalles, Oregon. As a Planetree hospital, MCMC espouses a model for improving patient care, including the provision of massage. Other elements for holistic, patient-centered care include patient education, spirituality, nutrition, and the arts, even architecture and design. Chapter 6 of *Putting Patients First* describes the massage program at MCMC. Every patient scheduled for same-day surgery is offered massage. Acute-care units offer massage through referrals by nurses, doctors, and patient self-referral.

Massage for cancer care is available for patients and family members. There is also infant massage and employee massage. When I visited the hospital in 2009, it seemed more like a friendly hotel than a hospital. Symbolic signs were ceramic numbers on patients' room and open nurses' stations. Further, doctors wore street clothes, not white coats, suggesting that they were *regular people*. When I stood in the middle of a circular labyrinth outside, I asked my host why the center was raised. She answered that the entire structure also served as the helicopter pad; the labyrinth had an embedded heater for melting ice and snow that needed to drain away. For these and other features, I thought: *how smart and caring!*

Planetree is a not-for-profit organization that works to improve patient care worldwide. As of this writing, it has 318 affiliates in the U.S. and 465 internationally. These include hospitals, medical centers, continuing care facilities, and others such as outpatient centers.

Sometimes the oncology service is the first area of a hospital to have massage, and other areas follow as the benefits become known. Patient satisfaction (and survival) go up, and doctors and nurses become aware of the values of the value of massage to patients, family, and medical staff.

I believe hospitalized patients should routinely be offered massage. They will be happier, heal faster, and leave the hospital sooner in better shape or better prepared for death. Family members will be happier. Also staff, as they enjoy the caring touch of massage. We have a long way to go for routinely providing massage in American hospitals. While 8 per cent is up from earlier amounts, it is still very low.

RANT 3: MASSAGE SHOULD BE ROUTINELY AVAILABLE IN HOSPITALS!

Massage relaxes people and makes them feel better about themselves and happier in general.

Massage lessens anxiety and boredom. Happy patients can put more of their resources into healing, so they heal faster.

Massage brings release of "happy chemicals."

Massage can get patients' bowels after surgery to move again, thus helping a quicker discharge from the hospital.

Massage comforts people; they feel less alone.

Massage brings people into the world of the well, even as they are sick.

Massage brings joy to patients' bodies and minds.

Massage is good for medical personnel; nurses love it.

Massage therapists are "safe," no bad news, no painful treatments.

Massage therapists symbolize the world of the well that cares about patients and includes them.

Massage therapists have time to listen to whatever patients want or need to say.

Massage is an excellent integrative therapy. Every hospital should provide it.

Mammals have done massage forever…they know it works, even if humans have forgotten.

Hospitals must provide food to keep patients fed. Massage is not as necessary, but it is also a form of nourishment. Some hospitals have coffee carts for outpatients as they are treated—not "necessary," but supportive.

Another comparison: visits from a hospital chaplain, or clergy from a patient's church, synagogue, or mosque. Although not "medically indicated," they usually provide a big boost to a patient's spirit.

If a *hospital* is, in fact, *hospitable*, massage is a most appropriate *hospitality* to offer.

Modern hospital design now approximates welcoming hotels, many of which hospitably offer massage.

Leading hospitals in the U.S. and abroad provide massage.

Why are we lagging?

Providing massage is a no-brainer! Let's provide it!

TOUCHING BRUISE: STORIES BODIES TELL

I shape the mood of my treatment room with soft lighting, music, and a comfortable temperature. I read over the charts for my next clients. Are they cancer patients or family members? Under treatment currently? Are there notes by massage therapists, our polarity therapist and/or a Reiki therapist; what can I learn from them?

If the client is new to me, I'll say, "Hi, my name's Howard. I'm a cancer survivor."

"Oh really, what kind did you have?"

"A leukemia. Six months of chemotherapy knocked it into remission."

"No kidding; how long ago?"

"That was in 2001."

"Wow, that's great. I hope I get something like that."

"I hope so too."

I ask them how they are feeling, any pain today, anything sore or tight? What about a concern mentioned in the recent chart notes?

Soon we have agreed on a treatment plan. Is starting face-down OK? If not, we start face up.

I leave the room so that they can undress. They take off their clothes—sometimes keeping underpants on, sometimes not—and place themselves between my sheets. I knock on the door. Ready? Occasionally clients don't want to disrobe. No problem; I massage them through their clothes.

People sometimes like to talk while being massaged. I've read articles in massage magazines about this: some therapists don't like it; others believe it's a client's right. I'm OK with talking. Indeed, as a newcomer to North Carolina, I learned from my clients where to get a good country breakfast, where to buy local produce, how to cook cressy greens, and where to buy heritage apples in the mountains. I learned that the Triad and the Triangle are two different arrangements of cities, but that the town of Apex has nothing to do geometrically with either one. Such conversation makes us social equals, not a well person and a sick person.

Sometimes my clients talk about their illness, their job, their family—whatever's important to them. As their therapist, I'm a safe person in a safe place. After a while, however, they relax and become quiet.

One day I pulled the charts on the two clients signed up for me. This was years ago, and I don't remember the first one at all. As for the second, I don't remember her name, so I'll call her Karin, and I've changed details to protect her privacy. When I greeted her and shook her hand, she seemed very tired. Her chart said she was caregiver for her husband, who had cancer. Such caregivers are almost always worn down and stressed. I remember the toll on my family when my father was dying of brain cancer.

I took Karin back to my treatment room and quickly went over her chart, which didn't tell me much. For a cancer patient, the chart tells the kind of cancer, the treatment, whether there's a port for chemotherapy, the number of radiation treatments, and the like. For Karin, however, there was nothing else to guide me other than her role with her sick husband.

"What can I do for you today?"

"Oh, just general relaxation."

"Any injuries? Anywhere particularly sore or tight? Anything I should look out for?"

"No, no, no, no."

I asked her to let me know if anything doesn't feel right or outright hurts; she agreed.

I left the room and gave her time to get ready. Out in front, the receptionist said to me, "Say, did you know her husband died about two weeks back?"

"Good Lord, no," I said, immediately feeling a burden. Whatever massage would be best for a grieving, young widow, a woman who had just lost her husband? Besides getting my strokes technically right, how might I care for this person who had just endured such a tragedy? Was I up to this?

I knocked.

"Come in."

Karin was lying under the sheet, face down on my table. I asked if she was warm enough. She said, "Yes" so softly I could barely hear her. "Would you like a bolster under you ankles?" She agreed.

I stood beside her and placed my hand on the middle of her back just to feel her breathing through the sheet and to let her get used to my touch. My breathing coordinated with hers; in the language of energy healing, we were *entrained*. I flowed my left hand—the more sensitive one—about ten inches above her prone body, picking up dark spots, blockages, or places that don't feel right, places I'd be sure to massage. Her body felt inert, her breathing shallow. She seemed restricted, even softly mummified. I never had a client like this. Would I know what to do? What would be the best approach? Furthermore, she didn't know that I knew about her husband. Neither of us mentioned him. Was he a ghost in the room?

I went to the head of the table, where her head was face down in the face cradle. I *thought this is my friend Karin; I care about her*. I do this with each client, because it's important to have the right intention, the right energy for massage. I rubbed her scalp, running my hands through her hair. She had wonderful blonde hair; her husband must have loved it. I send healing energy into her head through my hands.

I pushed the sheet down from her shoulders to her waist. As often, I found the taper of this woman's back touching…lovely. She had perfect skin, finely textured and almost white. Using my massage crème, I massaged her back gently, slowly sending energy into her body. Her muscles were quite loose, in contrast to many tight shoulders and backs.

"Pressure OK?"

"Yes," she said.

I covered her back and moved to the foot of the table. I pushed the sheet away from her right leg. It was a beautiful leg, with the same white skin. I massaged her leg, then her other leg. I affirmed again. *This is my friend, Karin; I care about her.* Her husband must have loved her legs. I loved her legs too, but in a different way.

Sometimes I do the glutes, the large muscles in the buttocks. For Karin, however, I skipped this area, and did some sweeps up and down the body through the sheet; I like a person to feel whole, reconnected from head to toe.

"You doing OK?"

"Yes." Her voice was lower, more resonant.

"Good. OK, Karin, let's move you down the table and turn you over." With care to the draping to keep her covered, we repositioned her face up. I adjusted the bolster under her knees, then bared one leg and foot and did the strokes. Then the other leg. I covered the second leg.

Moving up the table, I gently took her left hand—with its wedding ring—from beneath the sheet. When I bared her upper arm, I saw a large bruise on the bicep, the deltoid, and on around to the triceps. It was dark red in the center, purple further out, then just fading into brown in the edges—perhaps a week old, and representing one hell of a knock.

"You've got a bruise on her upper arm," I said neutrally; "is it too sore for me to touch?" Bruises are often from sports, but sometimes from domestic violence. (Massage therapists routinely look out for disease, for example skin cancers.)

"You may touch it…might even do some good."

"OK."

She added, "I fell down."

"What? Oh really?"

"At my husband's memorial service." *There, it's out. I picture a woman collapsing in grief.*

"I'm so sorry," I said.

"Thanks." She added, "I hadn't worn high heels for a while."

Like a young teen wearing heels awkwardly…except Karin was mature, and she wore them to honor her husband on the occasion of his death. High heels—a small gesture with much meaning.

"Yes. I see. My sympathy."

"Thank you."

I worked her hand and, carefully, this bruised arm. Was I touching it, or was it touching me? Karin and I were both quiet, having made some kind of peace with the dead. Her husband was no longer a random ghost in the room. Perhaps he was now present as a blessing spirit.

As I worked her hands and arms, I thought of the many things she must have done for her husband, well or sick. I remembered small kindnesses I did for my dying father. I remembered the many gifts my wife gave me when I was sick. I massaged Karin's neck, shoulders, upper back and head, sending love and honor through my hands.

Holding her head gently in my hands, I entrained our breaths again. Her chest rose and fell, now in slow, deep breaths. I slowly moved my hands away from her head.

"OK, Karin. Take your time getting up. I'll be outside."

I left the room and made my note on her chart. I was happy that, after all, I had known what to do. I drank a cup of water and prepared one for her too. When she came out, I handed it to her.

She took the cup and drank. Then she said earnestly, "Thank you," all the thanks I could ever want. "You're most welcome," I said, but I probably should have thanked her as well. Although I'm licensed as a therapist, someone who provides healing for other people, often it's me, as well, who feels that he's been healed.

NORTH CAROLINA CANCER HOSPITAL

I learned of an opening for a part-time massage therapist at the NC Cancer Hospital, one of the UNC Hospitals. I applied and was hired. I went through the multiple steps of becoming an employee, including a background check for criminal activity, a drug screening, and a careful session with my boss while we counted the number of letters and spaces that my new ID badge could accommodate. She was wanted me to display Ph.D. after my name (as well as my massage designation of LMBT) even though my field of comparative literature had little direct bearing on massage. The indirect bearing, however, is the literary concept of comedy in the largest sense, the good humor of people being happy together and seeing their commonality past divisions such as rich vs. poor and sick vs. well.

There was a two-day orientation of all new employees, maybe over a hundred, most of them nurses. The strongest memory I have is the speaker in charge of housekeeping. He said, "If you see a scrap of paper, pick it up, just like you'd do at home. Our hospital is a home, and we want people to be comfortable here. Patients are our guests. Also, of course, we want germs kept to a minimum so people don't get any sicker."

UNC Cancer Care has two thrusts: *research*, mostly at the UNC Lineberger Comprehensive Cancer Center and *patient care*, mostly at the NC Cancer Hospital. When I started, it was in a 1940s brick building, formerly a TB treatment area. In 2009 the new Cancer Hospital went up, a modern, airy, attractive building, with the feel of a luxury hotel. I found it a pleasant place to work and—when I came as a patient for my yearly cancer checkups—welcoming and easy to use.

As part of patient care, the Comprehensive Cancer Support Program helps patients, caregivers, and families with cancer treatment, recovery, and survivorship. Among other services there are counseling and mental health service, supportive care, a chaplain, nutrition support, and Patient and Family Resource Center (PFRC) that offers patient-centered support from wigs and head coverings to a large array of publications. There's also a relaxation room for massage, one of the places I provided massage. In front of the hospital is a large labyrinth. Now and then, I see someone making a slow, deliberate walk there.

After the formalities of hiring, including the hospital ID badge, I ordered a second badge from a stationery store. It read CANCER SURVIVOR in white letters on a green background. The green is my reference to the Shakespearean green world of comedy, Hildegard's healing *viriditas*, and the primal powers of fertile nature. It cost me $20 and worth every penny, because cancer patients see it on my shirt and routinely ask, "Oh, what kind did you have?" I tell them, and immediately they have an image of a well person who has passed through the same vale that they are currently in. I am now an analogue to the woman described earlier who lost her hair three times, the volunteer who provided cookies and drinks to chemo patients where I was treated.

Most of my clients want to talk, especially when they see my badge. If there's a spouse accompanying the patient, he or she is often glad for the diversion I provide. I've had conversations about gardening in North Carolina, black bears on the coastal plain, martial arts, academic research, and barbecue. As they talk, they relax. One man I claimed I had cured his Achilles tendon on my previous visit. One woman, reluctant to take a pill offered by a nurse, did so with no trouble while I massaged her hand. The nurse thought relaxation was a factor—or maybe it was the distraction I provide. Earlier, we saw "distract" as a term for moving flesh. In a medical setting, a conversational distraction can pull a patient back to the world of the well.

CANCER MASSAGE: A DOOR TO THE WORLD OF THE WELL

It's Thursday, my day to provide massage at the NC Cancer Hospital. Come along with me; there are so many interesting people to meet! Some will come for their very first day of treatment. They are usually frightened and nervous; I especially enjoy these, remembering my own experience. For all of patients, I help them connect to the world of the well, even as they travel in the world of the sick. As I talk with them, they often recall their ordinary worlds. One elderly man had farmed tobacco down on the coastal plain. He said, "You know, you could do a hell of a lot more in one day when them tractors came along." As I rub his hands with lotion, I can feel that they are still strong, although the calluses from working with mules are long gone.

Who gets cancer? Sad to say, it's almost anybody. Poor people and wealthy people. People with little or no English. People who have come to the Triangle area because of the Research Triangle Park or one of our universities—myself among them. People from overseas: Europe, China, Korea, Africa, Brazil. One cancer book has the title *One in Three*—the chances a person will have cancer over a long lifetime (Wishart, 2007). What they all have in common is cancer in one form or another…and the strong desire to be well.

Most patients, however, are from North Carolina. A woman tells me, "When I was a kid, I had to get up in the barn rafters to hang the tobacco leaves. It was so hot up there, sometimes I was afraid that I'd faint and fall down." I've met farmers, businesspeople, soldiers (both men and women), people in government, doctors, truck drivers, retired people, professors, artists, lawyers, fishermen, musicians, teachers, computer folks, and on and on. I've met people training for a new adventure, hoping their disease won't derail them. One man quit the business world to become a train engineer. He liked a particular route that went by a horse farm. He'd blow his whistle, and the horses would come to the fence and run parallel to his engine.

I've met famous people, names you'd recognize, but I can't name them here or anywhere because of ethics, even law—not to mention common courtesy. So I change identifying details and let folks speak for themselves.

Inpatients

I knock twice on a patient's door. I've had a referral from a nurse, the chaplain, or the family. I'm wearing street clothes, not scrubs. I've thought about scrubs: how official; how like a nurse or a surgeon! I decide that such protective coloration would be pretentious and symbolically dividing me from the patients. Street clothes are better, labeling me as a regular person from the world of the well.

I've already checked with the nurse for this patient. If there's any reason not to massage, she'll say so and I'll understand. Some patients are asleep; some are too sick; some are receiving a treatment.

"Come in." I hear the words but I also listen to the quality of the voice for any hints of the person's mood or vitality.

I enter, introduce myself, and offer massage. Perhaps they've never had a massage and my offer isn't acceptable for them. Perhaps they are too sick for a massage. When they say "No," for whatever reason, I reply, "I hope your treatment goes very well" and take my leave.

If they say "yes," I move toward them.

"How are you feeling?" I smile and look them in the eye. There are a variety of responses, of course, some of them clearly inaccurate, especially if they've chosen a standard social response: "I'm fine." I continue (as with regular massage clients), "How are you feeling *in your body?*" and then get specific replies about where they ache or hurt.

At the same time, I'm scanning them for clues. Posture in bed…facial color…approximate age…how sick do they seem to me? I'm also scanning the room. What have they brought in that tells me about them? Rural? Professional? Religious? Sports fan? The rooms are pleasant with large windows. Patients and their relatives often decorate them with photos, banners, and pictures of loved ones, friends, or pets. There are get-well cards, sometimes balloons, or paper flowers—no real flowers—too germy. Sometimes Bible verses or other religious symbols. Also images of sports teams, entertainers, or NASCAR (this is North Carolina). Window Crayons allow patients, family members, and friends to draw or write on the windows; the colors are lively and bright when the sun shines through them.

"Where are you from?" I ask.

They tell me. If it's a larger town, I probably know it. Anything smaller, I'll say, "Uhn huh, and what's that near?" I've visited most of North Carolina's 100 counties and can usually mention something about the area named. In this conversation I'm creating a social relationship based on the patient and me as social equals; this is a foundation for everything that follows. I always ask about their home to acknowledge their healthy life. Home is especially important to patients hospitalized for a month or more…and far away from home means that visitors are few.

In one room I recognize a photo of Hendersonville; the patient and I discuss the main street, the tourist bureau, and an inn where my wife and I have stayed.

Another patient has a large picture of Iona, Scotland, where there is a Christian community.

"That picture keeps me going," he says. "I intend to go there when I get well."

I'm also scanning the white board on the wall. It lists the doctor, the nurse, the CNA. Also the latest blood counts, a measure of sickness/wellness, especially for white cells, the infection fighters. I also look at platelets; a low level of them puts the patient at risk for bruising and I have to adjust my pressure. My check-in with the nurse is also a protection.

"Well, what can I do for you today?" This is a pivot because it gives the patient control. Some of them are surprised, thinking that I was going to be in charge and tell them what to do.

Nine times out of ten, they respond with, "Oh, it's my neck and shoulders."

"OK, that's pretty common," I say. Patients like it when I say that because it suggests they are normal. I continue, "Show me exactly where."

Again they are in charge. Their hands point to spot, often upper trapezius and/or levator. They point to some of the small muscles in the neck, or splenius capitus along the side, or all along the occipital ridge, the base of the back of the skull. These are old friends to me both as a therapist and as a human being; I get sore and tight there too.

"Where do you want me?" They ask, giving back control to me. I share it: "Well, we can see what makes the most sense, but first we need to do a little paperwork. This is a before and after survey on how you feel. It just takes a minute." Patients mark on 10-point scales for pain, nausea, fatigue, and the like. Fatigue is usually the big winner and, when I tell patients that, they are glad to know this is common. These numbers are called (in amusing medical terminology) SUDS Scores, the acronym for Subjective Units of Distress.

Let's make a plan

However slender, a relationship has been established, and we move on to create a plan for the massage: in a chair, in bed, side-lying—whatever is easiest for the patient. We make sure IV lines are OK. I might say, "I see you have a port, and I won't touch that." If the patient is in bed, I raise it so I won't hurt my back. If the TV has been on, I ask if we can turn it off during the massage; patients always agree. Almost ready, I ask, "Are you comfortable? Are you warm enough? Be sure to tell me if anything doesn't feel right or actually hurts." I check the room clock and put my hands gently on the patient. As described in the Portland training above, hospital massage is light, comforting touch. It's not about deep tissue work, lengthening muscles, or increasing range of motion. Sometimes I work through clothes, sometimes on bare skin—always with care of draping for modesty. Some massage is just "lotioning," using oil or crème with hands held still; this can be surprisingly effective, an invitation for the patient to relax. As the massage proceeds, I ask if the pressure is OK, usually twice, but more if I have any sense of a recipient's discomfort.

After the massage, patients report again on the survey about their pain, sleep, fatigue, as well as physical and emotional distress. The scores typically improve, sometimes dramatically. Patients also evaluate the massage overall, almost always highly. This is not news to me. Massage promotes relaxation helping body and mind to heal. Patients tell me directly, "Oh, that was great." "My hands feel better." "My headache is gone." Even, "Can I take you home with me?"

As I leave a room, I tell the patient, "I hope your treatment goes just great!"

Occasionally I hear an angry statement. One woman complained, "What is all this crap about breast cancer: pink this, pink that, foot races, yak yak yak? You know, there are a whole bunch of other nasty cancers, and I've got one of them!"

Outpatients: massage for hands and feet in the infusion clinic

I take the elevator down to the third floor, where the Infusion Clinic sees some 60, 70, even 80 patients a day for chemotherapy. Part of me feels right at home here because of my own treatment. Like my Florida clinic, there are bays for four patients in their Barcaloungers, but there are also curtains that can be pulled around any particular patient. There are views to the outside. On a second wing, there are small rooms with glass doors, but even there, patients can see across to large windows. Gentle lighting and good interior design make the space pleasant.

My job is to offer hand or foot massage. If a patient agrees, I pull up a stool and sit in front of them. We do some chatting and the SUDS intake form. If they look cold (or, as I touch them, feel chilly), I offer to get them a blanket from a large warmer.

"What kind of cancer did you have?" a woman I just met asks me. She has seen the badge on my chest.

"I had a leukemia," I say.

"How long ago?"

"Ten years."

"Were you treated here?"

"I was treated in Florida before I moved here. I come here for my check-ups. On Thursdays I do massage; once a year I'm a patient."

She nods and smiles.

"So you are healed?"

"So far, so good. And I wish the same for you."

She relaxes back in her chair.

"Is this pressure OK?" I ask her.

"Yes. It feels great."

When I was sick with cancer, it was important for me to have images of rejoining the world of the well. Now it's my turn to comfort and inspire sick people, but I often find that they inspire me with their stories, their doggedness, their faith in medical treatment, and their ingenuity in dealing with lengthy treatment. Many bring books, crossword puzzles, music, or movies. One man brought a chess set. I've seen women with various crafts: crocheting, quilting, sketchbooks, journals. A patient and a friend played cards. Many watch TV on screens overhead. Some people doze, nap, or fall asleep. Some put blankets over their heads.

I'm rubbing the hands of an African-American man. They are scarred and callused.

"Do a lot of work with your hands?" I say.

"You could say that," he replies. "I built three houses."

"Are you in construction?"

"Naw, those were for my family. I did streets for the city. If you learn one trade, you can learn most others." I recall that Ivan Denisovich said the same thing in Solzhenitsyn's novel.

Most of our outpatients come from, say, a 150-mile radius including the Piedmont, the coastal plain, the Outer Banks, and the Sandhills, but some day patients drive over from as far as the Appalachian mountains. Some come from South Carolina or Virginia. Some get up at 4:00 a.m. to drive to Chapel Hill. They must see a doctor, have blood drawn, et cetera, before they get upstairs for chemotherapy.

One woman says she's from the mountains.

"Near Asheville?" I ask.

"Further west and south."

"Anywhere near the Cradle of Forestry?"

"That's 20 miles from my house."

We have a chat about the area while I massage her hands.

Some men refuse massage. They say, "I'm doing *just fine*." For many Southern men, massage is outside of their experience. Given their serious disease, this affirmation of *fine*-ness is ironic, but I can understand the reticence. One man declared, with an odd smile, "It might make me gay!" Whatever their reasons, I say, "I understand. I hope your treatment goes great."

Now and then a man says, "Heck, what can I lose?" I give him the first massage he's ever had. He says, "Gosh, why did I wait so long?" Some of these men are over 60.

"Are you a volunteer?" some patients ask me.

"No, I'm not. I can't do this as a volunteer," I explain. "I have to be a hospital employee. This means I've had a background check, drug tests, and a two-day orientation to many aspects of the UNC Hospitals. You are, for sure, getting *quality*." People usually chuckle. In fact, they have little idea of the many provisions for their safety in modern hospitals, such as disaster planning, infection control, information management, confidentiality, insurance, and more.

Humor is a healing resource. I have to be careful about this, but you can usually sense who's open to it. Sometimes I say, "I can massage your hands or your feet while you are having a *wonderful time* with us." If the patient laughs, I know how to proceed. The nurses know my sense of humor. When they bring drugs for the patient, I sometimes say, "Hey, nothing for me? How about a gin and tonic?"

"Sure," they say, "And perhaps I'll join you."

The nurses are great. Hard-working, technically proficient, and kind to their patients. Sometimes they say to me, "Chair number 23 is new today. Can you see him?" "Sure thing," I say. Sometimes a patient has a high blood pressure, and nurses call on me. There could be many reasons for this reading, but conversation and gentle massage usually lower it. I like watching the digital display: 155 over 90, then 140 over 88, perhaps even 137 over 85.

Some patients are on curative treatment; they will get well and resume their ordinary lives—usually with a heightened sense of gratitude. Some patients will need continuing treatment for the rest of their lives. Some patients are on palliative care, damping down, as possible, the symptoms caused by their cancer. Some are headed toward death; these are referred to hospice. Eventually, of course, we all die. Spending time in a cancer hospital makes clear our shared mortality. Every morning I read the obituaries in the Raleigh newspaper *The News & Observer*. Now and then, I find an obit for someone I have massaged. One was a large but gentle man. Treatment extended his life, but could not ward off death. I went to the viewing at a funeral home, the only time I saw him in suit and tie.

Intent can relieve pain

Such variety! One day I offer massage to a woman. She accepts, indicating her hands. We do the paperwork. She reports no distress for every scale except PAIN, which she rates at a 9, just one level short of 10, the *worst* it can be. This is quite unusual.

"What? You are rating pain at a 9?"

"Oh yes."

"Oh, I'm so sorry. Where is it?"

"Well, it's my radiation burns, on my neck, on my chest, my armpit, and my breast."

Good Lord, I think. *That's terrible.*

"So how come you're sitting here, talking calmly to me?"

"I'm on oxycodone, for one thing." She smiles.

"Uhn, huh. But still…"

"And—even more important—I just had my last radiation treatment this morning. I know this is the *very bottom*; it can't get worse. I've done surgery. I've done chemo. I've finished radiation. I'm *moving up* from here."

"I see."

Do I see? It's hard to imagine all she's been through, but her psychology, her attitude, her *intent,* are all aimed, considered, and directed to be on a healing course. I believe she will do well. Some nurses say they have a pretty good idea about which patients will do well or do poorly, because of their attitudes.

Another day an elderly man rates pain, nausea, fatigue—everything at a zero. Zip. Nothing. *What?*

"Wow," I say, "You must be feeling pretty good."

"Yes," he say, "I surely am." As I massage his hands, I can see he's actually pretty sick, and his ID band says he is 79.

We chat for a while…then share silence as I work.

After a while he says, "You know, that tape is what keeps me in the pink." He gestures to a small black case in his lap.

"How's that?"

"My doctor at home had a hypnotist make a tape for me. I listen to it twice a day, sometimes before I go to sleep at night."

"Interesting idea."

"So I have no side effects, no fatigue, and my cancer is healing up."

"Hey, that's just wonderful!"

"Here, take a little listen!" He pushes a button and hands it to me. As I raise the device to my ear, I hear a voice droning *...are heavy, very heavy... your eyelids...are very heavy...they ...want to close now...you...are very relaxed...very relaxed...your jaw...is relaxed, very relaxed ...your mouth....*

"I like his voice," I say.

"Very relaxing."

"How long is it?"

"Forty-five minutes. I'm usually fast asleep before it finishes up."

I wonder: if this is so effective for this man, wouldn't it be for others, and, if so, why isn't this routinely offered as part of standard treatment?

These two patients illustrate how mind can be a powerful resource.

A joyous survivor

I'm sitting in the foyer of the Patient and Family Resource Center. I'm chatting with my boss Tina while I wait for a client. A lively woman looks in our door and calls out, "Hey, are you Howard?"

"Sure thing."

I don't recognize her. This often happens: they know me, but I haven't a clue. Now healthy, they look very different, unstressed, energetic, wearing different clothes and newly grown hair.

"Well, I'm Cora." She smiles.

Looking at her face, I can now see the person I massaged several times and many months ago.

"Oh my gosh. You look wonderful!"

"Don't I?" she enthuses. "I've got hair!" One hand pulls on her hair above her ear. "I've got boobs!" He hands bounce her breasts up her chest. "And I've got energy like before!" She smiles.

"That's just great," we say. "Congratulations!"

"Well, I've got to run," she says, and off she goes, back to the world of the well.

ON THE COLLEGIALITY OF MASSAGE THERAPISTS

I've yearned for expansion of the massage program, and one day there's good news. Another therapist will be joining, and she can do three days a week.

Soon I meet Stephanie Nussbaum, who has been working at Mount Sinai Hospital in New York; Mount Sinai has one of the top cancer programs in the U.S., and massage is available in oncology and several other services.

I'm thrilled to have a colleague…one with whom I can discuss therapies, the variety of patients, and whatever arises week in, week out. Through the magic of cell phones, we are usually able to meet for lunch to talk about our work and to support each other.

I have always found massage therapists to be collegial: sharing and supportive. At continuing education classes, a dozen or more of us, usually strangers, are able to work together on the material, touch each other, and share the lunch hour with interesting conversation. We work in different practices, have different specialties, and some have often traveled a long way. Nonetheless, we enjoy hearing from each other and gaining new insights into the art. They are, as we say in the South, "good folks."

Therapists specialize in many areas, for athletes, for spa clients, for infants, for pregnant women, or for the elderly. It's a field with many possibilities. Stephanie has techniques I don't, including the use of aromatherapy. She's very good at helping bowels wake up after anesthesia, indeed so good, one service calls her the "Peristalsis Princess." I don't know how she does it, but when I try myself, using Qigong, I get a family member's bowels working. The sooner bowels are awakened, the sooner patients can go home if all other signs are favorable.

"EVEN BETTER THAN SLICED BREAD." PRAISE FOR THE CONDUIT

After massages I give, I usually receive praise. I feel that I do a good job, but the praise should actually go not to me but elsewhere: to the traditions, techniques, and art of massage itself as well as to client's body and mind that are receiving and processing. I'm just a conduit. Besides, I've already gotten news directly from the body, various signs of response: looser muscles, slower breath, a voice that is lower pitch and slower in speaking, all signs of the parasympathetic response.

Some of the phrases I've heard:

That was just wonderful.

Muy, muy bueno,

I'm so relaxed now.

Massage is the only thing that truly relieves my pain.

One man said, with a wry smile, "I enjoyed that right much. It's even better than sliced bread."

The SUDS scores on the surveys are favorable: on a five-point scale, I get fives most of the time, occasionally fours.

I've had offers of tips of money, a southern custom. To these I say, "Aren't you sweet? Thanks, but I really can't accept them." Not quite satisfied with those words, I discuss this issue with Stephanie. She explains that she tells a patient, with a smile, "You wouldn't tip your nurse or doctor, would you? When I do massage here, I'm a medical professional."

I was massaging a woman's feet. After a while she sighed and said to her husband nearby, "Honey, you've got a lot to learn" and gestured to my hands on her feet. Apparently their style of conversation included various digs and teases. He sat still a while. Could I sense wheels turning in his head? After a while he came back with, "Say, this massage must be great stuff: she's been *quiet* for 5 minutes!"

And they both laughed.

SOCIAL ROLES IN MASSAGE FOR SICK PEOPLE AND CARE PROVIDERS

We looked at massage in general in "How Does Massage Work? Matter, Intent, and More" in the Section I. The seven dynamics discussed for all clients pertain as well to sick patients, although with some differences. Massage for patients is gentle, sometimes just laying on hands. It is also specific to areas agreed upon by patient and therapist, and of course, it is focused by medical information and the constraints of hospital beds, IVs, and so on. Also, massage therapists in hospitals never give health advice.

In the world of the sick, there are some other considerations. We can discuss these as an eighth area of inquiry, the social roles of therapist and patient in a medical setting. Massage is a common ground where the worlds of the sick and well overlap, a place where, for an hour, there is the green world of comic integration or reaffiliation.

Sick persons: marooned in the world of the sick

In the world of the well, the social roles of therapist and client are, more or less, as equals within the structures and rituals of massage. In a hospital, clinic, or sick room, however, patients are removed from ordinary life and from familiar patterns of activity, discourse, and environments. The English language has words for separation, some of them unpleasant: "exile," "disconnection," "segregation," "ostracism," and "banishment." Similar words have been used medically: "quarantine," "confinement," and "isolation."

Patients often feel emotional impacts of their removal: especially loneliness or abandonment, even guilt. They may feel that they acted wrong, that their body betrayed them, or that an injury came upon them unfairly... absurdly. Any of these emotions can produce stress. There are also discomforts from treatment, side-effects of drugs, and—from whatever origin—pain. Even moderate pain pulls us away from the world of the well. In Elaine Scarry's formulation, "Intense pain is world-destroying" (*The Body in Pain: The Making and Unmaking of the World*, 1984, p. 29).

Well people carry stress in their necks and shoulders. With heightened stress, sick people often carry stress there (and elsewhere), as if guarding themselves against threats. Pain, discomfort, worry, and all the novelties of the hospital setting contribute to stress that manifests in muscles and joints. Massage can be very helpful, reducing stress and pain and, even, giving patients tools of touch and imagery to control it by themselves. Such tools reduce victimhood ("Poor me!") and enhance agency, even control.

Massage therapists: emissaries from the world of the well

Massage therapists symbolize the world of the well. They arrive in a patient's room with good cheer, neat appearance, and trimmed and smooth nails. As they visit and chat, they represent the world of the well even before the massage begins. Their very presence suggests that the patient is still part of that world.

Massage therapists are "safe" in the sense that they do not represent any tests, medical treatments, or bad news; they don't draw blood or take vital signs. Thus they are similar to dietary workers who are also safe...and bring nourishing food. Of course, dietary workers can't stay to visit or provide treatment, and they come regularly and, we might say, non-medically.

A better comparison are the chaplains who come by specific referral and stay to chat, pray, and otherwise support patients. Chaplains and massage therapists both bring intentions of healing. They both excel in listening to patients. When I trained as a pastoral care volunteer at another medical center many years ago, listening was one of the most important topics discussed.

Arriving at a patient's room—and labeled both Massage Therapist and CANCER SURVIVOR—I represent both the world of the sick that I had passed through and, currently, the world of the well. I am a figure of reaffiliation with well people, an image patients may aspire to.

Massage as common ground for convergence of the two worlds

Massage therapists in hospitals treat patients as social equals; they make "small talk" that has large value; they listen well and take cues from the patients' appearance, demeanor, and personal items in the room.

The therapist and the patient create a plan together. The treatment will be "custom made," and proof that the therapist listened well. The patient will have a clear expectation and will be invested in the treatment, thus bringing possible placebo benefits. Therapist and patient will both be "on the same page." They will be in synch, in harmony, in material bodies and in their minds. Sometimes they breathe together.

Respectful, appropriate touch will match what the patient agreed to. This touch lowers stress and invites, allows, and promotes relaxation.

"COATS," LIKE NURSES, NO LONGER PROVIDE MASSAGE

One day I meet a new nurse. She's the only one at the nursing station, so I ask her if she's caring for a patient I'm looking for. She is not, but I should get to know her, so I explain who I am and what I do. She puts her hand to her chin and thinks a moment.

"You know, room 24 is pretty sick and has his birthday today. Any chance you could see him?"

"Sure," I say. She gives me details about the patient—a tough case, but OK for me to see.

"And…" She pauses.

"Yes?"

"Actually I've got another one, a young woman who got a very bad diagnosis today—right out of the blue."

"OK," I say. "I can see her next, if you say she's OK for me to visit."

"Yes, she is," she says but she furrows her brow.

I look expectantly.

"You know, the coats just don't do that much for them."

"What?

"The coats."

"I'm sorry, but I'm still not understanding you."

"The docs!"

I finally get it: the doctors in their white lab coats.

"They breeze in and out and are all technical. They just don't have time to give comfort to the patient."

She has a point…and doesn't. I've seen plenty of caring docs who relate well to patients. Sometimes I've been in the room doing massage when the group on rounds arrives. "Would you like me to step out?" I say. "No, you're fine," they say. The docs are competent and kind, it seems to me, but morning rounds are not a time for an extended visit.

"My time is flexible," I say. "I can listen if they want to talk."

"Good," she says. "I'm sure they'll enjoy your visit."

As I head to the first room, I recall the hospital in Oregon, the Mid-Columbia Medical Center, where doctors did not wear white coats or ties, just regular street clothes. With these and other changes, the building felt more like an apartment complex than a hospital.

I knock and identify myself. I add, "Your nurse said it was your birthday and maybe you could use a little massage." The patient nods agreeably. We do the all the usual readying routines. I check his cell lines on the white board. He's sitting up in bed but has little vitality. We shape a plan for neck and shoulders so he won't need to move at all. If I push my hand back into the pillow and bed behind him, I can work easily enough. As I give light massage, he closes his eyes. His breathing slows down. There's a smile on his face.

The next room is dark…lights off, shades drawn. A woman is in fetal position on the bed. After the preliminaries and her agreement, I press the button to raise the bed. I place my hands on her and say a prayer in my mind.

"I hear you got some bad news today," I say.

"Yeah...real bad."

"I'm so sorry."

I recall my Oregon teacher's sentence: "You can do a lot for patients just by putting your hands on them." And that's what I do for her neck and shoulders.

Later I'm in the back elevator and a coat joins me.

"Are you the massage guy?"

"Yes, I am."

"So you just saw the patient in 24?"

"That's right."

"Well, it's his birthday, and that meant a lot to him. Thank you so much."

MASSAGE FOR CAREGIVERS

Caregivers benefit from massage as well, whether they are strictly medical or family members. The latter are often worn down with the worry and extra efforts for travel, care at home, or just stressful waiting. For them massage is a form of respite care.

As for medical personnel at the NC Cancer Hospital, several times a year I take my massage chair upstairs and offer 10-minute sessions for nurses. They are their feet much of the day, and they must bend over to treat patients. While not all accept massage, those who do love it. Most have helping personalities; some would advance the next finger toward me when I was massaging their hands. One of the more jovial nurses liked to chortle, "Howard *does* nurses."

One day my boss says, "You know, the children's unit has had some tough times lately...losing patients. Could you take your chair up there for an afternoon?" After some planning in the morning, I take my chair there after lunch, set up in the Orangutan Room, and provide some relief and relaxation to nurses and a doc.

MASSAGE FOR THE DYING: FAREWELL GIFTS

We heal many cancer patients outright. For others, we extend life and quality of life. Some come back to us with relapses or new cancers.

Some go home to die, often on hospice care. Some die in our hospital. If only medicine could heal all cancer patients!

I've heard several nurses say resignedly, "We can't save them all." Nonetheless, they give competent and caring service, regardless of a patient's path.

But we can provide farewell gifts to the dying. "Comfort care," as it is called, can help with pain, breathing, temperature control, and more. Gentle touch, including massage, can help the dying person relax and feel connection with a loving caregiver. For a longer list of benefits, see the National Institute on Aging website given in SOURCES for this section.

About once a year I was called to provide massage for someone dying or for a parent waiting for an infant to die. Some of these were chaplain referrals. Sometimes family members asked for me because I had seen the patient at an earlier time. In one case, a patient was on life-support and scheduled to be disconnected the next day. The nurse said, "He can't say anything." As I massaged his shoulders, he pushed gently back against my hands, his way of telling me that we were connected.

One day a nurse asked if I'd see a woman who was actively dying. I had seen her before when she was lively and cheerful. I was sad that her life was ending but glad that it was in a safe and supportive place. She had no IV lines or other measures to extend her failing life. She couldn't talk or respond in any way. I massaged her feet, thinking *This is my friend; I care about her.*

If I were dying—when I will be dying—I'd be glad to receive gentle touch.

"WOULD YOU MIND IF I DID A LITTLE ENERGY WORK?"

Some years back, I had a cancer outpatient scheduled for a table massage. On the intake sheet she marked a 5 for pain, while the other marks are in the 0 to 1 range. She marked Nausea, for example, at zero.

I asked her, "Where is your pain?"

"It's in my pelvis."

"Can you show me where?"

She pointed to a spot on her lower back (the posterior superior iliac spine). It seemed a definite location, possibly from a muscle tear away from the top of her pelvis or some minor trauma, like a whack by a car door.

"Any idea what caused that?"

"Sure. That's where they took a bone marrow biopsy."

"Of course…that's where I had mine," I blurt out. Then: "Too bad I can't help you there. I do muscles," I heard myself saying. *What? I shouldn't give up that easily.* I regrouped. "On second thought, I do some energy work off the body. I think it might help you. Would it be OK with you for me to try?"

"Heck, I'll try anything. It's really annoying."

After other preliminaries, we began the massage. She was clothed and face down. I did my usual routine, then paused to consider her biopsy site. I took a couple of deep breaths and pumped up my energy. I felt my palms tingle. I scanned her back with my hands and felt something near her waist pushing them away. It felt *jangly* there or, *out of sorts*—hard to find the right language. Working above her body, I combed through the area and pulled away a handful of heaviness, taking it to her knee for an exit from the body. I dumped it to the earth. I did this three times.

"OK," I said. "I hope that helps. We'll go on with the regular massage."

When the massage was over I said, as usual, "How do you feel?"

"I feel very relaxed. That was great."

"And how's the pain in your pelvis."

She seemed surprised at the question. She put her hand to her lower back.

"Oh my gosh. It's gone!"

I was only a little surprised. I'd been studying Qigong, a means of healing energy.

THE BUBBLING FIRE IN THE BONE MARROW TRANSPLANT UNIT

Note: this first appeared in *The Intima: A Journal of Narrative Medicine.* (Spring 2016).

For several years, I was a part-time massage therapist in a cancer center. As my skills grew, I used more techniques than those I learned in massage school.

I remember a woman I'll call Ellen.

The rec therapist and a physical therapist are helping an elderly woman walk around the halls of the Bone Marrow Transplant Unit. As they go by, I say to the woman, "Good for you! Keep it up!" The rec therapist says to me, "If you're looking for a massage patient, how about Ellen, here? She's just about done with her workout."

I look at the woman for a sign. She's in a hospital gown and tethered to the IV pole on wheels.

She gives a small smile and nods.

"Sure," I say.

The P.T. and a nurse take Ellen into her room and get her settled. As I wait in the hall, the rec therapist tells me, "She's having a lot of trouble with her breathing. Anything you can do for that?"

"I imagine so," I say, and starting imagining.

The settling takes a while, as many things do in hospitals. When it's my turn, I knock, enter, and introduce myself.

Ellen is about 70, with messy gray hair and a rumpled hospital gown. She's lying on her back, and the bed is adjusted to a semi-sitting position. She looks at me doubtfully.

Her breath is rapid and shallow.

I extend my hand. We shake hands gently.

"Ellen…a little massage for you?"

"Yes," she says.

"Where's your home?" I ask.

She tells me and looks me over.

"Not too far away," I affirm.

"No."

I take a quick glance at the room, but I find no signs of her life that I can pull into our conversation. Often there are photos of family and pets, get-well cards, even insignia of sports teams or colleges. Here, nothing.

"Well, how are you feeling?"

"A little tired."

"I hear that all the time up here. And you were just out walking."

"Yes, I was." She seems proud.

"How about I massage your neck and shoulders gently so that you can rest?"

"That would be OK."

I move her overbed table out of the way, check her IV lines, and raise the bed to a good height for me. As a massage therapist, I'm thinking accessory breathing muscles. The scalenes can help raise the first and second rib therefore creating space for the upper lungs…but they shouldn't be doing the lion's share of work. As a recently trained Qi Gong therapist, I'm thinking Lung Meridian. Combining energetic and mechanical approaches can be effective.

The head of her bed is already raised at an angle that suits me. She has nasal prongs in her nose to deliver oxygen. A tube connects her to a port in the wall, where there's a humidifier that makes bubbling sounds.

"Your bed OK like this?"

"Yes."

I lay my hands on the back of her neck and the shoulder closer to me. She's skinny and frail. I remember one of my teachers for hospital massage saying, "Just by putting a kind hand on a patient can help them."

Ellen breathes rapidly and shallowly, just the upper parts of the lungs.

"Ellen, how about we take a couple of breaths together?"

We do that. I breathe loudly in order to pace her. As I slow down, she slows down.

"Can you use more of your belly to pull in air?"

She does just that, and I start some gentle work on her neck and shoulders.

"This pressure OK?"

"Yes."

"And, Ellen, you must tell me if anything feels bad or hurts, you must tell me right away, because maybe I couldn't tell. Is that a deal?"

"Yes."

I continue gentle work.

With no TV blaring and no other people in the room, it's very quiet. The humidifier bubbles merrily.

I take her hand and work the Yuan point on the Lung Meridian.

I continue with her neck and shoulders a while, listening to the bubbling behind the bed. Maybe I can use that.

"You know, that bubbling sound reminds of a fireplace."

She raises her eyebrows.

"Yes," she exclaims. "It sounds just like that!"

We listen for a while.

"I'm thinking of a winter evening," I say, "when it's dark outside and few lights are on in the house."

"Yes," she says.

"Maybe there's a dog asleep on a rug in front of the fireplace."

She says excitedly, "We have a Golden Lab. He likes to sleep with his head laid on his paws."

I move to the other side of the bed to work the other shoulder. Her muscles are relaxing. Her breathing has slowed considerably. I watch her belly rise and fall.

I continue to sketch the scene with the fireplace. She nods in agreement.

Over the door to the hall is a clock. I count her breaths against it: 14 per minute…now 12…now 10.

"You're doing great," I say. "And you know, when I'm gone, you can use the sound of that humidifier to remind you of that fire, that calm evening, that dog dozing on the rug."

"Yes," she says. She smiles.

As I leave, I am glad that I could help her during our time together but also satisfied that now she has a way of using her mind to calm and energize herself in the future. She can use imagery, the narrative we sketched, and her mind's intent to influence how her body works—all basic concepts in Qigong.

I report the improved respirations to the rec therapist; she says, "Oh that's great; tell it to her nurse

The next day, it dawns on me that the imagined scene parallels my childhood event with my Dad and family dog as we three lay on a sofa watching the red embers in the fireplace. What powerful urges were at work? Notions of safety…comfort…protection… warmth …quiet… rest…home…companionship—qualities often missing during a long hospital stay for a serious illness.

ESSAY 3: INTEGRATIVE HEALTHCARE AND INTEGRATIVE MEDICINE

Essay 2 focused on integrative medicine from the patient's perspective. In one sense, it can *only* be patients who do the integrating because it is their bodies and minds that respond to standard care plus other modalities, creating, for example, positive placebos.

I think of patients who had a pain in a shoulder, a knee, a hip. After some discussion with them (and maybe checking with the nurse) it seemed possible that there was tightness because the patient was placing her or his stress there. After some gentle massage and conversation about talking to that area, such pain diminished or went away entirely.

When hospitals and clinics offer only standard medicine, patients won't experience integrative healthcare there, but they often make their own private integrations, diet supplements for example. Historically, those two streams were separate: standard care versus *alternative* care. The 1993 article by Stuart Eisenberg et al. startled many healthcare professionals by describing the enormous amount of alternative care patients were using, typically without their doctors' involvement. This discovery was one impetus for professionals and institutions to learn about other choices and help patients chose among them. Ideally, such a combination would be *complementary*, as if—as geometry students learn—the two together created a vigorous 90-degree angle and maybe even included a pun on "complimentary." The word "complementary" also suggested that some of alternative medicine was not useful and even harmful. Patients are often eager for "*something else* they could do" and hungry for support beyond what professionals offer. For good or ill, they make decisions about other approaches, some with dangers such as delay or avoidance of needed medical care, dietary supplements that interact with medicines, even financial loss to patients for treatments or products that have no medical benefits. There can be misinformation in websites or passed between patients. To avoid such risks, medical professionals should be informed about other approaches, discuss them with patients, in order to make complementary medicine truly integrative.

Today, the term "integrative" is commonly used. Done well, it brings benefits to patients, relatives of patients, caregivers, even the spaces where care is provided. Caregivers benefit too, seeing better results for patients, working in a healthier environment, and using for themselves stress reduction, rest, good nutrition, and the like. As we saw earlier, the NIH institute evolved its name from the National Center for Complementary and Alternative Medicine (NCCAM) to the National Center for Complementary and Integrative Health (NCCIH).

The recent and robust growth of integrative medicine

The growth of integrative medicine in this millennium has been rapid and far-reaching.

For institutions, there is the Academic Consortium for Integrative Medicine and Health originating from a 1999 conference of eight institutions. There are now 60 members, typically universities, healthcare systems (including the Veterans Health Administration), or large clinics mostly in the U.S., but some in Canada and Mexico. According to their website (2016), their core values are as follows: "Every individual has the right to healthcare that: provides dignity and respect, includes a caring therapeutic relationship, honors the whole person, mind, body, and spirit, recognizes the innate capacity to heal, and offers choices for complementary and conventional therapies."

For individual physicians, there is the American Board of Integrative Medicine that certifies practitioners; this is one of some 20 national certifications, along with mainline specialties such as surgery, emergency medicine, and psychiatry. The website describes a rigorous examination process for Integrative Medicine; it covers nine domains, including nutrition, mind-body medicine and spirituality, as well as complementary and alternative therapies. The domain of "whole medical systems" includes Traditional Chinese Medicine, Ayurveda, Traditional Medical Systems (e.g. Native American, Shamanism), Homeopathy, and Naturopathy.

There is an increasing confluence of interest in integrative medicine from lay people (patients and their families), many medical providers (doctors, nurses, various therapists), and institutions, including hospitals, medical centers, and—very important—medical schools that increasingly offer courses in integrative medicine. Because the popular press often reports on integrative medicine, and because there is easy access on the Web to many aspects of it, it is important that medical personnel are informed and involved with patients making choices in treatment.

Integrative medicine has been influential since Andrew Weil's book *Spontaneous Healing* (1996) and growing continuously since then. Weil directs the Center for Integrative Medicine at the University of Arizona; their "Defining Principles of Integrative Medicine" give a good overview of the concept (see website: https://integrativemedicine.arizona.edu/about/definition.html, accessed July 4th, 2016).

1. Patient and practitioner are partners in the healing process.
2. All factors that influence health, wellness, and disease are taken into consideration, including mind, spirit, and community, as well as the body.
3. Appropriate use of both conventional and alternative methods facilitates the body's innate healing response.
4. Effective interventions that are natural and less invasive should be used whenever possible.
5. Integrative medicine neither rejects conventional medicine nor accepts alternative therapies uncritically.
6. Good medicine is based in good science. It is inquiry-driven and open to new paradigms.
7. Alongside the concept of treatment, the broader concepts of health promotion and the prevention of illness are paramount.
8. Practitioners of integrative medicine should exemplify its principles and commit themselves to self-exploration and self-development.

If we look at the two realms of standard medicine vs. integrative medicine with an eye to *matter, mind,* and *energy,* we can show contrasts.

The following figure is a force-field model that leaves out a lot, mental health for example. Further, this kind of model ignores overlaps and it is inherently extreme, leaving out areas of standard medicine that already partake of integrative medicine, for example team approaches. I apologize for this reductive polarization, but I hope that it allows for clarity, insight, and future progress.

	Standard Medicine	**Integrative Medicine**
Clients/patients	Bodies as *materials*	*Body/mind/spirit as a whole*
	Focus on disease	Focus on health
	Rescue model	Maintenance, healing model
	Cures are *tissue based*	Care is *whole person based*
	Death is a failure	Death can be managed, palliated
Treatment aim	Get diagnosis and treat	Support, heal patient as a person
Patient *Mind*	Secondary to disease and cure	A primary focus for care
Caregiver *Mind*	Rational, linear, reductive	Multi-layered, multiple inputs of emotion, instinct, intuition, spirit
Energy	Technical: X-ray, drugs, etc.	Aim to increase and balance energy of patient and caregiver

Figure 3: A Force-Field Comparison of Standard and Integrative Medicine

Integers, wholeness, reaffiliation

In mathematics, *integers* are whole numbers (1, 2, 57, -649, etc.) and never fractions. Metaphorically, we speak of *integrity* as a positive virtue for a person who has moral clarity, consistency, and the like: "Sam has integrity; you know where he's coming from," and *Steadman's Medical Dictionary* lists a meaning for health: "The state of being unimpaired; soundness or wholeness."

Wholeness, health, moral virtue—all good, we believe. If we push the word for origins and hidden dimensions, however, we find the Latin roots of *in* = "not" and *teger* = "touch," as in "tangent," "tangible," etc. The risk of *integrity* as "untouchable" is being alone, untouched by other people, cultures, or ideas. Some forms of integrity can be self-isolating, whether for people, neighborhoods, elites, or nations. The "integer" can be a sports team, a terrorist organization, a church choir, or a white supremacist group. So it's important to ask, what, at root, is the integer, and does it promote divisions with others or harmony and peace?

Beyond integration, the notion of reaffiliation (or affiliation), with its nuance of family, is helpful. In discussing affiliation as "a hallmark of healthcare," Rita Charon sees affiliation as a source of relief from the "contemporary bioscientific ethos that challenges the particular with the universal, the personal with the corporate, and the intimate with the mechanical" (Charon, pp. 157-158).

The antonym to integration is *disintegration*: falling apart from a unity. When people are reduced by economics, medicine, politics, racism, warfare, or violence…they are dis-integrated. When the cancer patient refused massage because "It might make me gay," he reduced himself through homophobia. In contrast, the article on hugs showed that as hugs went up, colds went down because of social support.

Cardiologist Bernard Lown's *The Lost Art of Healing: Practicing Compassion in Medicine* (1999) touches on some of these topics. The cover of my paperback copy has a quotation from Andrew Weil stating that Lown's book is (or was, at the time) required reading in the Integrative Medicine program at the University of Arizona. This book should be reread again today. Modern doctors, Lown writes, have emphasized curing (scientifically) to be neglect of caring. He cites the oft-quoted remark of Francis Peabody that "the secret of care of the patient is caring for the patient" (p. xvi). Listening carefully to patients and family is important (pp. 3-48). Touching patients helps to bond doctor and patient and to heal the patient (pp. 23-28). A patient's *mind* is a factor in illness, especially in cardiac illness. Although some "alternative" medicine may lack scientific proof, some of it can contribute to healing when, for example, there's a placebo effect.

A friend described to me his treatment for a medical emergency. The ER docs gave him, he felt, "assembly line service" and referred to a specialist. The specialist recommended immediate high-tech intervention, using "robotic new toys" that seemed to my friend like "an engineering approach," as if he, the patient, were a machine. Another specialist, however, spoke to him as "an actual, real human being" and treated his emotions as well as his body. He prescribed a low-tech treatment of pills that was simpler and effective, thus providing cure as well as care.

Is death disintegrative or integrative?

From the usual, Western perspective, death is disintegrative: the life of our bodies falls apart and eventually all the materials as well.

Various religious (including Christianity) and some philosophical perspectives understand that the body disintegrates, but the mind, psyche, or soul integrates with another reality, such as heaven—an ultimate affiliation, or reaffiliation, depending on your theology.

What are some of the obstacles to integrative medicine?

When I asked a North Carolina acupuncturist when might doctors be willing to use his services, he said, "When hell freezes over." Clearly, he felt excluded, *beyond the pale*. If he lived in California or New York, he'd likely have more opportunities, but in the South there's a cultural lag among both professionals as well as lay people concerning integrative medicine. Surveys have shown that physician resistance, budgetary restraints, and lack of evidence-based research are leading obstacles to offering complementary medicine.

Resistance (or downright opposition) to integrative medicine varies. Here's a list of obstacles in no particular order: traditions of standard medicine; reliance on mainstays of drugs and surgery; disease models versus health models; rescue medicine vs. health maintenance; beliefs that medicine is "done" to passive patients versus an understanding that all people have healing capacities; lack of research about integrative medicine—or ignorance or rejection of existing research for whatever reasons; lack of leadership in institutions, including lack of funding, and concern about possible legal exposure.

And, of course, there is "we've always done it this way." Professional and institutional change can be very difficult.

In the wider culture, we tend to believe that medicine is all-powerful and that rational approaches to illness will work and cure us. If one medical approach doesn't work, surely another will. Wise caregivers know, however, that medicine has limits, there are many uncertainties, and that death will come to us all one way or another.

Who can help? Everyone, from a parking lot attendant to a doctor

Anyone working at a medical facility can help patients by being kind, supportive, by treating them as fellow human beings: food service people, receptionists, cleaners, clerks, technicians, as well as, of course, nurses and doctors.

Another consideration: perhaps it is unreasonable to ask all doctors to be interpreters of a dozen or more integrative approaches in busy hospitals or clinics. Could others be others assigned for that task? Could there be integrative specialists offering a menu of choices, parallel to food service people?

For large places, top administrators are crucial to the delivery of integrative care, promoting it, rewarding it, and explaining why it is central to the mission and identity of the hospital.

The healing resource of witnessing

Jungian therapist Jerome Bernstein uses the term "witness" for person-to-person recognition: being present to another person and providing authentication of him or her. Current speech uses the phrase "being *there*" for someone who observes and acknowledges: "When my mother was dying, my friends *were there* for me."

As I worked in the world of the sick, I noticed that the chaplains, especially, but also effective nurses, doctors, social workers, and others served as witnesses to the personhood of a patient. They could offer support beyond technical medicine by signs of recognition, some subtle, some overt. They might say, "I understand you're going through a rough patch" or "Gee, your color is good today." A smile, a pat on the arm, eye-to-eye contact. Once again *small can mean big*, as the resources of mind come into play.

Medical personnel are often so focused on assessment and treatment, however, they may not act as witnesses in this way. In some cases, such as surgery or emergency care, focused technical care must be central, but even then, caring intent can play a part.

On the other hand, social workers, chaplains, and massage personnel have defined roles that exclude diagnosis or medical treatment: no bad news, no needles. Patients perceive them as safe and are often more open talking with them. Further, massage therapists have more flexible time than, say, phlebotomists, and typically try (at least in my experience) to create a social bond first, before any treatment. They use rituals of equality and create a plan with the patient. Although the basic roles of sick and well are still present, signs of witnessing acknowledge and honor the patient. Because hospital rooms or other sick rooms are often strange and inhospitable in many ways, a sick person can find much comfort in being witnessed by a friendly person. Bernstein quotes another Jungian analyst, Ann Belford Ulanov commenting on all persons, sick or well: "We need to be witnessed to feel real" (p. 150).

A witness can be technical, as in a legal witness, or simply a presence, as a witness to a wedding, the wedding party that "stands up" for the bride and groom. When Bernstein discusses the Navaho witnesses to a healing ceremony, he emphasizes that only the holy man is technical, while all the others witnessing are ordinary tribal members.

When we witness for anyone, we are present in body and mind, with our minds aware on several levels, perceiving, understanding, caring, and more. Even our modern sense of the related word "wit" may apply, as in the humorous or playful speech that entertains and creates social cohesion.

I remember a hospital team that I was working with had a patient who was actively dying, but a religious family member would not let go and demanded further medical interventions. The chaplain said, "I know you love her, but her work here on earth is over, and there are angels waiting for her on the other side. She is ready to join them. Can you wish her Godspeed and let her pass over?"

Such help, kindness, and witnessing can all contribute to paths for reaffiliation.

Integrating body, mind, and (?) energy

As medicine became more technical, more time-stressed, and in many ways more abstract, doctors stopped using massage for patients in treatment. Later, nurses also stopped providing it. This phasing out is sad in itself—a regrettable story. Nevertheless, that's where we are now, and there is partial restitution in some hospitals and also by patients who know about massage (and can afford it) and seek it outside of standard medicine. I believe we should make massage available in many healthcare settings and recommended generally for stress reduction (see Sources below),

First, there is a comforting social dimension of therapist and client as they meet as social equals, talk person-to-person, and agree on a treatment plan. Done well, the process is healing in itself as the client (or patient) is reaffiliated into the world of the well. He or she is no longer an outsider, someone alone and worried in a hospital room, someone now defined by illness or injury. Therapist and recipient are "on the same page"; they share intent and *mind*. The word "comfort" carries the Latin roots "together" and "strength."

Second, there is solace in the pleasures afforded by careful and loving touch. Massage, when it matches the needs of the recipients, affirms them in body and mind. The recipient *feels good*, feels kind attention, feels solace, feels worthy, cared for, and witnessed. One dictionary gives deep roots and analogues for the word "solace" as "good mood" and "happy."

Third, worry and stress go down, allowing the body and mind to relax and contribute to healing. From what I have seen, massaged patients heal faster, go home sooner, are more satisfied with care.

Fourth, patients who are actively dying (and their families) benefit from massage, a part of palliative care.

Firth, caregivers who receive massage (as breaks on the job or elsewhere) also benefit from massage. They can return to work refreshed and re-energized.

In my journey so far, I can better understand mind and body integration. As for energy and its relation to body and mind, I see increased energy in patients as they relax from massage. Energy that earlier made muscles and muscles and joints tight now flows better, relieving pain and promoting relaxation. Relaxation allows for better rest, which allows for healing.

What I understand less is the role of energy. I have an inductive sense from examples I have experienced as a patient and as a massage therapist. I have an intuitive sense without terminology or theory. I have my early successes with Ellen's labored breathing and the woman with the post-biopsy hip pain. In these cases, Qigong integrated into massage clearly worked. Section IV describes world(s) of Qigong and its linkage with a nature that, increasingly, is separate from modern life.

SOURCES 3

Braun, Mary Beth, and Stephanie J. Simonson. *Introduction to Massage Therapy,* 3rd ed. Philadelphia: Lippincott Williams & Wilkins, 2013.

Carter, III, Albert Howard. "The Bubbling Fire in the Bone Marrow Transplant Unit." *The Intima: A Journal of Narrative Medicine*, Spring 2016.

Center for Integrative Medicine, University of Arizona (seehttps://integrativemedicine.arizona.edu/about/definition.html accessed July 4th, 2016).

Eisenberg, David. *Encounters with Qi: Exploring Chinese Medicine*, rev. ed. New York: W.W. Norton, 1995.

_____, Ronald C. Kessler, Cindy Foster, Frances E. Norlock, David R. Calkins, and Delbanco, Thomas L. "Unconventional Medicine in the United States—Prevalence, Costs, and Patterns of Use." *New England Journal of Medicine* 28 January1993.328: 246-252. See: http://www.nejm.org/toc/nejm/328/4/DOI:10.1056/NEJM199301283280406

Field, Tiffany. *Complementary and Alternative Therapies Research*. Washington D.C.: American Psychological Association, 2009.

_____. *Massage Therapy Research*. Edinburgh: Churchill Livingstone, 2006.

_____. *Touch*. Cambridge, Mass.: MIT Press, 2003.

_____. *Touch Therapy*. Edinburgh: Churchill Livingstone: 2000.

Frampton, Susan B., Laura Gilpin, and Patrick A. Charmel, eds. *Putting Patients First: Designing and Practicing Patient-Centered Care*. San Francisco:

Jossey-Bass, 2003.

Lown, Bernard. *The Lost Art of Healing: Practicing Compassion in Medicine*. New York: Ballantine, 1999.

MacDonald, Gayle. *Medicine Hands: Massage Therapy for People with Cancer*. Forres, Scotland: Findhorn Press, 1999.

_____. *Massage for the Hospital Patient and Medically Frail Client*. Philadelphia: Lippincott Williams & Williams, 2004.

National Institute on Aging. Comfort Care. https://www.nia.nih.gov/health/publication/end-life-helping-comfort-and-care/providing-comfort-end-life, accessed March 5, 20016.

Remen, Rachel Naomi. *My Grandfather's Blessings*. New York: Berkeley, 2001.

Scarry, Elaine. *The Body in Pain: The Making and Unmaking of the World*. New York: Oxford Univ. Press, 1985.

Walton, Tracy. "Help for Hospital-Based Massage Therapy" See: http://www.massagetoday.com/mpacms/mt/article.php?id=14693, accessed March 5, 2016.

_____. *Medical Conditions and Massage Therapy: A Decision Tree Approach*. Philadelphia: Lippincott Williams & Wilkins, 2004.

_____. *Spontaneous Healing: How to Discover and Enhance Your Body's Natural Ability to Maintain and Heal Itself*. New York: Alfred A. Knopf, 1996.

Wishart, Adam. *One in Three: A Son's Journey into the History and Science of Cancer*. New York: Grove Press, 2007.

PART TWO

HEALTH IS CONTEXTUAL:
CONTRASTS OF HEALING QIGONG AND
THREATENING CLIMATE CHANGE

IV THE MANY WORLDS OF ENERGY, QIGONG,
AND THE UNIFIED WORLD OF NATURE,
HUMANS INCLUDED

WHAT IS QIGONG?

Reviewing our trio of matter, mind, and energy: Section I on massage emphasized the matter of the flesh, Section II on a cancer patient emphasized the mind. Section III on massage for cancer patients emphasized mind and matter, with an introduction to the energy in the ancient Chinese tradition of Qigong.

In this section, we focus on energy, and our discussion focuses on Qigong. Because Qigong brings matter, mind, and energy into a whole, it has applications for health and healing for humans and, even, the earth.

Qigong is an ancient Chinese art rediscovered and popularized in the 20th century. The word itself is a modern coinage. "Qi" (or "Chi" or "Ch'i") means energy, subtle energy, bioenergy, or universal energy. "Gong" can be translated as use, study, practice, or work. Qigong has spread world-wide now, but standard Western medicine is largely unaware of its resources that could help patients, caregivers, entire hospitals, even the culture as a whole.

Qigong understands energy as local as any given person and as large as the earth, even the surrounding universe. Qigong believes that mind can move energy within a body or across distance. Qigong understands that all matter has energy, from a snail to a distant galaxy. For Qigong, **nature**

and the universe are coherent in matter, mind, and energy, all in dynamic harmony.

Although I had some exposure to martial arts and Reiki over many years, the world of Qigong was new to me. Not only was I a beginner, I was skeptical of some of the concepts and claims for its efficacy. In section VII, we'll see how modern physics may explain how Qigong works.

ENERGY AND QIGONG

Have you ever felt that So-and-So *rubs you the wrong way*? Maybe it was the way he or she spoke or acted. Maybe the appearance, maybe a memory of a similar person, maybe tone of voice, maybe pheromones—these and more are possible factors. It is also possible that, at a deeper level, the energy from that person conflicted with your own energy. Hippies of the '60s and '70s spoke of "bad vibrations, man" to describe the harmful energy of a person, place, or anything else. I consider myself a pretty friendly and open guy, but now and then there are some people that just rub me the wrong way. I try to understand this reaction and to take steps to be fair with them, but I also want to understand such rubs better.

Or maybe it was a person who rubs you *the right way*? You feel an immediate attraction—sexual or otherwise—to this person. You must speak with him/her. You are sure you have something in common, experience, emotions, ideas, or something else.

Qigong is way of understanding these events; it sees humans as relating because of our energy that we send and receive…all the time, but Western culture is not much aware of it.

The study of Qi promotes our awareness of Qi energy that pervades the universe and especially in humans because of our active minds. We are like antennas, sending and receiving energy all the time and particularly with fellow humans, but also pets and favorites places, such as our homes.

Like most Westerners, I have lived in the standard, Newtonian world, but now and then I visited the Eastern and "psychic" (for lack of a better term) worlds, the other side of the same street we all live on. Over several decades, I had some training in Judo, Aikido, and Yoga. In the martial arts, we were taught to feel our opponent's energy. In Yoga classes, we were taught to "breathe into a tight area," meaning taking awareness there so that it would relax. My wife and I studied Tai Chi (a choreographed version of Qigong) and found it relaxing. We took a workshop in London that included "dusting" of the auras, concentric layers of energy around the body.

Energy has many meanings in other cultures, some of them ancient, some countercultural, some indigenous and active today.

QIGONG IN "THE PARIS OF THE PIEDMONT"

In 2010, my current massage therapist Joanne mentioned an afternoon workshop on Qigong in nearby Carrboro. She was the one who told me, "This may sound weird, but we're going to talk to these muscles." Because her strategy had worked, I trusted her instincts and opinions, even though I had never heard of Qigong.

Even the little town of Carrboro suggested promise. Nestled against the larger and more upscale Chapel Hill and nicknamed "The Paris of the Piedmont," Carrboro is home for hippies, bohemians, teachers, artists, professors, healthcare workers, and the like. It is known for good restaurants, places for music and comedy, and a large store for gardening, small-scale farming, even pet and livestock supplies. The original building of the Carr Mill processed cotton for many years; later it was urban-renewed into small shops and restaurants. A rail spur ended there; young men (no women in the old days) coming to the university by train were taken up the Hill by horse-drawn wagons.

In the relatively conservative state of North Carolina, Asheville and Chapel Hill/Carrboro are the main centers of liberalism and interest in new ways of looking at, for example, health. In the *Qi Journal*, advertisements for Qigong, Tai Chi, and the like have many listings for California and the West coast in general, some for the East coast (mostly in New York), but much less for the "fly-over states." Little or nothing for North Carolina, except for Asheville.

Joanne and I drive down to Carrboro and park at a physical therapy office. We climb a set of stairs and find seats. A tall brunette woman gives us a pleasant smile. This is Lisa B. O'Shea. Before long she's presenting to some two dozen people.

Lisa tells us that everyone has Qi, which is variously described as energy, bio-energy, and/or universal energy. She asks us to feel it as a ball between our hands, or warmth, or a color, even a sound. I feel it, but some people say that they do not. Next, we are to glide a hand above the opposite arm; can we feel energy from the hand on the arm? I think I do…or am I kidding myself?

"How did you get interested in Qigong?" someone asks her.

She replies, "I went with a friend to a workshop like this. I was having problems with asthma, so I decided to try a treatment or two. My asthma cleared up. I was planning to teach high school science, but Qigong was more interesting."

I like her. She is smart—a former chemical engineer—non-flakey, and seems to have a good energy about her.

Master Lisa B. O'Shea, her lineage, and Qigong

According to Lisa's bio on the Web, her parents and grandparents included physicians. She trained in Chemical Engineering at Rensselaer Polytechnical Institute and worked for some big companies in her field. Feeling urges to change, she studied Reiki and hypnotherapy and went back to school for a teaching degree. At the same time, she was taking the classes in Tai Chi/Qigong that alleviated her asthma. She began to study Qigong and Traditional Chinese Medicine in earnest. In 1995 She became a Certified Qigong Healer through the Chinese Healing Arts Center with Master T.K. Shih. She became a student of Dr. Yang, Jwing-Ming, learned his school's system, and became certified. In 1998 she created the Qi Gong Institute of Rochester, offering a variety of classes in Qigong, Tai Chi, Qi Healing, Meditation, and Nutrition as well as certification in Qigong Therapy. She continues to study and add skills and techniques. In 2012 Dr. Yang named her Master.

Lisa's lineage, as we say in this field, is through Dr. Yang. He studied Qigong and a variety of martial arts extensively in China, arriving in the U.S. in 1974. He earned a doctorate in Mechanical Engineering at Purdue University while teaching Kung Fu and Taijiquan to students. In 1982 he established Yang's Martial Arts Association in Boston. Now known as YMAA, it is a publishing house and a worldwide organization.

Qi = energy, Gong = practice

As I listened to Lisa, I found the concepts and history interesting. "Qi" means energy within each one of us and within the universe; Qi is like spirit, like breath. Indeed an early written character refers to steam rising from cooked rice. "Gong" means study, practice, or work. A perpetual student, I liked that as well. When I received two degrees of Reiki many years ago,

I respected the idea of healing energy but did not find enough intellectual richness to interest me or to explain how Reiki worked. By contrast, Lisa's presentation seemed nuanced and inviting.

Qigong is related to Tai Chi, the slow dance-like exercises people often perform in parks. The most well known version has 24 forms agreed upon in 1956 by the Chinese Sports Commission, but there are other versions, including a set of 108 forms. Tai Chi, which can be translated as "great harmony," may be understood as either a choreographed form of Qigong or as a stylized martial art. Related to both, Kung Fu is the martial art made popular by a television show of the 1970s. Qigong, Tai Chi, and Kung Fu all share the came concepts of energy, harmony, and the mind's ability to direct energy. (An earlier translation of Tai Chi included meanings of "fist" or "boxing.")

"The mind moves the Qi," said Lisa, a statement I would hear many times in my training with her. Was this possible? I found it a fascinating idea.

She also said, "Qi energy is universal and plentiful. We should use it."

She also said, "We use acupressure points to nourish and move energy… but no needles."

ANCIENT CHINESE ARTS AND MODERN RENEWALS

The modern term Qigong came into usage in China in 1953 (Harrington, p. 228). During Chairman Mao's Cultural Revolution from 1966 to 1976, however, Qigong was suppressed because it was part of the "Four Olds" of customs, culture, habits, and ideas. It is indeed very old, perhaps 4,000 or 5,000 years old. The deepest roots were the shamans of ancient China. In its long history there have been various emphases and names, including Dao Yin (guiding and stretching), Tui Na (breath work), and Jing Zuo (sitting meditation), and Xing Qi (circulating Qi by the mind).

An early and pivotal document is the *Huangdi neijing*, variously translated as *The Yellow Emperor's Book of Internal Medicine*, or *The Emperor's Inner Canon*. Scholars date it somewhere in the 5th century BCE to 3rd century CE, although it reflects traditions far older. It is detailed; my modern translation has 260 pages. Dr. Yang describes texts and archeological finds from early days up to the present in his *Qigong for Health and Martial Arts*, 2nd ed. (pp. 3-9). He also describes four schools of Qigong theory and practice with overlaps among them. The Confucians wanted Qigong to make people fit for their social roles. The Buddhist monks sought freedom from suffering through meditation and breathing exercises to quiet the mind; their Shaolin temple is famous for Kung Fu today, despite its suppression under Chairman Mao. The Daoists followed the Dao (the "Way") of harmony of the universe, seeking perfection and, even immortality; many of their symbols come from nature. The *Tao Te Ching* of Lao-Tse is a well-known Daoist text. Physicians (the fourth group) could be in any of the first three; they sought balance of Qi in persons.

Mary Beth Braun and Stephanie J. Simonson see a possible link between Kung Fu and modern massage. "The ancient Chinese medical reference Cong Feu was translated into French by Jesuit P. M. Cibot in 1779. It is very likely that this text served as the inspiration for Per Henrik Ling's Swedish Gymnastics" (*Introduction to Massage Therapy* 3rd ed., p. 11).

Qigong was often practiced within monasteries, such as Shaolin, or kept within families, so that widespread transmission and acceptance, even in China, did not happen until the late 20th century. Today there are hospitals and clinics using Qigong, some combining it with Western medicine. *Chinese Medical Qigong* is the official textbook for medical Qigong. (Tianjun Liu, 2010). A comprehensive text of 653 pages, it is used in Chinese colleges and universities. For an introductory overview of Qigong, see Christine J. Barea's *Qigong Illustrated* (2011).

In 1984, David Eisenberg published *Encounters with Qi: Exploring Chinese Medicine* that describes his time in China to study medicine, Qigong, and more. He found all these to be efficacious. "My observations in the clinics of a Beijing hospital suggest that acupuncture, herbal medicine, and massage may be highly effective therapeutic tools" (p. 134). He mentions more than once that Chinese medicine is more concerned with prevention than intervention and that "life-style is an essential part of health" (p. 153). Years later, Eisenberg became a professor of medicine at Harvard, and founder of the Osher Center for Integrative Medicine. In 1993 he and others published the article previously mentioned that startled many readers because it said that one in three Americans used "unconventional therapies," specifically some 36 alternative "interventions," including massage, relaxation techniques, chiropractic, energy healing, and acupuncture. Typically, these were not discussed with the treating physicians, so they cannot be called "complementary" from a medical standpoint, although they might be called "integrative" from the perspective of the patient. This survey showed how pervasive other therapies were, that they were outside standard care, and that they had large economic impact.

Today the National Qigong Association in the U.S. has some 450 members. There are teachers, seminars, and centers across the U.S. and around the globe. There is a large amount of information on the Web, including videos of exercises. In 2015, a "Qi Summit" at Omega Institute, Rhinebeck, NY, drew over 200 participants (see "Take-Home Treasures…" below).

"ENERGY HISTORIES" OF PEOPLE, MODERN SOCIETY

When we discussed "touch histories" in relation to massage, we focused on the variations among individual people, given their past experience. We can consider a parallel term of "energy histories" for individuals and even for our society at large. I believe a Westerner person's energy history is now dominated by four levels of contemporary "social energy history": Newtonian, kinetic/industrial, capitalist, and human. These pervasive social values in "developed" countries are so dominant that we have difficulty understanding other definitions of energy, including indigenous concepts such as Qi and also notions from Quantum physics (see Section VII).

Energy as Newtonian, kinetic, industrial, and/or nuclear

Our education stresses the Newtonian physics of forces, gravity, planets' orbits, and material objects in general. We learn that there are kinetic and potential energy, but it is kinetic energy that excites us, because it is active and "gets things done." We learn in school that "Energy is the ability to do work," Also: "Simple machines focus and transmit energy." Potential energy includes coal, gas, and oil in the ground, but these are "worthless" until extracted and *put to work*.

We associate atomic energy with military and industrial uses. We know something of the old-fashioned Bohr atom (nucleus surrounded by planetary tracks), the atom bomb, and nuclear fission in power plants. We are vaguely troubled by atomic weapons meant for war (and no other purpose!) as well as nuclear disasters such as Three-Mile Island in the U.S. (1979), Chernobyl in Russia (1986), and the Fukushima Daiichi Nuclear Power

Plant in Japan (2011). Typically we see or hear news of these but soon forget them.

Influenced by notions of kinetic and industrial energy, we are generally unaware that all matter—our bodies included—is made up of vibrating atoms and molecules that routinely interact with each other and, according to Quantum physics, with the entire universe.

Capitalist energy

More specifically, Western culture defines energy primarily as the power provided by coal, gasoline, electricity, or nuclear fission, although sometimes coming from renewables such as wind and sun. This energy powers our cars, lights, computers, and many other things. We use carbon energy without a thought to its origins of coal, oil, and natural gas. Some people find it hard to accept the notion of Climate Change, in part because of fear that humans have largely caused it by burning carbon fuels, in part because the projected changes are so frightening as to be unimaginable. These fossil fuels are the products of plants from long, long ago, and their energy came from the sun (our local star) via photosynthesis.

Before the industrial revolution, the power of humans or animals was the source of mechanical energy, and wood (or peat, coal, or dung) provided heat. With the mining of plentiful coal (making possible steam power) and then oil, the modern lifestyle evolved with trains, cars, planes, air-conditioning, movies, TV, computers, and much more. As a commodity, energy is now bought and sold; we buy and sell energy stocks. Some observers speak of a "post-industrial society," but most of us cannot imagine this. The capitalist meaning of energy is now separated from the origins of nature (carbon from the earth) with scant interest in other kinds of energy, such as wind, waves, weather, and, beyond these, no awareness of planetary orbits, stars, cosmic rays, etc. Energy in the capitalist sense is abstract because of economics: a strong change in the price of crude oil can take the numbers of stock markets up or down immediately.

Human energy, personal and social

We admire people with high energy, real "doers," "go-getters," "water walkers," and/or "movers and shakers." Companies seek energy within their employees, working groups, and executive teams. We admire zeal, zest, verve, *élan*—whatever the name—"to do work." This meaning of energy emphasizes the values of individualism, capitalism, and Neodarwinism discussed in Rant 1 above and often focuses on the making of products and profits. Energetic people often have trouble relaxing; many of them have stress-related illnesses. Typically they are so urbanized that they have little contact with nature or awareness of it, except for "bad" weather.

In summary, these notions of energy history, both social and personal, have eclipsed other meanings of energy, both ancient, as in Qi, and modern, as in Quantum physics.

OTHER ENERGY TRADITIONS, INCLUDING INDIGENOUS AND PSYCHIC

We recall that Tom Bender lists 65 indigenous cultures that routinely use energy, and it is safe that assume that there are more and have been many more ("The Physics of Qi," 2006).

In his *Atlas of Mind Body and Spirit*, Paul Hougham discusses ancient and/or indigenous cultures with holistic approaches to energy: the Jewish Sephiroth (spheres of light) in the Kabbalah, the symbolic body map for the Dogon Tribe of West Africa, and the LightWheels of human potential of the Mayans, Toltecs, and (later) the Navajo. These traditions stress energy as well as connections of humans and the universe (London: Gaia, Octopus Publishing Group, 2006).

From various sources, here is a brief list of energy concepts in other cultures: Prana (the Yoga of India and Tibet), Qi (China), Ki (Japan), Baraka (Sufi), nafas (Islam), Wakan (Lakotas), Orenda (Iroquois), Megbe (Ituri Pygmies), nilch'i or winds (Navajo), ha, breath of life (Hawaiian), Pneuma (the ancient Greeks and early Christians), Ruach (ancient Hebrews), and Spiritus (early Christians). The last three are found in the Bible.

For millennia around the word, many people have had larger views of energy. Are modern Westerners somehow the last to know...or remember...or be reaffiliated?

As for larger cultural perspectives, psychologist Richard E. Nisbett has explored the differences in Western and Asian thought. Among the differences, he finds the following. (We can see resonances with Figure 3 above comparing standard medicine and integrative medicine.)

WESTERN THOUGHT	ASIAN THOUGHT
Hierarchies, causality	Relationships
Independence	Interdependence
Persons as isolated	People as embedded in society
Debate, control	Harmony
Objects, things	Ground/Field
Logic, final outcome	Contexts of process
Seminal thinker: Aristotle	Seminal thinker: Confucius
Approach to health: atomistic	Approach to health: holistic

Figure 4: Western and Asian Thought

More specifically, Nisbett observes that Eastern medicine stresses health as a balance of forces, while Western medicine emphasizes illness and interventions to specific body parts or systems (*The Geography of Thought: How Asians and Westerners Think Differently...and Why*, 2003, p. 193).

As I thought about exploring Qigong, there seemed to be two separate worlds of perception and causality, Newtonian and ancient Chinese. Could Lisa O'Shea, trained as an engineer, make sense of Qigong in a way that would merge the two worlds for me?

Indigenous concepts of energy, contemporary and ancient

While on an ecology tour in Alaska, I met a native man, Larry Merculieff. He had the dual heritage from his Aleut tribe—he was an elder—and from Western science, in which had earned advanced degrees. He was, we could say, a citizen of both worlds. Like many natives, he was a quiet man, soft-spoken, and never, it seemed to me, in a hurry.

When he described his upbringing by another elder, he said they didn't speak much. His role was to watch and learn. One day he sat for a long time with hunters overlooking water. Suddenly all the men looked at the same spot...and a seal rose up. Apparently the men shared the same sense of energy within their arctic world and any immediate change. I've heard of desert men sitting still then suddenly leaping up and then running in the same direction because they all sensed rain in a specific place. Westerners of the dominant culture have lost such awareness; their only exposure to it may be the scene in the movie *Star Wars* when Obi-Wan Kenobi feels "a disturbance in the force."

Larry told us that Alaska's coastline now had much less ice, owing to Climate Change. Coastal communities were at risk from waves and habitats were changing for wildlife, including seals, walruses, and polar bears. Spring comes three or more weeks earlier, and the behaviors and needs of birds and other creatures are disrupted. Driving up to Fairbanks, we saw "drunken forests" of aspens that can no longer stand upright because the permafrost around their roots was melting. Further, there is methane released by this thawing also in vast Siberia, and methane is an even worse greenhouse gas than carbon dioxide. As polar ice melts and disappears, reflectivity of the sun's rays diminishes, and the revealed dark sea absorbs more energy. The poles have been called "the air-conditioning for the earth." Lose polar ice: you lose a lot. We'll turn specifically to Climate Change in Section IV.

Larry extended his native realm to the academic, Western realm (see Merculieff and Roderick, 20013). I'm extending in the other direction, from an academic career—including study at the University of Chicago, then a very rational place much influenced by Aristotle—to explore the worlds of massage, Qigong, and the changes in consciousness that inform those practices.

California insights
The cosmopolitan Bay Area
In 1954-55, my father had a fellowship to UC-Berkeley. The Bay Area was a revelation to my sister and me, who, raised in Arkansas, had never been out West. Our Berkeley grade school included children of many backgrounds, among them Asian, Latin, and Jewish. From our schoolyard we could see the Cyclotron perched in the Berkeley hills, although we had no idea then that experiments there dealt with nuclear and particle physics, proton therapy, and commercial production of radionuclides.

One day our family had a meal at San Francisco's famous Cliff House. In the lobby there was a small statue of a human with red, vertical lines that had no relevance to the body as I—a sixth-grader—roughly understood it. I asked my Dad about it. He said, "It's Chinese, and a different way of understanding how the body works." I was amazed; how could the Chinese model, looking so different, be correct? Sixty years later, I have my own charts of energy meridians or, more accurately, "channels," and have used them in Qigong.

Psychic aura readings at Esalen Institute in 1974

I discussed massage at Esalen in the Section I. Now I'll describe an event in the Bioenergetics Workshop at Esalen led by Hector Prestera, M.D. One morning he said that he was just gaining the ability to see auras. Another participant said, "You know, I could sit here and say nothing, but I think I should mention that I am a psychic who sees auras in detail and can do readings." We all stared at him. Hector asked him if he'd be willing to go around the circle with readings. "Sure," he said. For each person, he tilted his head and seemed to unfocus his gaze. He proceeded to tell all of us, one at a time, specific things about their lives, families, pets, and so on.

After a reading, most people said something like, "Wow, that's amazing. You're exactly right." Now and then, however, someone said, "No, I don't think so." The man said, "Don't worry about it; you'll think of it later. When I first got this gift, I didn't trust it, but now I know it's always right." For one man, he said, "I see a child with some kind of cards." The man said, "I can't think of what that would be." When the psychic turned to the next person, she said, "You know that child with cards? That's my daughter who does visual therapy with cards." The psychic said, "The image was between you two sitting there, and thought it belonged to him."

At lunch, I found my wife and took her to see him. Could he do another quick reading? He agreed. He cocked his head and unfocused his gaze. His left hand jumped to his right arm and he cried out. He had accurately located a bone graft done on her arm many years before. Covered by a sweatshirt, her arm appeared normal.

SPOON BENDING AND REMOTE VIEWING AT THE RHINE CENTER

I had heard enough accounts of spoon bending to believe that it might happen, but nonetheless my American-Newtonian mind dismissed it because I couldn't imagine how it would work. Such accounts are often called "anecdotal," a disparaging word used by some scientists to discredit anything they are not ready to study or even consider possible. "Anecdotal" suggests a casual story or a piece of amusing gossip about a singular occurrence, certainly "an oddity that we should ignore." Nonetheless, oddities can be true, actual parts of experienced reality that our culture does not currently recognize, measure, or even consider a useful area of future inquiry.

The Rhine Research Center in Durham, North Carolina, has long studied so-called Psi phenomena. The Greek letter psi (ψ) starts the word ψυχή or "psyche," meaning "mind" or "soul." In modern usage Psi means "paranormal," an adjective describing such phenomena as ESP (extrasensory perception) and "psychokinesis" (moving objects with the mind). One Friday in May of 2015, my wife and I went to a Rhine open house because spoon bending was to be the final activity. As a doubting Thomas, I needed to see with my own eyes, but as a student of energy, I wondered whether I, myself, could bend spoons.

About 20 of us listened to an introductory talk then split up into three groups.

One group went to experience Zener cards, each card displaying one of five designs. Our leader held up cards, the design toward him, and we were to guess/sense/intuit that particular design and mark on our papers. At the end, I came out "according to chance," that is, no demonstrated psychic recognition of the hidden designs, although I had more success near the end of the demonstration. Could this faculty, I wondered, be learned with practice?

In another room we looked at clock face of unlit bulbs except for one blinking at 12 o'clock. We were to move the light round around the circle with our minds. I started in very well, moving it three-quarters of the way around before my ego became pridefully involved and my influence abruptly ceased. Apparently some capacity of my mind had worked well, a capacity that I didn't know I had.

Remote viewing

My group moved to another room for remote viewing. I had read enough accounts, including military uses, to believe this ability to be possible, although, again, rational understanding escaped me. Scientist Dean Radin discusses instances of remote viewing, scientific studies that validate the phenomenon (including military uses) as well as possible theories in his *The Conscious Universe: The Truth of Psychic Phenomena* (1997, pp. 101-115).

Our leader asked a member of the group choose from several envelopes, each containing a photo. What was on the photo of the chosen envelope? We had paper and pencils to draw whatever we sensed. We were told to relax and breathe slowly. I immediately imagined a large, white church with a prominent steeple and quickly sketched it with a dozen lines. My ego returned and chided me, "That really isn't possible; you are totally full of crap."

For the rest of the five minutes I sat in some turmoil, inviting another and perhaps correct image to come to me. None did. When the photo was revealed, it was some kind of large church with a tall steeple, not exactly what I had sketched, but close enough to impress the group and the leader. In the realm of intuition, something had worked, but I had not idea what.

Spoon bending

All groups reconvened for the celebratory ritual of spoon bending. With excitement, we took our seats around tables. Our leaders stated that we might have 100 per cent success, and they hoped for this result, but typically about half the group bent spoons. They brought out a large, red plastic tub with a huge number of ordinary spoons. They said we should empty our minds, hold the spoon upright on the table with two hands, and picture energy from the universe coming to the very top of the spoon, then moving down. Soon there were gasps and "wows" as spoon bowls bent over. The woman across from me sat there with a big grin on her face and a dramatically bent over spoon on the table. I picked it up and felt the bend; it was not warm. I tentatively tried to unbend the spoon. It didn't budge.

During the three trials the leaders urged us onward—which I didn't like. After all, I was consciously focusing my energy and didn't want any distraction. Evidently they knew that distraction was good because it took rational minds (like mine) away from the "task."

At the end, some 25 to 30 spoons were crooked over, thoroughly bent. This form of psychokinesis (mind influencing matter) clearly works. Of the six people at my table, three bent spoons; I bent none and wondered why. The woman across from me kept asking for larger spoons. The last one she bent was a heavy serving spoon.

Scanning the Web, I've found two theories to explain spoon bending. First, Van der Waals forces holding the metal together dissipate and the spoon is sent back to an earlier time in its life, perhaps when molten. Second, Quantum energy of mind is directed to Quantum level of metal, where mind's intent makes subatomic particles increase energy so that metal atoms come apart. These two accounts may be saying they same thing in different language, but neither gives me a clear picture. Therefore, I understand the reality of spoon bending as an event that is so far unexplained, therefore a mystery but, nonetheless, also a reality. I imagine science will eventually catch up.

"WEIRD," "TWILIGHT ZONE," THE "WOO-WOO FACTOR": REJECTING NONSTANDARD NOTIONS OF ENERGY

Popular responses

Spoon bending, remote healing, and telepathy are so bizarre in comparison to our schooling and typical mass media that it is sensible for us to ignore them or reject them without a second thought. These are usually "non-starters" because they would disrupt our received opinions and filters. Common responses include:

"That may be possible for other folks, but surely not for me."

"That's the Twilight Zone, you know, the woo-woo factor."

 "Weird, man, and so much bullshit that I cannot be bothered."

If we are enmeshed in the dominant paradigms of reality and causality, it is very difficult for us to consider other approaches to reality. On the other hand, there are people who accept such topics without critical thought and many other notions such as aliens, UFOs, and past lives. (Yes, these are beyond my current acceptance; maybe that will change?) These provide believers order and understanding, even though science gives no substantiation. Some fanciful ones are useful as entertainment in fiction, art, and film, where we suspend disbelief and enjoy our esthetic imaginations.

How do we discriminate plausible models, possible models, and outright fanciful ones?

Scientific knowledge and evolution; lags of social acceptance

Psi phenomena have been studied by scientists with much care for two or three decades. Phenomena such as psychokinesis, remote viewing, and telepathy are real, but mainstream scientific theories have not caught up to them, nor have education or mass media given them attention, except as oddities. Science itself is, however, always changing, and people have a wide range of attitudes about science. Some people are skeptical about the word "theory," even though the gravity that holds them on the ground is described and quantified by a theory.

Martin Gardner described himself as a scientific journalist and a "fringe watcher." Some of his collection of columns from *Skeptical Inquirer* are collected as *Did Adam and Eve Have Navels?: Debunking Pseudoscience* (2000). He is a good example of 1990s skepticism and rejection of Psi phenomena. In Chapter 6, "Reflexology: To Stop a Toothache, Squeeze a Toe!" (pp. 83-91), he lumps reflexology ("preposterous") with Zone therapy ("as imaginary as the energy lines on acupuncture charts"), and myotherapy (removing "sharp pain"). Further, he writes dismissively of the twelve meridians also "a mysterious form of energy the Chinese call "ch'i energy," and "yin and yang energy." He criticizes two reflexology books from 1938 and 1951, probably easy targets to attack. He has more difficulty dismissing Sea-Bands, a commercial product for nausea developed in the 1980s that presses on P-6 (near the wrist) of the meridian system. He writes, "tests…were inconclusive." However, some earlier studies and several after his book appeared have shown efficacy for morning sickness, travel sickness, post-operative nausea, and nausea from chemotherapy. The manufacturer of Sea-Bands lists citations on a website (http://www.sea-band.com/for-medical-professionals, accessed July 14, 2016).

There are lags between the specialists, popular media, and a wide public—some of whom will never believe in, say, evolution or Climate Change. I've heard that "energy work" can be construed in North Carolina as dangerous, the influence of the Devil or other malign spirits.

Any social change of basic attitudes is difficult and slow. Although science and clinical evidence showed the danger of tobacco usage, tobacco companies (having the same information) nonetheless fought the truth for a long time. Eventually common sense and fair play won out.

Ancient traditions of energy have been around for millennia, but these are often regarded as *primitive* or *savage* and not worth any serious thought.

FROM MOTHER'S HAND TO A TEMPORAL ARTERY SCANNER: PARADIGM SHIFTS ABOUT MATTER AND ENERGY

But we do change some of our views and habits. The first users of telephones shouted into them because they didn't understand that speech could efficiently change from sound waves into electrical signals varying in frequency and amplitude and, at the other end, change back into sound waves. Not long ago we walked up to our TVs in order to change the channel. Now we use a remote that uses infrared energy as an information carrier; we neither see this energy in the air nor understand the physics, but we accept that it works. We slowly change our awareness and acceptance of some things like microwave ovens, cell phones, and GPS devices that would have seemed bizarre/magical/woo-woo a few decades ago.

What factors are involved in such paradigm shifts? First, would be our personal experience. Second would be general social attitudes. Third would be the state (or states) of the science of the day.

Personal accounts by scientists

As for personal histories, Lisa O'Shea became interested in Qigong when Qigong treatments cured her asthma. Bernard Lown became interested in complementary medicine when acupuncture in China took care of a back injury he had suffered (Lown, p. 129). In Section V we'll meet Imke Bock-Möbius, a Ph.D. physicist doing research in China. She describes being healed from "heavy symptoms" of coughing and blowing her nose after one treatment of acupressure and application of Qi (*Qigong Meets Quantum Physics: Experiencing Cosmic Oneness*, p. 8).

Even allowing for a placebo dimensions, such personal accounts by scientific people are impressive.

Shifting paradigms in taking a temperature

When I accompanied my wife for a surgical procedure, I was startled to see a nurse pass an instrument near her forehead and say—as if magic had occurred—that that her temperature was normal. "Hey, what's that?" I asked and got a brief explanation.

The device is called a temporal artery scanner; it is available in retail stores for about $50. A pyroelectric crystal in the device senses the infrared energy emitted from the artery in the forehead. As that name suggests, it converts the radiated heat (pyro) into an electric (electric) signal that is further manipulated to produce a displaying number. Because that artery brings blood from the heart, the registered heat is assumed to be near or at the core temperature of the body. The TA scanner measures infrared ("below red") energy just below the visible spectrum; humans can't see it, although spiders, some insects, and snakes could (if present) and this scanner definitely can.

When I was a kid, my mother's hand felt my forehead to check for fever. Her hand (or a glass and mercury thermometer) measured my heat by convection: material-to-material. Changing from this older method to energy measurement by the TA scanner illustrates a paradigm shift in technology linking the energy of the body's matter to the scanner that converts energy—first thermal, then electrical—to a material, numerical display.

Like a TA scanner, a Qigong healer's hand feels Qi energy and can radiate energy as well.

Medicine, like many fields, has a history of changes that were radical in their day but routinely used today: vaccinations, sterile technique, heart-lung bypass, green (not white) scrubs, and, today, slow acceptance of nutrition as a central component of health.

"MIND MOVES THE QI" AND MORE: STUDYING WITH LISA B. O'SHEA

Over the next four years I took five weekend seminars, some in North Carolina, others in Rochester, N.Y., Beginning, Intermediate, Tai Chi, Advanced, and Nutrition. Fortunately these counted hours toward my massage CEUs.

Some features were similar to the massage training, such as presentations by Lisa followed by pairing up to practice on tables. There were also differences. My massage class had the same students for every meeting. Lisa's seminars, distributed over time, had participants of much variety, even people from other countries. On an average, they were two decades older than the 20-something massage students I studied with and typically quite intelligent and highly motivated.

For students interested in certifying with her, there was another difference: reading assignments in *Between Heaven and Earth: A Guide to Chi-*

nese Medicine (1991) by Harriet Beinfield and Efrem Korngold, pioneers in American acupuncture, The book presents Yin and Yang, meridians, the five elements, and an excellent overview of Traditional Chinese Medicine (TCM) in this 430-page book. In Essay 1 above, I cited Victoria Sweet's prose about the organic and holistic system of Hildegard of Bingen's medicine as opposed to the industrial and mechanical framework for modern medicine that replaced it. Similarly, Beinfield and Korngold contrast the modern, Western philosophy of the human body as a machine and the doctor as a mechanic with the eastern worldview of the body as a microcosm of nature and the doctor as gardener. This discussion (Chapters 1 and 2) is the conceptual basis for all that follows in the book, especially the notion of *balance* in health.

Lisa's seminars covered such topics as the nature of Qi, Organ systems and meridians, meditations, healing sounds, exercises, treatment methods, and energy points used in treatments.

As we did work at tables, we became more aware of energy in our partners and in our own bodies. Some of the classic Chinese terms were poetic, for example an exercise called the Microcosmic Orbit, suggesting that humans are microcosms of the entire universe.

Throughout the basic goals were these:

- To be aware of energy in bodies,
- To help energy move freely and in balanced ways,
- To understand that such energy pervades the universe and that there is always plenty of it.

In Qigong, the activities of our *material body*, our personal *energy*, and our *minds* all interact.

WHAT DO YOU HAVE IN MIND? SOMETHING...TOO MUCH...NOTHING?

In a choral rehearsal, the man next to me pulled his cell phone out of his pants pocket whenever we stopped singing, probably looking for emails or texts. Whatever the reason, it seemed to me he could only half listen to the conductor, could not study or mark his part, nor learn from hearing other singers rehearse—which is to say, he left the shared intents of the rehearsal because he needed some other stimulation.

"The monkey mind" is an Eastern term centuries old: it suggests the chatter, buzz, and frenetic activity of a restless mind. (See https://en.wikipedia.org/wiki/Mind_monkey, accessed April 20, 2016.) Many "New Age" (or, often, "old age") traditions promote ways to quiet the mind, including varieties of meditations: sitting, standing, or moving. In one Buddhist meditation Lisa taught, we sat still and imagined our clothes disappearing, next our skin, then our muscles, our internal organs, and finally our bones. We imagined ourselves as nothing, totally effaced. Then we imagined putting the layers back on one at a time to reconstitute our bodies. Years later, I see this as a mental ritual in reaffiliation of self. Meditations help calm our *minds*, guide our intent, calm the *matter* of our bodies, smooth out our *energy*, and help us reconnect our energy to larger energies.

YING AND YANG, HUMANS AS TREES, AND OTHER BASIC CONCEPTS

In the lobby of a restaurant in Beijing, my wife and I saw a large platter, most likely ceramic, on a low stand. It displayed in beautiful colors, a large, dramatic dragon descending from heaven and, below, a lovely phoenix, gentle and tame. The two figures represented Yang, the male sky energy, and, Yin, the female earth energy. In Chinese philosophy, these are the two basic energies of the universe, cosmic forces that bring the actual world, humans included, into being. They are interactive, each including a portion of the other, as shown by the two dots in the symbolic representation of the dark, descending Yin, and the light, rising Yang.

Figure 5: The Yin Yang Symbol

This design is also called the "Tai Chi symbol," meaning "Grand Harmony," or "Grand Ultimate." For the ancient Chinese, the energies of Yin and Yang exist in dynamic, creative, and balanced harmony.

The variations of Yin and Yang can be seen in the history of the written Chinese characters. Yin has the meanings of the shady side of the hill (dark, hidden, female), while Yang means the sunny side of the hill (bright, manifest, male). These two dynamically change—even reverse—as the day goes on. In Chapter 42 of Lao-Tse's *Tao Te Ching* we read:

> All things bear the shade on their back
> And the sun in their arms;
> By the blending of breath
> From the sun and the shade,
> Equilibrium comes to the world.
> (R. B. Blakney's translation, 1955)

Some accounts describe how the progression of shadows from a vertical pole (like a sundial) create on the ground the S-shaped curve between Yin and Yang, thus linking heaven and earth.

In terms of Western neuroanatomy, there's a parallel to the interactions of the two branches of the autonomic nervous system, the sympathetic, energizing (Yang) branch, and the parasympathetic, relaxing (Yin) branch. Other alternations and interactions include inhalation and exhalation of breath, systole and diastole of the heart, sleep and wakefulness.

Physicist Neils Bohr was fascinated by the Yin Yang symbol and put it into his personal coat of arms with the motto *contraria sunt complementa*, Latin for "opposites are complementary" (Bock-Möbius, p. 69). Indeed, Bohr introduced the "complementarity principle" of Heisenberg to Quantum Mechanics.

Trees and humans: both microcosms of the universe

One meditative form is called Standing like a Tree (different from Yoga's one-legged Tree Pose). There are many variations of it in name and form in martial arts and meditative traditions; Kenneth Cohen devotes all of Chapter 10 to it in his book (*The Way of Qigong*, 1997). One version is called Hugging a Tree because the meditator stands and holds arms out in a circle, palms facing the chest. For Qigong and TCM in general, this pose helps energy flow up and down the body. A tree, the largest land plant, has roots in the earth but branches, leaves, flowers, and fruit in the sky, all facing the sun. Thus it symbolizes the vertical connection of earth (Yin) and heaven (Yang). The pine tree often shown in Chinese art is evergreen, symbolic of life even through the winter. In myth studies, the tree, the Cross, mountains, and other vertical structures are known as axis mundi or world pillar. In the American spiritual, the words proclaim, "We shall not, we shall not be moved…just like a tree that's standing by the water, we shall not be moved," words that echo Psalm 1 (1.1-3). A human meditating in Standing like a Tree can feel, consider, and celebrate earth energies and sky energies, even becoming a channel for them, a microcosm of the energies of the universe. Energy sensitive people can feel the energies of individual trees. One summer day I tested several trees to feel their differences. When one gave me nothing, I was alarmed; was I losing my touch? I looked up and saw no leaves because it was dead.

Five Elements, five Seasons, five organs

The ancient Greeks considered the four basic elements (earth, air, fire, and water) the basis for material reality, a tradition that continued until the English Renaissance. Indeed Shakespeare used "humour characters" in his plays, persons who had a strongly predominant basic element, such as love or melancholy (see Carter, *Clowns and Jokers Can Heal Us*, pp. 148-56).

Similarly the ancient Chinese saw basic elements (or phases) for reality, and they were five in number: wood, fire, earth, air, and water. These elements also related to five organ systems: Liver, Heart, Spleen, Lung, and Kidney. As bodily systems, these were not simply the physical organ named, but a complex of values and meanings. Spleen, for example, implies nutrition and growth, as well as the fluid elements of the body, such as saliva, mucus, and blood.

Like the classical four elements of the West, the five Chinese elements also depend on each other and need to be in balance.

For each organ system, there is a further pairing. Kidney, the Yin element, has the Yang partner of Urinary Bladder—which makes physiological sense because Kidney removes waste from the blood, sending it to the bladder for excretion.

Further, there are five seasons of the year, each correlating to an organ, element, and color. Summer, for example, has the partners of the Heart, fire, and red.

Further still, each organ system has a sound, a vibration that can be used for calming or energizing.

A circle puts the five organs/elements/colors into a wheel, one feeding the other consecutively in a clockwise fashion, B feeding C, but the precursor to B, namely A, limits C, all to create a dynamic balance in persons, the year, and all things. The circle runs: Liver, Heart, Spleen, Lung, Kidney in parallel with Spring, Summer, Late Summer (Harvest), Autumn, and Winter.

Channels (meridians) and vessels

Each organ has a nourishing energy channel. Although these are often called "meridians," the word "channel" is a more accurate and evocative for how energy flows. While meridians on a globe are external and rigorously geometrical, channels of the body are internal and vary in size and direction like streams and rivers. They run up and down the body, but with curves, zigzags, and connections to an organ, hence their names like Lung, Heart, Liver, Bladder (urinary), and Gall Bladder. These are Yin/Yang pairs with turn-around points in the head, hand, or foot. Many acupuncture points are on these channels. Still larger conduits are vessels. While these channels appear to have no relation to Western anatomical structures such as arteries, veins, or nerves, some investigators believe they may relate to fascial planes (Eden, p. 111). For a detailed account, see Giovanni Maciocia's *The Channels of Acupuncture* (2006).

ALEC'S QIGONG TREATMENT

How does a treatment go? Let's imagine a client named Alec. Since it's his first visit, I've scheduled him for 90 minutes. Following sessions will fit in an hour, but a 90 minute session has its own magic.

The magic of 90 minutes: talk and silence

When my bone marrow was damaged by the chemo, I went to an acupuncturist. I was impressed when she took a long time on my first visit to interview me and hear my story. There were even times of silence, with no rush, no pressure to speak on either one of us. I realized later that some things I said could come up only with that use of time. Further, she created a relationship that was supportive on a human-to-human level, the kind of witnessing discussed in Essay 3 above. How many of us ever feel *completely heard* by another person? When it does happen, supportive energy linkages begin.

Ninety minutes gives the client to ask questions, to review current health, and to remember concerns that have been hidden. Some people have no experience with healing energy, while others have extensive experience in the martial arts, Yoga, or Qigong itself. If Qigong is totally new to them, a slow, gentle time is important for gradual understanding.

As with massage, I start with social questions. "Where are you from?" "How was traffic today?" "How are you feeling today? ("Fine, thank you. And you?") "What are you looking for in a session today?" "What medical treatment are you receiving treatment currently…for what…how's it going?" And "What do you know about energy therapies, Qigong in particular?"

During our chat, I get an impression of Alec's energy and his level of relaxation or stress. Also, his notion of medical treatments, be they alternative, complementary, integrative, or even a last-resort trip to another country for unusual treatment.

I prefer that sick or injured clients are receiving standard care so that my massage or Qigong is complementary, even integrative, and not alternative, but I have to accept whatever they have decided upon.

An overview of Qigong for Alec

I have a three-ring binder with basic information—just nine pages.
What is Qi Energy?
Bioenergy, subtle energy, universal energy, Quantum energy…perhaps more.
Mind moves the Qi.
This is a basic tenet; consciousness can change the materials of our body.
Yin, Yang….heaven and earth.
Yin is female, earth energy, the shadow side of the mountain.
Yang is male energy, sky or heaven energy, the sunny side of the mountain.
A stick figure shows humans as an energy link, like a tree, between earth and heaven.
Tai Chi, Qigong, Kung Fu
These are all cousins, so to speak, using the same energy for different purposes.

Energy Channels...with Yuan points

The body has pairs of energy channels; special points to increase and manage energy.

Moon (Yin)	Sun (Yang)
Lung	Large intestine
Spleen	Stomach
Heart	Small Intestine
Kidney	Bladder
Pericardium	Triple Warmer
Liver	Gall Bladder

Relaxation

Energy management increases energy efficiency and allows the body to relax and heal.

Basic Principles of Qi

Qi is natural and universal.

There is plenty of Qi now and forever.

Nature and the Universe make sense: neither is chaotic, nor are they running down.

A Qigong therapist's hand is like a magnifying glass focusing the sun's rays. Mind moves the Qi.

Coherence can order incoherence.

The "three adjustments"

These are ways we recognize and use energy...as easy as A B C:

- Align the body
- Breathe deeply
- Calm the mind

I allow time for Alec's questions or comments about connections he sees with other traditions...or anything else in his world.

During this introduction, I try to recruit Alec's imagination, so that he can actively participate in our session. I want his mind engaged so that it can move Qi by himself in the future. As opposed to typical massage therapy, Qigong does not rely on the passivity of the client.

I believe this: if people just did versions of the three adjustments most days, they'd have less stress, more energy, and better health.

These three actions fit well with the three universals: our bodies are made of *matter*, breath suggests *energy* in Qigong, and *mind* is, of course, mind, both personal and universal.

From the general concepts to Alec's specific health and a treatment plan

I explain that I have a set of 74 questions to ask about symptoms. These come from Lisa's teaching; many are in the Beinfield and Korngold book mentioned earlier. These may appear strangely disparate to Westerners, especially for Liver, but they have their bases in Traditional Chinese Medicine (TCM), where, grouped by five organ systems, they make sense. I'm not focusing on individual questions as much as on their groupings. For the Heart, for example, I'll ask, "Do you have vivid or disturbing dream?" and "Do you sweat, with no apparent reason?"

As we go, I make check marks next to the five organs. I add them up and show Alec the results:

Lung	||	2
Spleen	||||	4
Heart	|||	3
Kidney	|||||	5
Liver	||||| |||	8

Next we make a plan based on Alec's initial interview and the organ systems with the most checks. I describe how he'll sit on the table then lie down, and I'll do work off the body and with light touch on his clothed body.

I'll have my client sit on the table, ideally with something green in view, like a bush or a tree out the window. I stand in front of Alec and ask him to place his hands about a foot apart.

"Do you feel any energy there? A pressure, a ball, some heat?" Some people do. Some don't. I crank up my energy with deep breaths and activate my Microcosmic Orbit a few times. Then I run the energy down my arms and place my hands outside my client's hands.

"Feeling anything now?"

"Hmm. Yes, there's something there," says Alec. Clients can usually feel the difference between their own hands.

Standing behind him, I scan his back with my hands, combing out blockages and filling the energy back up. I do energy circles on his Kidneys. "Everyone needs Kidney," Lisa said.

With Alec lying face up, I scan from head to toe, again combing out blockages and filling energy back up in those places.

There are other techniques, depending on what he needs and/or what I sense. It would take pages to describe them.

Much of this work is intuitive. A therapist feels energy, creates a path to universal energy, and then gets out of the way. I love this philosophy for two reasons: (1) the universe knows what it is doing and will help anyone (therapist or client) who tunes in and (2) the therapist's personal power (or ego) is not involved; as s/he works, the universe takes over to create a new harmony: Tai Chi (Grand Harmony.)

I make one more scan down Alec's body. Troubling areas now feel smooth or much improved. If there are a few stragglers, I'll treat them again. Then we are done for now. My hands nourish his Dan Tian (a central energy center) until it feels full.

I sweep my hands above and below him to close up his energy field and separate it from mine. I close my field and center down.

I have Alec sit up. He may feel sleepy, dizzy, or alert. (One set of head and neck treatments often puts clients to sleep.) I'm careful that he doesn't fall to the side.

"So, Alec. Take a moment…how do you feel?" This request gives him a moment to sense his energy, and this is important in a society that has, by and large, lost this ability.

If there was a particular ache or pain, I ask how that feels now. Usually it feels better or is entirely gone.

He replies according to how he feels. We head back to our interview area.

Self-care strategies to take home

Therapists like cooperative clients who will continue their care. My half-sheet form lists the three intentions mentioned (Align, Breathe, Center), a list of exercises that I choose from, a healing sound, and six meditations (for which I have handouts).

I state that I'll ask him on his next visit which ones seemed helpful.

EXERCISES AND MEDITATIONS FOR QIGONG PRACTICE

Two basic ways to practice Qigong are exercises and meditations. Meditations can be moving or sitting. They can be a form of gentle exercise or a mental exercise that imagines the body moving in space. Either way, they are synthetic of matter, mind, and energy.

The Exercise of Cloud Hands

There are hundreds (some say thousands) of exercises, forms, or other versions of moving meditation.

I'll sketch two of them; fuller descriptions are easily found online; some advise the guidance of a teacher.

The clouds in Chinese art are traditional as symbols between and linking heaven and earth. They can stand for dragon's breath (Yang) and simultaneously the sources of life-giving rain (Yin). A moving form of meditation is Cloud Hands, part of Tai Chi sequences or a standing exercise by itself. Basically, one hand makes a slow clockwise circle from midline outwards in front of the right shoulder then, with no break, the other hand makes a slow counter-clockwise circle outward from the midline in front of the left shoulder. (The plane of the circles would be on a wall you face about a foot away.) The two circles may be felt as a sideways figure 8, or an infinity sign in front of you. There are many variations including how breath is used, the size of the circle from very small to large sweeps of the arms, and the speed. In one variation, the descending hand symbolizes rain, the rising hand evaporation from land or sea—in short a miniature, hydrological cycle.

Exercise sequences

There are sequences of exercises, some with delightful names: Eight Pieces of Brocade, Crane Flying (also known as Soaring Crane) and my favorite title, Five Animal Frolics (also known as Five Animal Play). These are much older than the 1956 version of Tai Chi commonly taught.

Spontaneous Qigong is not a strict sequence; standing, you shake, twist, and move about anyway you like, loosening up tight places.

Meditations

Besides the Buddhist meditation mentioned above, there are many others. Here are two in brief:

Dan Tian meditation

Dan Tian is usually translated as "elixer field," meaning that you grow energy like a crop. It is located about two inches below the navel, similar to the third chakra in Hindu belief or the "hara" or center in Japanese tradition; there is no similar structure in Western anatomy.

Here is a basic version; there are many variations. Seated, you take long deep breaths. Turning you hands down, you picture Yin energy rising from the earth, through you feet and legs, and settling in your Dan Tian. Picturing your Dan Tian as coals, you see them glow brighter as you draw up energy and breathe into them. Turning you hands up, you draw Yang energy from the heavens, down through your head, neck and torso, and gathering in your Dan Tian; again the coals brighten as you breathe into them. You can choose the ratio between Yin and Yang nourishment depending on what you feel you need.

Microcosmic orbit

The sitting meditator moves energy up the spine, over the head, and centrally down the front. The details are more complex, but the basic idea is that the person is a microcosm of the Yin and Yang energies of the universe. The circle is like the Yin Yang symbol turned sideways (back to front, sagittal plane), and the meditator rests within the circle, feeling both calmness and nourishment. I've done it in line at the supermarket, waiting in my car at a red light, and lying in bed calming down to go to sleep.

Much of Western life assumes that our mind and energy goes outward…in order to get things done! In meditation, however, we quiet mind and energy, so that other energy and mind can *come to and into us*. In one scriptural formulation, "Be still and know that I am God" (Psalm 46.10).

A STUDY GROUP

I learned that there was a study group for Qigong in Chapel Hill and joined it. We regularly meet each month in the evening. Members took turns reviewing a concept or a technique for 30 minutes or so, and then we set up tables and exchange treatments. Sometimes at the end we asked if there was anyone needing energy sent remotely. This was new to me…and fascinating.

One evening I said that my wife had an eye problem that was bothering her. Without hesitation, the group—most of whom did not know her—was quiet with eyes closed and sent energy. When I got home, I asked Nancy how her eye was. She blinked a moment and said, "It's fine!" I asked if she knew, by any chance, when it got better. She said, "Oddly enough, I do know. I looked at my watch when it felt better. It was 8:50." That was the exact time we went her energy.

On other days, a study partner and I met to prepare for Lisa's exams, written, oral, and practical. We made lists of what we felt we understood and, at more length, what we did not know. We divided up the latter and reported to each other at our next meeting.

Before I could take her exams, however, Lisa required her students to provide 150 treatment sessions. Where would I find cooperative clients beyond my study group and family members? I offered sessions to members of the Church of Reconciliation, a progressive, Presbyterian church where my wife and I are members. Over six months, adventurous souls gave me the gifts of their bodies, minds, and energies to work with. Lisa said that we could ask for payment for such work, but I did not.

WHO IS SENSITIVE TO ENERGY? (1) DOGS AND OTHER ANIMALS, ANCIENT AND/OR INDIGENOUS PEOPLE, AND SOME WESTERNERS

In 2004, an earthquake in the Indian Ocean caused a tsunami that killed some 250,000 people. In contrast, many animals, both domestic and wild, escaped death because they headed away from the shore before the disaster. Among them were elephants, monkeys, dogs, cats, flamingoes, and bats. Why did they do this? Perhaps they sensed subsonic sound waves or vibrations in the earth, but this has not been determined. What is clear is that they felt something that humans could not (Sheldrake, *Dogs That Know When Their Owners are Coming Home, and Other Unexplained Powers of Animals*, 1999, 2011, pp. 277-78.) Scientist Rupert Sheldrake has studied some 1,500 cases of animal behavior, including awareness of impending earthquakes, storms, and avalanches, even air raids. He believes that theories of physical energy are inadequate, and that premonitions and precognitions among animals and humans may have deeper and as yet unidentified energy. Quantum theory, morphic fields, or combinations of known energies such as radiation and electromagnetic fields may be factors. What is clear, however, is that in some situations dogs and other animals know more than we do.

Many years ago I spoke with a nurse who had energy sensitivity. At the time she was a night nurse in a neonatal unit. She said, "Another nurse asked me, 'Lib, how come no babies ever die on your shift?' and I said, 'Gosh, I don't know!' although of course I did know. I just have a way of holding them, talking to them, and keeping them safe and well."

A fellow massage therapist said to me, "I just met a man and felt that our energies overlapped just right, I mean it was *Bam*! *And you know*! That ever happen to you?"

"Yes, indeed," I said. "You feel a match, a congruence."

"*That's it*, even though you know it won't be followed up in any way, romantic or otherwise."

"Yeah,…maybe a nourishing friendship…or just a wonderful event for that day…but you *know* it happened for sure."

"And that person may recognize it also and then you can really feel the vibe…or they don't."

These are examples of energy awareness our culture does not understand and often, therefore, ignores. Perhaps "love at first sight" is another example.

Material proprioception

"Proprioception," the dictionary says, came into usage around 1905; it means *reception* of the material self, one's own ("proper") body, that is, how it moves or is positioned in space. If police officers think a driver has been drinking, they may conduct a field sobriety test. In one part, they ask the driver to close the eyes and bring the forefinger to touch the nose. With more impairment comes less proprioceptive accuracy that allows for matching the two.

Proprioception has been much studied through its muscle and nervous system components. The awareness can be conscious or unconscious, so the *mind* is variously involved, but the information is, as understood now, *material*: where is my hand right now as I dribble this basketball?

An understanding of proprioception is routinely applied in a variety of physical therapies. A common example is the relaxing effect of vibration.

This works on a simple principle called "proprioceptive confusion." If you move or shake the body at random, the brain gets a deluge of unusual proprioceptive data. The nervous system, overwhelmed by the random stimuli, effectively "gives up" and stops resisting the movement and allows for relaxation of the muscles used. "Spontaneous Qigong" is similar: practitioners stand, walk, shake, dance, and/or look goofy in random ways, loosening up body, mind, and energy.

While *bodily* proprioception is the usual meaning, we can extend in two other directions.

Mind proprioception: what are you thinking? Too much?

Mental busy-ness is not, however, new; we looked at monkey mind earlier. Yoga teachers may say, "Let's take some slow, deep breaths; let's center down; let's quiet the monkey mind." Awareness of the monkey mind is the first step in controlling mind rather than having it dominate emotion and thought. We take stock: how is our mind working, feeling, perceiving today? If it's racing in different directions, how can we resume a calmness, focusing as needed? Meditation, counting breaths, chanting, and prayer are well-known centering devices that can calm our minds.

In Qigong, "mind moves the Qi." A quiet and focused mind can send energy to places in the body.

Energy proprioception

Yoga practitioners feel tightness of material in the body and send breath (energy) there to relax the tissues. Similarly in Qigong there is this material awareness but also another sort of proprioception: how is *energy* moving within the body. As a beginner, I found this a strange concept. As a student, I found that this awareness could be developed. Scanning a client, we find areas where energy is blocked, stagnant, or weak, and we can send energy there.

We can also treat ourselves. I like doing energy sweeps from my feet to the top of my head on an inbreath, then from top to toes on the outbreath. Then I reverse the directions. It is relaxing and a good way to prepare for sleep. Even untrained persons banging a knee on a chair will typically put a hand directly in the hurt area, directing *mind* and *energy* for pain relief and healing of the *material body*. Proprioception can involve all three.

Seeing auras and other perceptions of energy

I attended a seminar with James Oschman, a cell biologist who has studied energy in humans. By chance, there were two aura readers out of our class of 24 people. One afternoon he took all of us outside. We stood in a circle with a massage table in the middle. A male volunteer lay on the table, and the rest of us took turns passing by him and sensing his energy; then in a circle we all sent healing energy. The two aura readers said they saw energy changes after our treatment. One described waves of orange and red around him. Part of me was jealous because I couldn't see this, but part was glad for the different gifts I was developing. No one, it seems, has all gifts, and people feel energy in different ways: a warmth, a color, pressure, tingling, a sound. Later, I spoke with the man on the table, eager to test what I had felt. I had felt blockage in his upper lungs. I asked him if he was having any trouble breathing. He said yes, he had asthma. Because I felt a block in his right knee, I asked him about that. He said he had a basketball injury there. This experience boosted my confidence in what I could feel.

In presenting Qigong to a group, I've asked people how they feel about someone reading over their shoulder. They grimace and some say they don't like it. I think there are at least two reasons for this: (1) there is a distraction of our attention away the text we want to read, and (2) our energy field detects the other person's energy field and considers it a possible threat because it comes from behind us, out of our visual awareness and, therefore, mental control. (For further examples, see Rupert Sheldrake *The Sense of Being Stared at and Other Aspects of the Extended Mind*, 2003.)

In my experience, many people can feel the energy ball between the hands; it may take a few minutes, but they feel something, even if that can't describe it. Training and practice can make an enormous difference. In China, Qigong experts can discern 27 different pulses in the forearm because they have learned the concepts and believe in the categories, and they have practiced a long time to discern them.

"Extraordinary" powers

For some people, so-called extraordinary powers are in fact ordinary to them because they grew up in families where energy was routinely used. I think of Edd Jones from rural Georgia. From time to time the Rhine Institute brings him up to Durham to look at his psychic abilities. I heard him speak there. He said that his grandmother had healing abilities and that people in his town would come to her for healing. These were accepted values for his local culture. He was born with sensitivities like hers and thought it was normal until he learned that other people didn't feel energies. I asked him if he always knew which way north was. He said "Sure," and pointed exactly that way.

More famous nationally, Donna Eden is also such a person. As she grew up, she was surprised to learn that not everyone used energy. In her book she describes how she can feel energy nine different ways, using nine energy systems in and around the body. The first two are Chinese: the five organs and the channels (*Energy Medicine: Balancing Your Body's Energies for Optimal Health, Joy and Vitality*, 2008). She writes about her rocky health history. In her 30s, she was so sick that doctors told her to "get her affairs in order," but she survived and taught energy healing to many people.

For some people, extraordinary powers follow a near death experience. Eden writes about her father's heart attack, including nine stoppages of the heart; afterwards he was a much changed person with different sensitivities. Other people who have had near-death experiences have found abilities for clairvoyance, remote viewing, reading auras, and so on, as if circuits in their brain had been remodeled to work in a new way or to reawaken abilities they had all along. I've met some of these; they are extraordinary people and fascinating to talk with.

Another well-known energy healer is Caroline Myss. She explains her work as a medical intuitive in *Anatomy of the Spirit: The Seven Stages of Power and Healing* (1996), drawing on several traditions, but emphasizing the chakra system. See also Barbara Ann Brennan, *Hands of Light: A Guide to Healing Through the Human Energy Field* (1987).

WHO IS SENSITIVE TO ENERGY? (2) EVERY PERSON, EVERY LIVING CREATURE, PERHAPS MORE

Who is sensitive to energy? In ordinary usage, this question is shorthand for "Who is *very* sensitive to energy?" because all humans (and many other living creatures) are routinely sensitive to energies such as sunlight, heat, noise, and wind. Sighted and hearing people are sensitive to light and sound, but only within certain ranges of wavelength. Spiders and bees can see ultraviolet light that we cannot. Dogs can hear higher pitches than we can, and elephants hear lower pitches.

So, we need to ask more specifically, "who is sensitive to *what* energies?" because our world is awash with energies, including ones we aren't consciously aware of, such as gravity (a constant, more or less), as well as cosmic rays that penetrate our atmosphere and our earth (to a certain depth), and even our own physical bodies.

So when Qigong therapists ask, "Are you sensitive to energy?" they actually mean "very sensitive," and the range of sensitivity can be surprising.

Tota and I were working together at one of Lisa's seminars. I was sitting on the table and she was scanning my back.

She paused and said, "Are you a religious man?"

"Yes," I said, wondering *what the heck?*

"Well," she said, "I just felt that prayer you people say, something about 'The Father.'"

"Wow," I said, "You're exactly right. My wife and I say that every night before we go to sleep. I'm amazed you could feel that in my energy field."

"Well, it was just *right there*," she said.

While some people are more sensitive to energy than others, the amount can change. Lisa said some people start off with one hand more sensitive than the other, but the other catches up. This was true for me, and both of them gained more sensitivity. Research scientist Melinda H. Connor has a book on learning to see auras; I've tried some of her directions and made some progress, having started at no ability whatsoever. Such development takes practice and effort and, in some areas, guidance; I was not willing to work further.

"Calibrate your intuition!"

In the Advanced Seminar, Lisa introduced us to see the colors of organ systems. This seemed odd to me, because I know from the grocery store that the livers of mammals and fowl were reddish-purple. For the Chinese, however, the symbolic color for Liver is *green*, reminiscent of the chlorophyll of plants and trees, springtime, and symbolizing the primal powers of nature in general. We may recall Dylan Thomas' "The force that through the green fuse drives the flower" or Hildegard's *viriditas* (greenness). I remarked to Lisa that green for liver seemed odd to me, and she exhorted: "Calibrate your intuition!" This startled and amused me. After all, she trained as an engineer, where calibration had specific meanings. In Qigong however, calibration has other senses: how we can perceive energy in other non-numerical ways.

Here's my best guess: I believe all humans are born with a potentially full set of so-called "psychic" gifts, but any of them only develop if they are nurtured. This notion parallels speech acquisition; babies make an enormous range of sounds, but they narrow to only the ones that appear to communicate. I believe that "psychic" is used approximately to include mental faculties we don't (yet) understand, although some (many?) of them could be useful, even central to our lives.

In my Qigong classes, people gained in sensitivity to energy, even if they had little experience with it beforehand. These were the majority. Some people, like Tota, came with unusual sensitivity. All of us were glad to have explanations for it as well as ways to increase it and apply it to other people.

Fairies, leprechauns, sirens, and saints: liminal figures combining matter and mind

Humans love to create, consider, or even discover figures that are bridges or portals between matter and mind. These figures are sometimes called "liminal" (*limen* is Latin for "threshold") and found worldwide and throughout human time. Some are guides, totems, or familiars in the form of an animal (dolphins, eagles) or forms that are animal-like (dragons, centaurs). There are also saints, angels, guardians, deities, and gods. Some are in forms of matter (paint, clay, stone, etc.) or language, whether written or oral. Whatever the medium, they focus minds that believe in them, and they raise and focus energy. In the Eastern Orthodox tradition, an icon is a window or portal into heaven, and energy may flow through it from either side. I've seen worshippers at a Shinto shrine clap to summon the *kami*, elements of nature and also spirits of the deceased.

In various traditions, a person never dies completely, just loses the physical body while the soul/energy/mind continues on. In some traditions (Jewish, Irish, and the American South) some people open a window when there is a death so that "the soul may go free."

Such notions are foreign to most Westerners because of rationalism, materialism, and even condescension toward indigenous, "primitive," or different people. These notions may offer insight into Near Death Experiences, dreams, trances, as well as past lives, familiars, guardian saints, and the like. Can Westerners *recalibrate their intuition* toward these concepts?

Are trees, rocks, and oceans sensitive to energy and mind?

Material objects react with the energies described by physics and geology: gravity holds the earth together, the sun warms all it shines on, wind can blow down trees, and all tectonic plates of the earth slowly move across molten rock. *Matter* and *energy* make up the earth and all of the universe. Do they work together in anything we might call *mind*?

At a national meeting of the Society for Values in Higher Education, I had two questions for David Hoekema, Professor of Philosophy and African Diaspora and African Studies at Calvin College (see Sources).

I asked, "Can rocks have mind?

He replied, "In America, no. In Africa, yes."

We attribute qualities according to our cultural heritages, and these differ across the world.

At the Australian Museum in Sydney, I was introduced to "The Dreaming," the Aboriginal tradition 50,000 or more years old. A website explains, "Once the ancestor spirits had created the world, they changed into trees, the stars, rocks, watering holes or other objects. These are the sacred places of Aboriginal culture and have special properties. Because the ancestors did not disappear at the end of the Dreaming, but remained in these sacred sites, the Dreaming is never-ending, linking the past and the present, the people and the land" (http://www.australia.gov.au/about-australia/australian-story/dreaming, accessed Aug. 3, 2016).

In such traditions—and there are many—the earth is alive with energy. The Chinese have practiced Feng Shui for some 5,000-6000 years. Stonehenge, Tara, Chaco Canyon, Nazca Lines, and Ley Lines are places considered to have energy. David Furlong lists over 200 sacred sites around the world; some of these are natural such as mountains or lakes, while others are human-made, such as monasteries, nunneries, temples, or burial grounds (*Working with Earth Energies: How to Tap into the Healing Powers of the Natural World*, 2003). The ancient Romans had scores of genii loci, or spirits of place. Tourists have felt awe at Yellowstone Park, Fingals' Cave (on the Isle of Staffa), Ayers Rock (Uluru), the Ganges River, the Temple of Heaven in Beijing, as well as shrines, mosques, and cathedrals, just to name a few. We can imagine (absent means of measurement and rational proof) that people's experiences vary with the personal beliefs they bring to such places.

I believe a church or temple or any other sacred site gains energy and mind as people visit, meditate, sing, and pray over centuries. If the energy and mind are bi-directional, like the icons, the site and the humans exchange energy and mind.

Naturalist E. O. Wilson has traveled the world looking at landscapes and animals, including his specialty, ants. He believes humans have a built-in love for the life of animals and plants. He calls this "biophilia," and so named of his book (*Biophilia*, 1984; see Essay 4.) Is this quality part of our souls? Can the material world sense when we perceive it with pleasure …or with fear?

Is panpsychism real?

I asked David Hoekema my second question: does panpsychism exist in reality. He said, "It depends who you ask."

A long article on Wikipedia describes panpsychism as "one of the oldest philosophical theories" and defines it as "the view that consciousness, mind or soul (psyche) is a universal and primordial feature of [all] things." This view, previously widespread, has been declining in the 20th century with the rise of logical positivism and the popularity of philosophy of science (see https://en.wikipedia.org/wiki/Panpsychism#Quantum_physics, accessed Aug. 3, 2016). Our loss of nature as *en-spirited* has other factors: urbanisim, materialism, living in buildings, traveling in planes and cars, light pollution that blocks out stars, other pollutions that keep us away from air, water, or earth, and a capitalist understanding that earth is just raw *material* to be mined and converted into products such physical things or carbon-based energy.

ANOTHER THOUGHT EXPERIMENT: UNIVERSAL MIND AS PANPSYCHISM

In this thought experiment (reminiscent of Atlas and Atlanta) we imagine that *the Universe knows what it is doing at multiple levels and actually wants humans to join in.* Perhaps we already join in sporadically, as when we see an inspiring mountain range or a beautiful sunset, when we marvel at a flower or feel closeness with an animal, or when we experience sex and all reason and logic fly out the window. The figure is another force-field, opposing Universal Mind with the typical Western Secular Mind.

UNIVERSAL MIND	WESTERN SECULAR MIND
HOME	
All space and matter and energy	House, apartment, town, city
TIME	
All time from the Big Bang 13.8 B years ago to a presumed end of time	Focus on, roughly, last 300 years with progress to the present and, surely, beyond
REALITY	
Complex and wonder-ful, in part beyond human imagination	Largely comprehensible, a puzzle being solved, especially by science
Matter = mind = energy and perhaps = more	Mostly matter, some energy (especially carbon-based)
The universe is coherent and orderly	Man the center: reality exists for man.
KNOWLEDGE	
Innate, complex, eternal; all levels of human consciousness, all living creatures and plants and matter, and more	Domination by logic, planning, strategy
	Values of individualism, Neodawinisn to promote wealth, status, progress
Omniscient, loving, eternal	Omniscience is just around the corner Technology and economics can fix anything
EARTH/NATURE	
A lovely planet with extraordinary life	Raw material for human consumption

Self-regulating, despite five extinction events; our Magnificent home	Unlimited resources are our birthright
	Matter converts to products and energy
HUMANS	
Made up of matter = mind = energy	Largely in control of physical reality
Large brains and focused intent could do much for the future of earth and beyond	Excellence shown through technology, market share, trade, cities, fashion, etc.
Wasteful and polluters of air, water, land	Dominant culture humans live in "luxury," uprooted from nature, unaware of
Beloved, along with all creatures, by Universal Mind	Universal Mind

Figure 6: Universal Mind vs. Western Secular Mind

For the most part, we live in the dominant Western mode that is detached from Universal Mind. Some of the results include loneliness, confusion, and a sense of being thwarted rather than supported.

A notable exception is Jungian thought: the collective unconscious, seen in stories and dreams, holds archetypal images shared across the world and throughout human time.

GURGLES AND, FINALLY, THE NEEDLE MOVED

In making passes with my hands over the stomach and intestines of my partners, I began to hear gurgles. At first these seemed like any random gurgles anyone might make, but soon it was clear that there was a cause and effect: when my hands crossed the area again, there was a gurgling response. I mentioned this to other Qigong students, and they said they heard them as well. We made jokes about going by people in crowds and surreptitiously waving our hands around their bellies so that they'd make large noises. Later, I helped a hospitalized family member get bowels moving again.

A fellow student brought a gaussmeter to our study group. This hand-held instrument measures magnetic fields. Could we measure Qi output? We tried various ways of holding our hands over the instrument but got nothing. "Is there a more sensitive setting," someone asked. There was, and we selected it. I stacked my hands over the instrument and directed energy through the P-8 of my palms. The needle moved...not a lot, but enough to show it registered energy. Magnetic energy is probably just one of several energies involved in Qi.

As opposed to this amateur effort, there is a rigorous overview of scientific measurement of energy fields in Beverly Rubik's article "Measurement of the Human Biofield and Other Energetic Instruments," in Lyn Freeman's *Mosby's Complementary and Alternative Medicine: A Research-Based Approach*, 3rd ed. (St. Louis: Mosby-Elsevier, 2009, pp. 555-573). (See http://www.faim.org/measurement-of-the-human-biofield-and-other-energetic-instruments, accessed Feb. 18, 2016.) Rubik has a Ph.D. in Biophysics from UC Berkeley.

QIGONG CERTIFICATION

Slowly gaining confidence, I scheduled my exams with Lisa. I competed a written exam (some 20 single-space pages), an oral exam (via Skype), and a practical exam, a treatment on her as a client in real time and space. I passed and became the first Qigong therapist certified by her in North Carolina.

RANT 4: DISHARMONIES AND PRISONS OF MODERN LIFE; ALSO, QIGONG CAN HELP

In Rant 1 we looked at social values of competition, Neodarwinism, tribalism, and hard work.

In this rant we consider more personal impacts of ideologies we scarcely realize that we hold and the resultant stresses of a living a high-wire life, stresses that diminish our well-being and can make us sick.

Living and liking high-stress

Modern life is high-stress for many reasons: multiple demands in time and space, as well as noise, crowds, economic pressures, and more. Much of the day we sit on chairs or in cars, and the upper body slumps. Our feet walk on concrete, not the earth. Our minds deal with a barrage of inputs and our own personal demands for high performance, often with time constraints. As our attention focuses on our duties, we forget about our bodies. We breathe shallowly and tense up: *ever alert, ready, on task*! We come to consider an adrenalized body and an over-active mind as normal. Indeed we get high from our monkey minds…and become addicted to stressful living as well as to caffeine and/or other drugs.

The over-active brain enjoys calculation, planning, criticizing, and other rational activities, at a cost to reflection, intuition, emotions, and other mental faculties. We disconnect from the slow rhythms of nature.

Many moderns are proud of having "a full calendar," a cocoon (or prison) of sorts, limiting our time for reflection, listening to others, or being open to new adventures. Perhaps we are afraid of silence and "doing nothing," times when we might get to know ourselves better and become ready (as from a respite) to meet other people or anything else, even new ideas.

Being stuck in the head often means the body suffers. Our nation suffers from obesity from overeating and lack of exercise, also addictions to tobacco, alcohol, drugs (street and/or legal—lately and tragically opioids), as well as "screen addictions" from TV to hand-held devices. The phrase "retail therapy" has gained currency; it is both satiric and accurate for people buying clothes, jewelry, cars, houses and/or expensive travel in order to feel *better* and/or to have a sense of *control and power*.

Qigong and other practices can bring peace and calm in awareness (proprioception) of body, mind, and energy.

I recall a fellow martial arts student saying, "Practicing Aikido is like learning a philosophy with your body." For him, the forms and focus provided calming and coherent meaning.

Emotions take time to assess, plumb, evaluate, heal, share with others.

Perfectionism

The current state is never good enough; there's always something better to strive for.

Fulfillment always lies ahead.

Bigger, better, and faster are performance goals. Enlightenment progress has run amok, creating illusory goals: *make more, do more, be more!*

Because *perfection is just around the corner*, we are always *a day late and a dollar short*, permanently thwarted by circumstances. We can't *measure up*, no matter how hard we try!

A more healthy approach was on Sesame Street when Big Bird said, "Everyone makes mistakes." At an Omega Center conference (described later) an instructor talked about Master Morihei Ueshiba, the founder of Aikido. When praised for the perfection of his form, he said, "Oh, I make mistakes all the time; it's just that I am able regain form very quickly."

Setting aside the "perfect," we can enjoy the fullness of the present, the wonderful now. We can appreciate today's leisure, grace, mystery...and welcome serendipity when it occurs.

Imagination considers different approaches, perspectives, and paradigms.

Exceptionalism/Nativism

Self-esteem is one thing. Self-privilege (hubris, pride, swelled heads) is another.

Neighborhoods can be self-isolating, also social classes, especially the very rich.

An entire nation can be self-isolating and fear immigrants.

Expressions of "Number One" (with a raised index finger; a crowd chanting) suggest that "Two"—or any other number—represents abject failure. "*Win or go home!*" is a current motto.

Better than being "Number 1" is being part of all of us as One.

Wealth concentration and the disenfranchisement of middle classes and the poor

The very rich live in a different world from everyone else. Many of them focus on maintaining their status and yearning for an earlier, "freer" time (fewer taxes, less regulation) when Neodarwinism, they assume, routinely rewarded them, certainly the fittest and best people.

Many young people can't find the bottom run of a career ladder or even a "make-do" job. Many, regardless of age, are in financial debt.

Some disaffected youth join a gang or go on Jihad, at last finding a role that appears to have meaning.

Many find their dreams blocked and unfulfilled.

In some neighborhoods, selling drugs is the only job available.

In some mountain areas, growing marijuana is the only cash crop.

The rich invest and become even wealthier. Some buy things to hoard in homes or bank vaults, or they send their money to Switzerland. Most want more wealth…showing *progress*.

The poor, even the middle class, pay for goods and services that are transient, often with no long-range gains; or, if there are plans, there are no dollars to make them happen. Living paycheck to paycheck, they have no money to invest, so "economic recoveries" don't include them.

Urban and rural spaces are more and more privatized…the notions of "the commons," public space, and "free range" are falling away. Our nationals parks are underfunded and threatened with "review" by the Trump administration and consequent plunder and despoilment by lumber and mining interests.

We need to restore the commons, create meaningful cities and jobs, and imagine economic policies that can benefit all people, all creatures, all of nature, and the earth.

Loss of nature, abuse of nature

Modern people live much of their lives in buildings, dwellings, cars, malls, and city streets, missing the beauty of hills, water, plants and trees, moving air, clouds, and sun. We are less and less aware of nature as mother, teacher, and exemplar of balance and harmony. We are *deracinated*, cut off from our roots in the materials of nature, the seasons and cycles of nature, the ecologies and energies of nature. These are personal losses for many people, families, and children.

Indigenous people typically understand that natural resources are limited and should be protected for the use and enjoyment of future people— the Seventh Generation Principle. This notion of sustainability was held by the Algonquin, Anishanabe, Cherokee, and the six Iroquois nations, among others.

Daoist origins of Qigong stress the beauty and balance of nature. Many Qigong exercises and meditations use natural imagery (clouds, trees, the lotus flower) and invite humans to be part of the energy, mind, and matter of the earth and, even, the heavens.

Losses of mystery, spirit, the sacred

Reason displaces awe and wonder.

Reductionism displaces wholeness.

Utility displaces joy.

The mundane displaces the cosmic.

The profane displaces the sacred.

No wonder many feel lost, adrift. We have diminished connections vertically with the heavens above and the earth below, and we have diminished connections laterally with nature, and many, perhaps all of our fellow creatures, other humans included.

Health maintenance is displaced by rescue medicine....

We don't do enough with preventive medicine. Instead, we put money, manpower, and emphasis on injury and disease. Medicine is dominated by Frank's "restitution narrative," meaning we seek to restore what is failing or broken. Industrial medicine stresses repair of injury and illness.

We love TV shows about emergency rooms, trauma centers, and "difficult cases." As a younger man, I trained as an EMT and spent many hours in an ER/Trauma center as a pastoral-care volunteer. Ambulances and helicopters brought in patients. The staff was great; we helped people. It was exciting!

But there are no TV shows about well baby visits, TB screening, or primary care. Too boring...but very cost-efficient in the long-run.

Qigong (and other practices such as massage, prayer, meditation) focus on the health and well-being of patients and care-givers within a meaningful world. Such values and actions allow people to stay healthier longer and, when sick or injured, heal faster. A wider view of nature allows for healthier views of illness and, even, death.

HOW DOES QIGONG WORK? (1), ACCORDING TO ANCIENT CONCEPTS AND MODERN SCIENCE, RELAXATION ALLOWS THE BODY TO HEAL ITSELF AND PROFIT FROM HEALING INTERVENTIONS

How does Qigong work? Once again, it depends on who you ask.

For the Chinese, Qi is complicated, with at least three sources of Qi (your ancestors, the food you eat, and the air you breathe) plus various chains of development to produce the Qi in any given person. Although

Qi can become blocked or stagnant, work with the body's energy field and energy channels can help it flow freely once again, within the person and in harmony with the Qi of the universe.

In a few words: Qigong helps the body relax by rejoining the coherence of the universe with the result that the body heals itself.

The Three Treasures

Most authors, whether Chinese or Western, see Qi within "The Three Treasures," a grouping of (1) Jing (Yin, a sexual energy) (2) Qi (life energy, both Yin and Yang), and (3) Shen (Yang, a spirit energy). This trio has a tantalizing parallel to our trio of universals: matter, energy, and mind, with the difference that *energy is already distributed to matter and mind* in Qigong.

In Qigong practice, therefore, energy pervades exercises (primarily body or matter based) and also meditations (primarily mind based). In the most basic theory: Qigong seeks the plentiful and balanced flow of energy in humans (as in nature) without blockages or stagnations.

In so-called Medical Qigong, the main aims are relaxation and self-healing. Relaxation means that the body and mind can maximize the flow of harmonious and balanced energies. The opposite of relaxation includes imbalance, lack of one or more energies, blockage or stagnation, or heightened stress, even routine stresses of daily life. Self-healing means that a person can use the mind to remove stress, balance and/or increase energy, and allow the body to function in a healthy way. Sick or injured persons can especially benefit from these practices at any time, and especially upon going to bed and preparing to sleep. Further, to be in tune with the harmony of the world

Three Treasures	Jing (Yin) Sexual Energy	Qi (Yin, Yang) Life Energy that pervades all	Shen (Yang) Spirit Energy
Practices	Exercises Diet Channels, Vessels **A**lign body	Mind moves the Qi **B**reathe deeply	Meditations Awareness of energy **C**alm the mind
Three Universals	Matter	Energy	Mind

Figure 7: The Three Treasures: Overlapping and Interchangeable Realities

helps to relieve doubt, sorrow, even the fear of death.

To understand that our bodies and minds are *treasures* is a rare and wonderful idea. We are products of millions of years of development, true wonders that we should appreciate, cherish, and treat with care. When Roger Jahnke led Qigong at a conference, one exercise he led was Cloud Hands. As we moved our hands slowly, he suggested that *we look at them attentively and appreciate all their complexity and ability.*

Mechanisms of Pain relief

A 2013 review article "Mechanism of Pain Relief through Tai Chi and Qigong" in the *Journal of Pain & Relief* describes how the three treasures help relieve pain in the short term and contribute to overall health in the long term (C. J. Rhoads, 2013). Several dynamics interrelate: relaxation, release of endorphins, open focus of the mind (alpha and gamma brain waves, not the beta waves of the "closed" mind), stimulation of the vagus nerve that activates parasympathetic functions (slower breathing and pulse also lower blood pressure), and, especially, the desensitizing of the dorsal horn (in the spinal cord) that, otherwise, will magnify sensations of pain as perceived in the brain and increase overall stress. (The last dynamic agrees with David J. Linden's explanation we saw earlier in RUBBING THE RIGHT WAY.) Many of these dynamics occur also in practitioners of Yoga and recipients of massage. In brief, relaxation leads to homeostasis and well-being, rest, and repair. I believe standard medicine should make these available to patients.

Word study: "idiopathic" but never "idiotherapeutic"

Some physicians love the word "idiopathic," by which they mean "an illness the origin of which is unknown." Right away we should be suspicious because this strange, multi-syllabic word really suggests that "the doctors don't have a clue." Worse yet, the word is made up of Latin and Greeks roots for the "self" (as in "idiom," or "idiot") and illness as in "pathology"), which really means "self-diseasing" or: *the patient caused the illness and is, therefore to blame.* So, we might sarcastically argue that doctors have made up a fancy word to cover their lack of knowledge and, further, to blame patients for dilemmas.

Further, where is the word "idiotherapeutic," which would mean: "self-healing" or: *the patient causes healing to occur?* This process is, in fact, normal and routine. Many minor injuries and illnesses resolve by themselves, with no medical assistance at all. When the doctor sets broken bones, it is the patient's body that repairs the defect. When the doctor prescribes an antibiotic that kills or inhibits a bacterium, it is still the patient who takes these as prescribed, whose immune system contributes to healing, and who must recover strength after the illness.

In a time before modern treatments, Voltaire wrote, "The art of medicine consists in amusing the patient while nature cures the disease." While satirically overstated, the witticism does credit a patient's harmony with nature and the healing resource of amusement.

American experts seek links between East and West, scientific understandings of Qi

Kenneth S. Cohen sees seven kinds of Qi in the Chinese tradition (p. 34). He reports that Western medicine has been able to measure correlates to Qi from electricity, biochemistry, bioluminescence, and consciousness (pp. 44-56). His book dates from 1997, but even then there were scientific studies to support these correlations. He stresses, however, that the proven activity of Qi over millennia is more important than how it is operationally defined. Further, he urges a "healthy respect for the unknowable, the mystery" in seeking a "knowable science of Qi and Qigong" (*Qigong: The Art and Science of Chinese Energy Healing*, p. 44).

Roger Jahnke's *The Healing Promise of Qi* (2002) takes an inclusive approach as well. He discusses the Chinese mythology: Qi existed before the world, the primordial energy that, later, infused all reality (p. 8). I'll summarize some of the points in the 30-page chapter "The Light of Science on Qi" (pp. 237-266). Jahnke sees current Western science limited in approaching the mystery of Qi. (We could add other limitations: reliance on numbers and linear causality, reductive isolation of factors, insistence on replicability, as well as funding sources such as drug companies.) Jahnke mentions the scientific work of the Institute of HeartMath and William A. Tiller, and The Qigong Institute, Menlo Park, California, active when Jahnke wrote, which collected many scientific studies of Qigong.

Jahnke reviews physiological equivalents (similar to Cohen's correlates) of Qi, mind-body connections, and bioenergetics equivalents, including conduction systems in cells (the work of James Oschman), and biofield equivalents of energy and light (the work of William A. Tiller). Throughout, he emphasizes *coherence* within and between systems. Coherence, including balance and harmony, is an essential quality of Qi that permeates the universe. A theoretical step beyond Cohen's correlates, these equivalents are just that, interchangeable versions of energy, much like our interrelating *matter, mind,* and *energy.*

His final section discusses the Quantum equivalent, including the famous equivalence from Arthur Heisenberg of light as either particle (matter) or wave (energy) depending on the observer (mind). Jahnke provides paragraphs on, among other topics, holonomic theory (although not Einstein's student David Bohm), group Qigong, prayer and nonlocal healing, and a Global Consciousness project, and love, in all its forms. Jahnke, like Cohen, believes that we should cultivate Qi to the highest levels, even though its ultimate mystery will probably never be solved.

Reports from science: Qigong and Tai Chi both improve health

In 2010, Roger Jahnke, Linda Larkey, et al. published "A Comprehensive Review of Health Benefits of Qigong and Tai Chi" in the *American Journal of Health Promotion.* This long article is a meta-analysis of 77 articles that described randomized controlled trials (RCTs) assessing outcomes of Qigong and Tai Chi. These articles were published in peer-reviewed journals from 1977 to 2007. In the Abstract conclusion we read, "Research has demonstrated consistent, significant results for a number of health benefits in RCTs, evidencing progress toward recognizing the similarity and equivalence of Qigong and Tai Chi" (p. 1).

Specifically, Jahnke and Larkey found nine outcome category groups with positive results: bone density (n=4), cardiopulmonary effects (n=19), physical function (n=16), falls, balance, and related risk factors (n=23), quality of life (n=17), self-efficacy (n=8), patient-reported outcomes (n=13), psychological symptoms (n=27), and immune- and inflammation-related responses (n=6). They also state that Tai Chi and Qigong are effective and, according to a study on older adults, have no adverse effects. The authors conclude, "The preponderance of findings are positive for a wide range of health benefits in response to Tai Chi and a growing evidence base for similar benefits for Qigong." Grouping the two together, they write: "The substantial potential for achieving health benefits, the minimal cost incurred by this form of self-care, the potential cost efficiencies of group delivered care, and the apparent safety of implementation across populations, points to the importance of wider implementation and dissemination" (Application to Research, p. 17).

C. J. Rhoads, cited above, writes that continued practice of Tai Chi, Qigong, and Yoga lead to long-term health benefits of stress reduction, better digestion, and healthier lifestyle in general.

Other sources of research

The Qigong Institute database lists 7,000 abstracts of studies from around the world since 1984. The topics vary widely, from biophysics to telomeres, from balance to stress reduction. A 2016 multi-center study found efficacy for Qigong in treating PTSD in American World War II veterans. More selective, PubMed lists 475 abstracts, including that PTSD study, "Qigong in Injured Military Service Members: A Feasibility Study" (A. M. Reb et al., *Holistic Nursing*, 2016).

In their section "Health Topics A to Z, "The National Center for Complementary and Integrative Health website has an entry on "Tai Chi and Qigong," which are described as "safe practices." The website provides "Key References" (including the Jahnke article just discussed). (See https://nccih. nih.gov/health/taichi, accessed May 17, 2016.)

One more dimension: faith

Qigong's heritage draws on both Daoism and Confucianism; indeed, Fritjof Capra suggested that Daoism was largely intuitive (female, Yin) while Confucianism was largely rational (male, Yang), and the two are complementary (*The Tao of Physics*, 1976, pp. 14, 105).

I see another parallel from the notion of the Christian Trinity where three merge as one, and Jesus and the Holy Spirit combine Ying and Yang. Accordingly we may expand the previous figure:

Three Treasures	Jing (Yin)	Qi (Yin and Yang)	Shen (Yang)
	Sexual Energy	Life energy that pervades all	Spirit Energy

Practices	Exercises Diet	Mind moves the Qi	Meditations Awareness of energy
	Channels, Vessels **A**lign body	**B**reathe deeply	**C**alm the mind
Three Universals	Matter	Energy	Mind
Awareness	Proprioception of matter	Proprioception of energy	Proprioception of mind
Eastern thought	Daoism Intuition (Female, Yin)	Dynamic balance	Confucianism Reason (Male, Yang)
Christian Trinity	Jesus (Yang and Yin)	God pervades all	Holy Spirit (Yin and Yang)

Figure 8: Expanding the Three Treasures: Overlapping and Interchangeable Realities

Is it possible that we have been too unimaginative, too unspeculative with regard to these important topics?

DAFFODIL IMAGERY HELPS A MASSAGE CLIENT

A man comes for massage at Cornucopia House. When he complains of pain in his chest, I am alarmed, thinking a possible heart attack.

Should I call 911?

I ask him what he thinks might be the cause.

"Oh that's easy, I got a port put in my chest." He gestures to his upper chest.

"When was that?"

"Two weeks ago."

"Have you checked with your doctor or nurse?"

"Yes. They say it looks fine." So…apparently no infection or internal bleeding or other life-threatening dilemma. With such risks out of the way, we finish all preliminaries and get him on the table for massage.

When I gently work on his chest, back, and shoulder, there are sore, tight muscles limiting his range of motion. Left upper trap is tight, also outside his scapula, the latissimus dorsi, and the "helper" muscle teres minor. His pec and pec minor are not so tight but touchy. I imagine his body has reacted to the "insult" of the surgery by splinting and guarding the area subconsciously with these muscles. Avoiding the port and its catheter—and mentioning that to him—I make some progress loosening his muscles, but I also want to recruit his help.

"I think some energy work can help you. OK to try?"

"Sure."

"What's a healing color for you?"

"Yellow."

"Great, yellow…let's see, a golden yellow…a light yellow…a strong yellow?

"Like those flowers in the spring, you know, whole bunches of them outside."

"Daffodils?"

"Yeah, that's it. Daffodils."

I ask him to picture a ball of this healing color swirling in his upper chest, then draining out his left arm and hand.

"OK. Let's give it a try."

We do this three times.

"How does your shoulder feel?"

"Much better." He rotates his shoulder, which now moves freely.

"Do you think you might do this meditation at home?"

"Well, I guess I could."

"Super! Picture that bright, shining, energy flowing through your shoulder. Three times a day would continue to help you, I believe, so it won't tighten up again. You could do it at mealtimes, just to have a schedule."

"I'll give it a shot."

What has happened? He has replaced the unconscious messages from his mind to his chest that caused tight muscles that, in turn, fired sensory nerves that registered pain. When his mind loosens the muscles, the pain stops. The daffodil visualization allows this man to be aware of his body and to work with it right away...and in the future.

It seems that the material work of massage and the mindful work of Qigong fit together very well. "Mind moves the Qi" for the well-being of the body. The specific imagery of lovely flowers involves our imaginative mind and heralds the rebirth of spring; these recall the Daoist roots of Qigong: the "Way" of nature is harmonious, beautiful, and healthy.

Later that month, I help a woman with a painful and stiff hip. She chooses the color blue from a tropical bay she remembers from a happy vacation. We swirl this color in her hip, then down her leg and out her foot. The hip loosens up and most of the pain is gone.

Modern medicine and healthcare in general should provide patients with ways they can contribute to their own healing.

A DISTANT RACEHORSE SHOWS THAT "REMOTE" HEALING IS ACTUALLY RIGHT NEXT DOOR

I had heard of various stories of "remote healing" and took them with many grains of salt, maybe even handfuls. My experience was that healers—such as surgeons, prescribing doctors, therapists, and the like—were always *physically present* to the patient. Healing, I assumed, had to be hands-on, scalpel-in, face-to-face, etc., which is primarily "material-to-material." Because remote healing was not part of my culture, it linked to the "woo-woo factor," or— cue the music—"The Twilight Zone," or just plain old "no way Jose."

At the same time, I knew that radio waves could go from a city to my radio at home or in my car, even "skip" off the ionosphere over hundreds of miles. Radios, including short-wave, were part of my history and culture, and I took them for granted—but only after I read about them, talked to other radio people, and went to stores with such equipment. So...openness to distant healing depends on information, experience, and, for many, a theoretical framework to place it in.

Healing a distant racehorse

Before my introduction to Qigong, I met a neighbor with a South American background; he said he was a shaman. I don't hear such a claim often, but I took his word for it. Further, I knew from my massage practice that more happened for patients than just my touch, so I asked him about his experience. He told me about some distant work he was doing with a racehorse.

"What?" I said, not resisting a small joke: "Do you phone him up?" I pictured the horse holding the handset to his pointed ear.

"Yes, in a certain sense," he said matter-of-factly. "I look at a picture of him and send him coherent energy. When he fills up, I say goodbye."

"How do you know when he's full?"

"I can feel it."

"Do you have any idea that it works?"

"Yes, and it's easy. His owners tell me his blood pressure comes down, his time trials are faster, and he's happier in general."

I wondered whether there were other factors to explain this, regardless of what this man did, but as we talked then and later, it seemed clear that there was a correlation between is efforts for distant healing and the horse's health and performance. Indeed, there were verifiable measures with numbers that I could accept according to my Western view.

While the horse (I assumed) would not have any direct knowledge of its remote healer, the owner and others would; perhaps they also sent healing energies because they had confidence in the process? But why rule out the horse? Maybe it could perceive—by equine proprioception—the sent energy without knowing the exact source.

MY REMOTE CLIENT LYNNIE

I received an email from a woman who wanted Qigong for her hospitalized mother in another North Carolina town. I didn't know either of them, but I wrote back that I might be able to help and gave her my phone number. She called immediately. I learned that the mother was sensitive to energy, and that the daughter was eager for my help. I said I wouldn't travel to provide Qigong but I could send energy if I had a photograph of the mother, a description of her, the name and location of the hospital, and the specific agreement of the mother. Part of me hoped that these restrictions would be daunting enough to end the project, but I received everything in two hours. The daughter also said the mother was in the ICU.

I studied the photo of a smiling, older woman and the written description. I'll call her Lynnie. I looked up the hospital on the Web and used a map and a compass to orient the direction I would send energy. The geographical direction may or may not be important, but it helps me focus my mind and stay in the "zone" of sending. I calmed my mind and did deep breathing with the Microcosmic Orbit. When I felt energized, my hands tingling, I faced the hospital, looked at Lynnie's picture, and held out my hands. I said, "I love you, Lynnie, and I am sending energy to your mind and body." I could feel her energy. She needed all Yin energies, and especially Kidney. I felt blockages in her torso, combed them out, and sent energy. When she felt full and balanced, I wished her good healing and said goodbye. I closed her energy field and separated it from mine. I closed mine and sat still a minute, wondering what effects there might be. The next morning I energized and sent to her again. She felt different and much better...or was I kidding myself, somehow answering my own desires?

That afternoon, I received an email that Lynnie was out of the ICU and that the nurses and doctors were surprised by how quickly her incisions from surgery had healed up.

HOW DOES QIGONG WORK? (2), MIND, MATTER, AND ENERGY ARE ALL ENTRAINED ACROSS THE UNIVERSE

I have also sent energy to family members with good results. My routine varies, depending on what I find. If the "target," as we call it, has mentioned a particular problem, I concentrate on that. I sometimes use Yuan points, channels, and other techniques, but the basic approach is sending universal energy to the entire person. Sometimes I consider the organ system involved, the precursor organ that nourishes it, and the farther precursor that can limit or restrict it. Throughout, I'm saying out loud such things as, "[Name], I love you and send you coherent energy from the universe...I am removing a block from you and sending it to the earth...I'm sending energy to [problem]...this is coherent Qi that will allow the area to relax so that blood can flow through there and the body can heal itself." These comments help me stay focused. I don't often get more specific about tissues or diagnosis. The main point is to make the connection and get out of the way so that universal energy flows through me to the person.

While I have reports of improvement from the small sample of people I've sent energy to, I have no scientific, quantified evidence that energy helped or how it helped. I believe, however, that scientists will someday be able to define this with a precision that allows medicine to accept and use remote healing.

Here are my thoughts, using the three universals. First *mind*. Because the targets know that energy is being sent, they have a placebo effect that changes cells and tissues in their bodies. Various aspects of their minds (imagination, hope, awareness, intention) allow their bodies to respond and heal. Further, my focused and energized mind is sending energy to this person who may, at some level, receive it. If other family members or senders (as in my study group) are involved, there is a further and deeper application of mind and energy. Wider circles of friends, neighbors, fellow patients, and religious communities also have influence. Perhaps some aspects of medical treatment became more effective because of a multiplier effect of received energy.

As to the *matter* of the body, focused energy removes an energy blockage or stagnation and allows for freer movement of energy. This helps circulation of blood and lymph but also allows for all cells to vibrate more freely at a molecular level, including water molecules in the ground substance and in structural elements such as the investing fascia that covers all muscles and organs. Because of thixotropy (discussed earlier: changes of gel to sol), body fluids can move better to transport nutrition and drugs) and take away wastes from cellular respiration.

And *energy* itself, at a Quantum level, comes from the sender(s) to the targets causing more coherent energy in the target, especially as sender and receiver are entrained. My focused and well-defined application of energy may reach the target. Entrainment can come from relationship of family or friends, from share experiences, or perhaps simply as fellow humans. The most comprehensive and exciting suggestion is this: *we and every aspect of the universe are already entrained because all of our elements and energies (and mind?) originated together in the Big Bang.*

WHAT CAN QIGONG OFFER TO SOLACE US?

For Qigong—and for many people still attuned to the nature world—our human nature and the world of nature are alike. We take pleasure in seeing trees, clouds, crops, and fellow animals. We enjoy sunrises and rainbows. If a night sky is sufficiently dark and clear, we look with wonder on myriads of stars, even the Milky Way, our home galaxy seen from the side. Even at noontime, we know stars circle above us and surround us, although we cannot see them because of the brilliance of our local star, the sun.

Permanence is not guaranteed. Our bodies change as we age, yearly, even daily, like trees, clouds, crops, and fellow animals. We are microcosms of the earth (and the universe) that is always changing. By paying attention to nature, we see patterns of change, even destruction and renewal. We should enjoy our bodies, minds, and energies, and take care of them in the limited time we have them.

Human nature and "nature nature" have overlaps that provide solace

Daoism may be considered a steady-state philosophy: the earth is in a dynamic harmony of a constancy that includes change. Any and all such changes are normal in the long term. The earth has always been in flux, even in danger; this is normal. Plagues and famines come and go. Glacial ages come and go. There were the five major extinctions before today. Climate Change will probably cause the sixth major extinction, but the earth will evolve and continue, with or without humans.

Qigong celebrates and uses various kinds of harmony, and these are calming in a world that is sometimes violent and chaotic. Qigong understands balance in each person, in all surrounding nature, and the entire universe. Humans are microcosms of the macrocosm of the universe, each reflecting the other, and the universe makes sense: it is harmonious and eternal. This is The Way, the Dao, the Grand and Dynamic harmony, the Eastern sense of balance that Western science calls, for human health, "homeostasis."

Qigong understands continuities between humans, all nature, and the universe. All are made of the same energy, mind, and matter. We are never alone or lost; we just need to tune in by meditation, exercise, or awareness.

Qigong can help us understand natural disasters

Because nature is dynamic, storms, earthquakes, floods, and the like are natural and expected events. Materials will re-arrange. If lightning causes a forest fire, the next plants will profit from the nutrients from the ashes and the opened space that allows the sun's energy to reach deeper into plants and the earth. (In next section we'll consider whether Climate Change is natural.)

Qigong can be a practical resource for health and healing

I am convinced the Qigong healing is real, efficient, practical, low-tech, and low cost; I believe this because of what I've read, seen in group sessions, experienced as a recipient of Qigong energy, and experienced as a Qigong therapist.

Qigong should be available widely and especially in hospitals and clinics. It should routinely be part of integrative medicine.

And, if not Qigong…what? Other forms of energy healing that calm and energize us: massage, Touch for Health, Reiki, Johrei, therapeutic touch, healing touch, acupressure, acupuncture, Quantum touch, vortex healing, Bio-Touch are all examples of disciplines that develop energy healing.

More generally there are all sorts of exercises, meditations, and beliefs that provide for celebration of our own bodies, minds, and energies. Even hobbies, sports, lunch with friends.

TAKE-HOME TREASURES FROM THE "SUPER QI SUMMIT" AT OMEGA INSTITUTE, MAY 2015

Omega Institute, Rhinebeck, N.Y., is as famous for the East Coast as the Esalen Institute is for the West Coast. When a Qigong client told me there was a "Qi Summit" at Omega, I chuckled at the unlikely phrase because the world of Qigong does not stress hierarchies, such as levels of rank or colored belts, let alone "summits." Nonetheless, I looked it up on the Web. Sure enough: nine Qi leaders were scheduled, including Ken Cohen and Roger Jahnke, whose books I had studied.

A weekend at the end of May, some 240 people gathered to experience Qi and learn further. We were a wide range from across the U.S., also Canada, Brazil, and Greece. Because we were, in some ways, outsiders to our dominant cultures, we were thrilled to meet each other and enjoy common ground. We were a tribe, a cohort, a family on the frontier, the borderlands.

I'll not give a full account here but list some nuggets I took home from the four-day worship.

Robert Tangora spoke of *tuning* the Chong Mai, the main energy channel in the body; a simple movement can be used almost anywhere, even on airplanes.

Robert Peng taught us "flower smiles." These arise from the heart and blossom in the face. I have used these for nurses' retreats; we smile at each other and break into laughter.

One master was quoted as saying "Laughter is the best Qigong."

Minke de Vos taught us a walking chant, "Chu Chu, Ha Ha" to energize the Kidney and the Heart.

Michael Winn (from Asheville, N.C.!) taught us a visualization, "Stroking the Long Beard," an oval in front of us that we could step into. This is a parallel to the Microcosmic Orbit mentioned above.

Roger Jahnke affirmed, "We all share the same mind with all people, all creatures, the earth, and the universe." I believe this is worth thought, meditation, and acceptance.

For four days we were outside of ordinary time. We were all kind to each other. The late spring landscape was alive with fresh green leaves. It all seemed a wonderful example of Hildegard's *viriditas* and Shakespeare's green world of fertility, joy, and love. I felt many affirmations to what I had studied and experienced.

ESSAY 4: REAFFILIATING WITH NATURE AND, THEREBY, HEALTH

As a beginner in energy work, I sometimes wondered whether I was self-deluded, waving my hands and looking like an idiot. Although there were traditions of three or more millennia behind me, the modern West has very little to explain or justify this approach. We are educated in the traditions of Euclid, Aristotle, Descartes, Newton, and similar others, all rational, orderly, and materials-based. For many (including me), religion was also a meaningful tradition, but it was a separate realm from all I learned in school or universities or saw in mainstream media.

What an array of choices we have to model reality!

The Existentialists suggested that the world, the universe, all reality is absurd, chaotic, unwelcoming. Therefore, angst, anguish, dread are appropriate modern responses, along with the heroism of individual humans—like Sartre's Orestes—finding their own ways in the world. At an opposite pole, the Daoists understood that the Way was a coherent path for all persons, all nature, even the Universe. For them, the universe makes sense—always and everywhere—in its repeating patterns, materials, and mystical Oneness.

Yin and Yang may sound exotic or arcane, but they may shed light on modern times. We recall that Yin was the shadow side of the mountain, with female, quiescent values, but Yang was lit by the sun, all masculine and active. Paternalistic medicine that emphasizes cure, can be very Yang, while supportive, integrative medicine that emphasizes healing (even a healthy death), would be Yin. At a longer reach of time, we now live in an inter-glacial period, a warm Yang time, as opposed to sleepy glacial Yin times of the past—another harmonic rhythm. As for Climate Change, we might consider that petrochemicals are a disruptive modern Yang capture of the sun energy, thus overheating the earth, all out of balance with Yin qualities of rest and renewal—the way night and day have functioned for millennia.

Sure, I'm an enthusiast for Qigong, and I know most Americans don't know about it nor will likely to study it. If we take my account of it as an *illustration*, however, we can think of parallel ways to recapture our sense of nature and our connections with it and the universe. The Romantics did it before us. Various faiths have viewed nature as the Deity's creation, therefore holy and to be worshipped, celebrated, and cared for by humans.

Indeed, we haven't forgotten entirely. We like nature. We like fresh green leaves on trees. We like flowers in different forms: cut flowers to take inside, flowers in a garden, and the flowers on trees, bushes, or plants that promise fruit, nuts, or vegetables. We like scenic vistas of mountains, valley, rivers, the sea. We like wildlife. We like pets. We remember natural places from our youth, a tree house, a bend of a creek, and special trips to a parks or scenic areas.

Edward O. Wilson, a scientist working worldwide, defines biophilia as "the innate tendency to focus on life and lifelike processes" (*Biophilia*, 1984). He describes both the complexities of nature and his love for nature in his studies of ants in Surinam, birds of paradise in New Guinea, plants and species revival in Krakatoa, and snakes along the American Gulf Coast. He writes that humans across the globe have enjoyed views of the savanna, with nearby cliffs from which to view it and to enjoy safety. Further, there should be clumps of trees indicating water, even lakes or shorelines promis-ing fish: "it seems that whenever people are given a free choice, they move to open tree-studded land on prominences overlooking water" (p. 110). This is just one example of biophilia, an innate human love for life. I believe biophilia is an important part of the human mind.

He argues that "to explore and *affiliate* with life is a deep and compli-cated process in mental development." Further, "[t]o an extent still under-valued in philosophy and religion, our existence depends on this propensity, our spirit is woven from it, hope rises on its currents" (p. 1, emphasis mine). Indeed, even the world is "congenial" to biophilia, and "to the degree that we come to understand other organisms, we will place a greater value on them, and on ourselves" (p. 2). If, we might add, our values are only economic or political or military, we will misuse natural resources and endanger ourselves.

He acknowledges the Darwinian notions of competition but sees also cooperation among species. "Coexistence was an incidental by-product that accrued from the avoidance of competition" (p. 9). Nature is replete with systems of symbiotic relationship, from a colony of leafcutter ants that func-tions as a "superorganism," to local ecosystems, to the earth itself, where millions of species interact in many—and often—subtle ways.

For a long time, humans have paved parts of the earth, and we show no signs of stopping. He writes, "'Push the forest back and fill the land' remains a common sentiment, the colonizer's ethic and tested biblical wisdom (the very same that turned the cedar groves of Lebanon into the fought-over-desert they are today" (p. 88).

He proclaims that "we are human in good part because of the particular way we *affiliate* with other organisms (p. 139, emphasis mine). We need, therefore, a conservation ethic that will slow species loss. In sum, "we need the most delicate, knowing stewardship of the living world that can be devised" (p.140).

Wilson's majestic, wise, and nuanced book is over 30 years old. Sadly, it appears we haven't followed his advice. Species loss has increased and Cli-mate Change is well underway (see the next Section).

We have not, however, entirely lost our biophilia, but we need con-cepts, vision, and imagination to see and protect the nourishing coherence between the world and ourselves. Qigong is one avenue. Religion another. Panpsychism, another. Also biological science. Also Quantum physics (See Section VII). Beyond such basic concepts, we need governments to put them into action. If governments fail, homebodies will be our hope (see Essay 6).

Romantic poet Wordsworth saw connections among nature, humans, and time:

My heart leaps up when I behold
  A rainbow in the sky:
So it was when my life began;
So is it now I am a man;
So be it when I shall grow old,
  Or let me die!
The Child is father of the Man;
And I could wish my days to be
Bound each to each by natural piety.

SOURCES 4

Barea, Christina J. *Qigong Illustrated*. Champaign, Ill.: Human Kinetics, 2011.

Beinfield, Harriet and Efrem Korngold, *Between Heaven and Earth: A Guide to Chinese Medicine*. New York: Ballantine Books, 1991.

Bender, Tom. "The Physics of Qi." tombender.org/energeticsarticles/qi_physics.pdf. accessed March 9, 2916. This is a "brief summary" of his DVD "The Physics of Qi" (2006).

Blakney, R. B., trans. *The Way of Life* [The Tao Te Ching]. See Lao-Tse, below.

Bock-Möbius, Imke. *Qigong Meets Quantum Physics: Experience Cosmic Oneness.* (St. Petersburg FL: Three Pines Press, 2010). Published in German also in 2010.

Book of Changes (*The I Ching*). Many editions available; also online.

Bohm, David. *Wholeness and the Implicate Order*. London: Routledge & Kegan Paul, 1980.

Brennan, Barbara Ann. *Hands of Light: A Guide to Healing Through the Human Energy Field*. New York: Bantam Books, 1987.

Braun, Mary Beth, and Stephanie J. Simonson. *Introduction to Massage Therapy*, 3rd ed. Philadelphia: Lippincott Williams & Wilkins, 2013.

Carter, III, Albert Howard. *Clowns and Jokers Can Heal Us: Comedy and Medicine*. San Francisco: Univ. of California Medical Humanities Press, 2011.

Capra, Fritjof. *The Tao of Physics: An Exploration of the Parallels Between Modern Physics and Eastern Mysticism.* New York: Bantam, 1977.

Cohen, Kenneth S. *The Way of Qigong: The Art and Science of Chinese Energy Healing.* New York: Ballantine, 1997.

Connor, Melinda H. *See Auras.* Marana, AR: EarthSongs Holistic Health, LLC, 2009.

Eden, Barbara. *Energy Medicine: Balancing Your Body's Energies for Optimal Health, Joy and Vitality,* 2nd ed. New York: Jeremy Tarcher/Penguin, 2008.

Eisenberg, David. *Encounters with Qi: Exploring Chinese Medicine*, rev. ed. New York: W.W. Norton, 1995.

_____. Ronald C. Kessler, Cindy Foster, Frances E. Norlock, David R. Calkins, and Thomas L. Delbanco. "Unconventional Medicine in the United States—Prevalence, Costs, and Patterns of Use." *New England Journal of Medicine* 28 January 1993. 328:246-252. http://www.nejm.org/toc/nejm/328/4/DOI: 10.1056/NEJM199301283280406

Freeman, Lyn. *Mosby's Complementary and Alternative Medicine: A Research-Based Approach,* 3rd ed. St. Louis: Mosby-Elsevier, 2009.

Furlong, David. *Working with Earth Energies: How to Tap into the Healing Powers of the Natural World.* London: Piatkus, 2003.

Gardner, Martin. *Did Adam and Eve Have Navels?: Debunking Pseudoscience.* New York: W. W. Norton & Co., 2000.

Harrington, Anne. *The Cure Within: A History of Mind-Body Medicine.* New York: W. W. Norton, 2008.

Hoekema, David. "African Politics and Moral Vision," *Soundings* 96:2 (Spring 2013), 121-144.

_____. "Faith and Freedom in Post-Colonial African Politics," in *Jesus and Ubuntu: Exploring the Social Impact of Christianity in Africa,* ed. Mwenda Ntarangwi (Trenton, NJ: Africa World Press, 2011), 26-45.

_____. "African Personhood: Morality and identity in the 'Bush of Ghosts.'" *Soundings* 91:3-4 (Fall/Winter 2008).

Hougham, Paul. *The Atlas of Mind Body and Spirit.* London: Gaia, Octopus Publishing Group, 2006.

I Ching or *Book of Changes.* Numerous editions available. There's an online version.

Jahnke, Roger, Linda Larkey, et al. "A Comprehensive Review of Health

Benefits of Qigong and Tai Chi." *American Journal of Health Promotion* Jul-Aug 2010; 24 (6): e1-325. Accessed March 12, 2016.

_____. *The Healing Promise of Qi: Creating Extraordinary Wellness Through Qigong and Tai Chi*. New York: McGraw Hill, 2002.

Kuhn, Thomas S. *The Structure of Scientific Revolutions*, 3rd ed. Chicago: Univ. of Chicago Press, 1966.

Lao-Tse. *The Way of Life* [*The Tao Te Ching*], trans. R. B. Blakney. New York: Signet, 1955. Numerous editions available.

Lown, Bernard. *The Lost Art of Healing: Practicing Compassion in Medicine*. New York: Ballantine, 1999.

Liu, Tianjun. *Chinese Medical Qigong*, eds. Tianjun Liu, and Kevin W. Chen, trans. Tianjun Liu, 3rd ed. (London: Singing Dragon Press, 2010).

Maciocia, Giovanni. *The Channels of Acupuncture*. Philadelphia: Churchill Livingstone, 2006.

Merculieff, Larry (Ilarion) and Libby Roderick. *Stop Talking: Indigenous Ways of Teaching and Learning and Difficult Dialogues in Higher Education*. Anchorage: The Univ. of Alaska-Anchorage, 2013.

Myss, Caroline. *Anatomy of the Spirit: The Seven Stages of Power and Healing*. New York: Three Rivers Press, 1996.

Nisbett, Richard E. *The Geography of Thought: How Asians and Westerners Think Differently…and Why*. New York: Free Press, 2003.

Qi: The Journal of Traditional Eastern Health & Fitness. A quarterly magazine, now in its 26th year, based in Temecula, California. See https://www.qi-journal.com/.

Radin, Dean. *The Conscious Universe: The Scientific Truth of Psychic Phenomena*. New York: HarperCollins, 1997.

Reb, A.M., et al. Qigong in Injured Military Service Members: A Feasibility Study. *Journal of Holistic Nursing* (Mar 27, 2016) pii: 0898010116638159. [Epub ahead of print]

Rhoads, C. J. "Mechanism of Pain Relief through Tai Chi and Qigong." *Journal of Pain & Relief*, 2:115 (2013). DOI 10.4172/2167-0846.1000115

Rubik, Beverly. "Measurement of the Human Biofield and Other Energetic Instruments." In Lyn Freeman's book cited above.

Sheldrake, Rupert. *The Sense of Being Stared At and Other Aspects of the Extended Mind*. London: Arrow Books, 2003.

_____. *Dogs that Know When Their Owners Are Coming Home*

and Other Unexplained Powers of Animals, rev. ed. New York: Three Rivers Press, 2011.

Tegmark, Max. "Consciousness as a State of Matter." https://arxiv.org/pdf/1401.1219v3.pdf, accessed May 8th, 2015.

_____. *Our Mathematical Universe: My Quest for the Ultimate Nature of Reality.* New York: Vintage Books, 2014.

The Yellow Emperor's Classic of Internal Medicine, trans. Ilza Veith. Berkeley, Los Angeles, London: Univ. of California Press, 2002. Originally *Huangdi neijing* and also translated as *The Emperor's Inner Canon.*

Wilson, Edward O. *Biophilia.* Cambridge, London: Harvard Univ. Press, 1984.

Yang, Zwing-Ming. *Qigong for Health and Martial Arts*, 2nd ed. Boston, YMAA Pubs., 1998.

V HARD NEWS ABOUT CLIMATE CHANGE: WORLDS OF DECAY AND DISASTER

THREATS TO OUR EARTH AND ALL OF ITS INHABITANTS

Omega Institute was edenic: beautiful, peaceful, calm. My cell phone did not work there. I heard no TV or radio news; I saw no newspapers. Because cars were parked away from buildings, we walked on paths with green grass and trees all around us. It could have been a glorious spring in any recent decade.

Although I profited from the new information, the inspiration, and the retreat in general, when I returned to the ordinary world, it had not changed. Indeed it seemed worse and continuously worse over the next year: TV, radio, and print media all described terrorist attacks, a bizarre presidential campaign in America, refugees from Syria and elsewhere stuck in camps or outright dying in transit, and a bombshell of a book, E. O. Wilson's *Half-Earth: Our Planet's Fight for Life* (2016) that recalls Al Gore's *An Inconvenient Truth: The Planetary Emergency of Global Warming and What We Can Do About It* (2006) a decade earlier. Wilson's book and other authoritative publications warn that Climate Change will disrupt many patterns of life on this earth, destroying some or even many of them. In peril would be all life, including humans. This is hard news to receive, and many people currently choose not to believe or even consider it.

Informative sources include:

Naomi Klein, *This Changes Everything: Capitalism vs. the Climate*, 2014.

Elizabeth Kolbert, *The Sixth Extinction: An Unnatural History*. 2014.

Pope Francis. Laudato *Sí: On Care for Our Common Home*, 2015.

David Orr, *Dangerous Years: Climate Change, the Long Emergency, and the Way Forward*, 2016.

The movie "Before the Flood," 2016.

See also "Facts" in Sources 5.

In Section II we followed the mental turmoil of a cancer patient confronted with questions about death. In this section, we may look ways humanity might understand *death writ large*, a Sixth Extinction on the earth, killing many species, possibly including humans, or making existence for them very, very hard. Appropriate emotional responses include shock, abandonment, fear, even terror.

In Essay 2 we reviewed Atul Gawande's concept of "hard conversations" for families with a dying member. Now we need to have hard conversations about the possible deaths of many ecosystems and species across the globe.

In 2015 195 nations joined in the Paris Agreement on Climate Change. President Trump later withdrew the U.S., but various mayors, governors and business leaders have said they'll go ahead with the written values nonetheless.

How ever will this all evolve?

E. O. Wilson's Book Half-Earth: Our Planet's Fight for Life (2014)

E. O. Wilson affirmed in his *Biophilia* that love of nature is a basic human trait (1984). Some 30 years later, he writes again about nature, now emphasizing the progressive and accelerating dangers to the earth: *Half-Earth: Our Planet's Fight for Life* (2016). This book blames humans: "all the available evidence points to the same two conclusions. First the Sixth Extinction is under way; and second, human activity is its driving force" (p. 55). The Fifth Extinction was the Yucatan asteroid impact that killed the dinosaurs 65 million years ago. The Sixth would be *ours*. Wilson adds that with the five earlier extinctions it took, on an average, some 10 million years for the earth to recover (p. 8). Clearly, that is a long, long time.

We can only guess at the sadness he, a lover and student of nature, must feel as species losses mount and ecosystems degrade.

Whatever responses humans make—and however successful they are—reading his book invites us to consider *the end of humanity's successful (more or less) run on earth*. Will we all die outright, or, more likely, will we struggle on in terrible conditions of war, starvation, searing heat, and various diseases?

He proposes setting one half of the earth aside to protect species and biodiversity in general. Is this possible? I doubt it. Barring some miracle—or changes in governments—it appears that we are doomed to an enormous die-back.

Whatever responses humans make—and however successful they are—reading his book invites us to consider the end of humanity's run on earth.

OUR BATTERED BUT SURVIVING (SO FAR) EARTH

Certainly our earth has been battered and bruised before. Before humans came on the scene there were five extinctions that wiped out many, many species. There have been wars, plagues, famine, and deaths—to evoke one version of the Four Horsemen of the Apocalypse described in the last book of the Bible (Rev. 6.1-8). There have been ice ages and, much more recently, nuclear accidents, tsunamis, and harsh weather and wildfires, some already linked to Climate Change. The earth has always been under siege and—while nuclear destruction is still possible—Climate Change appears now to be the clearest threat to humans and many living creatures.

Nonetheless, it is easy, indeed normal, for many of us who live comfortably to give scant attention to Climate Change. It seems abstract…distant…and just too bizarre to imagine.

CONTEMPORARY ORDER…ALSO CHAOS

Many aspects of modern life in the West seem regular, efficient, and dependable—at least in wealthy countries and for the wealthy people there. For such people, food and water are available, transportation works, and businesses provide products and services. Electronic media (including cell phones) allow for quick connection. We have entertainments galore. For many, this "surround" feels good, a cocoon of order, predictability, and certainty, but it's not the whole story, even within wealthy countries.

What else is happening?
- Wealthy people in the US and other places have ways to avoid impacts, but poor people usually do not. One example: cooling or heating bills. Others: medical care, rent, local pollutions.
- Thousands of refugees leave drought-ravaged Syria and northern Africa, heading for a better life in Europe, many die on the way. Thousands come toward the US and Canada from Latin and South America.

- Terrorists of many sorts (Al-Qaida, Isis, Boko Haram, "lone wolves") strike big cities, kidnap girls, make soldiers of youths, murder innocent people, and put nations on alert and on edge.
- There is chronic air pollution in most large cities, especially Beijing, New Delhi, and Greater Mexico City—all cities of over 20 million souls.
- Indeed human population grows at a brisk rate—but many people do not want to consider the topic. According to Worldometers, the current population on earth is 7.5 Billion, which is *three times* the 1951 value of 2.5 Billion (see Worldometers.info, accessed July 10, 2016) and appears to be headed for 9 Billion.
- Many cities, public spaces, roads, highways and airports are more crowded than ever.
- By some estimates, we have already exceeded earth's carrying capacity for humans.
- The U.S. has seen chaotic politics in 2016 and beyond, and fracturing of the body politic, news media, and civil discourse in general.
- With economic difficulties, many American Millennials are often not "doing as well" as their Boomer parents.
- Year by year some carbon fuels are harder (and more expensive and often more dangerous) to extract from the earth and thus more expensive for at least some consumers. Further, some nations are slow to find replacements for oil, gas, and coal.
- Sects and splinter groups speak of Armageddon, the Rapture, survivalism, "Doomsday prep," and so on.
- Disaster movies symbolize our fears; they show threats of asteroids, extreme weather, volcanoes, aliens, zombies, as well as threats of artificial intelligence, plagues, and nuclear war.

Many threats relate specifically to Climate Change:

- According to the EPA, "all of the top 10 warmest years on record worldwide have occurred since 1998" (https://www.epa.gov/climate-indicators/weather-climate, accessed January 22, 2019).
- Each of the years 2013, 2104, 2015, and 2016 have registered, progressively, the warmest temperatures ever recorded.
- Forests are being cut down for fuel and construction, also to clear

land for crops and human habitation. As wood is burned, more carbon enters the atmosphere, and less oxygen is provided by vegetation.

- Many animal species are in decline, owing especially to loss of habitat.
- Wildfires increase, especially in Australia and the U.S., owing to very dry vegetation; the 2017 and 2018 fires in California were especially damaging and frightening.
- Glaciers (especially in Newfoundland) and polar ice are melting; ice shelves in the Antarctic are endangered. Rates of melting are faster than predicted.
- Rising seas threaten some island nations and coastal cities. The seas are also acidifying so that corals bleach and die and fish stocks (that depend on corals) shrink. The seas are also warming and, in turn, warm the atmosphere, thus making weather more variable and violent.
- There are more extreme hurricanes in the U.S. (Harvey and Irma in 2017, Michael and Florence in 2018) also tornadoes, now possible at any time of the year.
- Droughts occur more often and for longer periods of time around the world.

Many of us feel uncertainty, disappointment, anger, frustration, and general malaise, spoken or not…conscious or not. Others deny the difficulties. Some, rich or not, escape into work, daily routines, material comforts, personal obsessions, even drugs and alcohol, all because the forces arrayed against us are overwhelming.

We may wonder: can nations govern themselves for the benefit of all inhabitants and the environment? Is government worthy of trust? In the U.S., the obstacles include the corruption by carbon industries to influence governments as well as wide-spread human lust for power and control. In the winter of 2019, Donald Trump took credit for the partial shutdown of the federal government for 35 days; during this time—to pick just one example—our national parks suffered various kinds of damage.

Whatever the turmoil of the moment, all humans must live on the same planet, "spaceship earth," as R. Buckminster Fuller styled it, and we all share—sooner and later—the same enormous risks.

Can we cooperate to save ourselves…and much of the nature that makes all life possible?

CLIMATE CHANGE IS HERE, AND IT'S WORSE THAN GLOBAL WARMING

In 2015 Nancy and I visited Alaska on an ecology tour. We saw the impacts of melting permafrost, loss of sea ice, and disrupted migrations of wildlife. Climate Change hits Alaska first and harder than the rest of the world that will soon be similarly threatened.

The phrase "global warming" didn't carry much threat. The "globe" is an abstraction, something diffuse or domestic, like my childhood globe on a stand. "Warming" is reminiscent of pleasant objects, such as slippers, toast, or a bed, and it suggests a slow pace, something gentle. Furthermore "global warming" can be construed to mean somewhere—anywhere—else and not in my country or backyard.

"Climate Change" is, however, more definite, and change can be scary. While "weather" means variable conditions now or next week, **climate is long-term**—human lifetimes and more. Climate occurs around the entire world, acting dynamically in the thin shell of our atmosphere, bringing rain or snow, gentle winds or hurricanes, chill or cold, warmth or heat. Every day and forever, climate impacts everyone, every human, animal, and plant. How much change? At what speed? We've liked what we've had so far; we take it for granted. We deeply feel that it should not change!

In the decade of the 2010s, it became all to clear that change was underway, even though deniers—too many of them now in government—still speak of a "scientific hoax."

"Before the Flood"

In 2016 the National Geographic Society released the movie "Before the Flood." This 96-minute documentary follows Leonardo DiCaprio, designated by UN president Ban Ki-moon as a "UN messenger of Peace with special focus on Climate Change." DiCaprio visits the Arctic, Canada, Greenland, Miami Beach, Beijing, New Delhi, Argentina, Kiribati (an island nation threatened by sea rise), and Indonesian rain forests. He interviews experts who report on the changes and possible responses. He visits Pope Francis and attends the Paris conference of 2015, where the Agreement was adopted. (For detailed background about this important movie see www.beforetheflood.com.) The film's images and information are impressive and can be disturbing to watch. Viewers must consider a possible extinction of the human race, not in a disaster movie with an dramatic asteroid, but in a relentless strangling by Climate Change with rising of the seas, flooding of coastline (cities especially), loss of arable lands and fish stocks, extremes of weather, tropical diseases (some as pandemics), migrations of desperate peoples, and more. After one showing I attended, the audience seemed stunned. After a few minutes, one man said, "I thought it was bad, but I didn't know it was that bad."

HEALTH IMPACTS OF CLIMATE CHANGE, HAPPENING NOW AND WORSE IN THE FUTURE

At the website for "Before the Flood" websites, there's a section on deniers of climate change and the myths that they foster. "Myth 5" is "Climate change isn't harmful." In rebuttal, a chart shows how the higher temperatures will impact human health in eight ways. These impacts are occurring now and will become worse in the future *even if changes are made now*…and will be much worse if improvements are minimal or absent. In a simplified version:

CLIMATE CHANGE CAUSE	IMPACT ON HUMAN HEALTH
Extreme heat	Heat-related illness and death, cardiovascular failure
Environmental Degradation	Immigration, civil conflict, and mental health impacts
Water and Food Supply Impacts	Malnutrition, diarrheal disease
Water Quality Impacts	Cholera, cryptosporidiosis, campylobacter, leptospirosis, harmful algal blooms

Changes in Vector Ecology	Malaria, dengue, encephalitis, hantavirus, Rift Valley fever, Lyme disease, chikungunya, West Nile Virus
Air Pollution	Asthma, cardiovascular disease
Increasing Allergens	Respiratory allergies, asthma
Severe weather	Injuries, fatalities, property damage, mental health impacts

Figure 9: Health Impacts of Climate Change

This is a frightening list. Any steps to lessen climate change will mean a healthier world and fewer health problems for all living creatures, humans included.

VARIOUS VIEWS OF TIME AND ATTENDANT RISKS

How do we situate ourselves in time, our future, our past history? Some models are cultural, and we take them for granted, with no thought to their assumptions or implications, some of which may be misleading or even harmful.

Some models of time are cyclical, as in the Mayan calendar or notions of reincarnation; these typically suggest that wheels of time, long or short, that will go on forever. Other models suggest a steady-state of dissolution and repair, as in some versions of the Daoist worldview. Still others are linear— an arrow of time—some with an absolute end-point, such as an Armageddon or the Omega point, as in the book of *Revelation*.

Is there a single, contemporary Western worldview? Or a combination of views, including ghosts from the past, such Neodarwinism (*over time: may the best man win*) with an overlay of Enlightenment progress (*a man on the moon, democracy for all, tomorrow will be a better day*).

Linear time that progresses forever

Westerners—and Americans in particular—tend to see (or assume) an arrow of progress going upwards forever. We believe in improvements in life and material success. Except for wars, the last two centuries appear to have borne out these beliefs. Many Westerners live longer and better. Material comforts abound. Technology improves. Some of these values come from the past. In his oration *On the Dignity of Man*, Renaissance philosopher Pico della Mirandola exulted that men could become like angels. Enlightenment philosophers sought liberty, progress, tolerance, and freedom. Reason and science would lead the way, giving control over the material world.

In America, we assume there will be progress forever and an endless timeline for the generations of humanity. In John Winthrop's formulation, The "city on the hill," was originally Boston but symbolically a place of ideals to inspire the world. The phrase has been used by politicians such as Walter Mondale, John F. Kennedy, Ronald Reagan, and Barack Obama. Manifest Destiny would expand our control from "sea to shining sea" because the Western frontier was "ours for the taking." Later, frontiers shifted to outer space, a War on Drugs, a Moon Shot for Cancer, the Internet, and more. Progress and expansion are always possible! We imagine wonderful adventures are just around the corner and will always be available to our imaginations and can-do efforts.

These are deep values taken for granted, also permanent and predictive for next year, next decade, and so on because *the best is yet to come*. There is also a Neodarwinist dimension: individuals should improve their individual lots and compete with others. For decades "self-help" books have been best-sellers, emphasizing how intent and strategy can improve individual lives (Carter, "Self-Help," 1995).

I believe this is the default position for most Westerners: that we will go onwards predictably to tomorrow, to next year, the year after, and the next decade with continuous improvements. The stock market will go up; a company's market share and profit will increase, and each [wealthy] person's net worth will grow. That's the American way!

Linear time as a trap

This ideology pleases and encourages many of us, but it also traps us into a false optimism and blinds us to present and future realities. Because earth's resources are limited, limitless growth is not possible. Nor is it equitable across a population or between nations. Nor are we in control of many forces on earth...even less in control of our sun, galaxy, or universe.

LINEAR WORLDVIEWS LEADING TO FINAL DAYS, SOME IMPLYING *CARPE DIEM*

Buddhist, Hindu, Islamic, Jewish, Zoroastrian, and Mayan beliefs...all have visions of final days, end times, the end of humanity. In many belief systems, heaven as a second, all-encompassing world for an afterlife. In some heavens, humans are re-united with family and friends, even united with persons of other times and places.

For Christianity, God is in charge and will take all of us home to a New Jerusalem at the end of time; this is described in *The Revelation to John* that concludes the Bible, a bookend to *Genesis* that describes the beginnings. Many (perhaps most) Christians, however, give little thought to such an end and assume continuous progress. In *Laudato Sí*, Pope Francis does not speak of end times one way or another. Various fundamentalists, however, look for "the Rapture" that will take 144,000 of them (and not others!) to Heaven. Some believers of final days—Christian or not—see a violent Armaggedon of one sort or another, and survivalists store food and make plans for the *very bad times*.

So, besides progress, there is another, darker strand in our understanding of time. Some people, consciously or not, understand that time and resources are running out so we should *seize the day* (carpe diem) and enjoy it to the fullest...because today—this day now—is already the acme, a zenith, the peak, and therefore a party should be enjoyed *now*. The end of the world will be the problem for other, future people, certainly not us.

Carpe Diem: Levels of Meaning from Hard-Core to Gentle

The phrase comes from an Ode by Horace. The final line runs "Carpe diem quam minimum credula postero," meaning "seize ["gather," "enjoy"] the day, trusting as little as possible in the future."

Hard-core carpe diem

Modern usage intensifies the meaning: *go for the gusto, get yours now*, or *grab all you can get*. A good example is a satiric *Far Side* cartoon from many years ago. Two men are fishing in a boat. In the background, there are several mushroom clouds suggesting nuclear war. One man says, "*I'll* tell you what this means, Norm...no size restrictions and *screw* the limit."

This hard-core sense fits with current usage of earth's carbon resources: rip the oil out of the tar sands, "frack" gas with dangerous chemicals, drill deep into the ocean beds, and remove mountain tops in Appalachia to collect coal. In "Before the Flood," we see helicopter views of devastated landscapes in Canada where the tar sands are sacrificed for oil. We see acres of Indonesian forests cut down to allow for growing palm oil (widely used in snack foods, such as potato chips). We see American feedlots for cattle, large contributors to methane pollution of the atmosphere, but *Americans want beef*. Near the end of the movie we see climate simulations on a large wall display: ocean currents and jet streams are disrupted; tropical temperatures rise dramatically, but for many investors today, maximizing immediate profit makes sense, even if it harms the earth and its creatures now… and even more in the future.

Gentler carpe diem

There is a gentler way to understand "carpe diem": enjoy today as today and be grateful for it. Gratitude can expand our experience of any day or of any hour, our surroundings, and our fellow creatures, be they human or animal. We may be realistically skeptical about the future because nothing is guaranteed, but—departing from Horace—we may also *have intent* to influence the future for the better.

CLIMATE CHANGE DOESN'T FIT WITH OUR LONG-HELD CULTURAL VIEWS

Many modern Westerners dislike the concept of Climate Change because it does not fit with our current values and sense of history.

Our instinctive dislike of an Omega Point

As for a final Omega point, our human instincts go against any possible end of our race; our DNA wants us to live and reproduce as long as possible. We want our offspring, our towns, our countries to survive forever. We think of ways to deny the threats: *surely science isn't correct.* Even if science is more-or-less correct, there are some proposed (if desperate and impossible) solutions, such as space colonization or technological fixes. Or the next ice age might—in some thousands of years—exactly match the temperature rise. Some may imagine a positive Omega Point, but they are in no hurry to arrive there, and especially not by a disastrous path.

Regrettably, Climate Change is happening now

The documentary "Before the Flood" shows streets of Miami Beach flooding from the manholes on a bright sunny day. Cars splash the seawater as they drive through. The city is raising some roads and using pumps to return water to the sea. Not only is it happening now, it will get worse, and some of the rises in temperature have been faster than projected. There will be problems for decades if not centuries, especially if there are tipping points as feedback loops kick in and nothing can stop further and accelerated heating.

In 2009 Nancy and I visited the Great Barrier Reef off the coast of Australia. It was then in moderately good health, with only some bleaching that we could see. In 2016, however, reports showed enormous damage to the northern sector, bleaching caused by warm water that made the algae (normally symbiotic) to emit toxins that kill the coral. While some bleaching events in the past were repairable, second or third events (that seem likely) will kill these corals and many others around the world. Corals are called the "rainforest of the sea," rich and varied ecosystems that are nurseries for the oceans' health and fertility of the food chain that eventually feeds humans. Corals also protect shorelines from stormy seas.

LINEAR WORLDVIEWS WITH NO CLEAR DIRECTION, NO TIMELINE AND INEVITABLE DISASTERS

Worldviews can still be linear but with risks that are uncertain and always possible, such as the 2004 Indian Ocean earthquake and tsunami, as well as the nuclear accidents such as Three-Mile Island, Chernobyl, and Fukushima Daiichi in the last 40 years—all events that seemed to come out of nowhere.

In the very long run: ice or fire

According to astrophysics, the earth—in the very, very long view (billions of years)—is already doomed because the sun will become a red giant star expanding to first desiccate then roast the earth. This scenario is based on observations of many other stars, but the time span is so huge, that, even if we've seen the scientific projection, we give it no thought. In a shorter long view (thousands of years) there may be another glacial age. Scientists examining ice cores and other evidence see past interglacial (warm) periods of 10,000 to 30,000 years. We are some 11,000 years out from the last one. Do we have another 20,000 in the bank? Or much less? No one knows, especially with the current, unprecedented heating of the globe. Again, the number of years involved and the uncertainty of scientific projection, make it easy to disregard the image of the earth half-covered in ice and all the attended dilemmas. (For discussion of these and other risks, see https://en.wikipedia.org/wiki/Future_of_the_Earth, accessed July 30, 2016.)

In a still shorter time frame, *even this current century*, we may lose much of nature because of Climate Change. Human population growth, scarcity of resources (food, water, fuels), sea rise, warfare, and/or spread of diseases—all these may cause considerable impacts on humanity; no one knows the extent, but they could be extreme. Some projections are dismal…truly frightening. They are also uncertain enough that we are glad to avoid considering them as predictive.

Other possible disasters…at any time

Over geologic time, the earth has always been at risk from tectonic plates shifting, super volcanoes, earthquakes, asteroid strikes, and bizarre solar radiation. These seem aleatory, according to chance…unpredictable rolls of the dice. Ice Ages have been a little more orderly in cycles, but the time scales are so enormous that we cannot plan for the next one.

Modern anthropology understands that the slow migrations of humans out of Africa could be possibly only after the melting of the last ice age and changes in climate that allowed for agriculture. Over thousands of years many cultures arose, peaked, and fell into ruin, including the Mayans, the Aztecs, and the Puebloan cultures of the American southwest, also the cultures of Egypt, China, Greece, and Rome…and many others. Modern scholars have sought the causes in the past: water or food shortage, disease, warfare, lack of slave labor, and so on. Many of these causes are possible today, with the addition of worldwide economic collapse and chaos, nuclear war, or artificial intelligence run amok (as in the *Terminator* movies). We tend to dismiss most of these as "long shots."

Climate Change, by contrast, is here, now, and with *clear and present dangers*.

Linear views that grind to a halt

According to Newton's Second Law of Thermodynamics, closed systems experience progressive and increasing disorder. Even with energy input from the sun, friction, materials fatigue, and more lead toward chaos. Natural processes are irreversible, and there's an asymmetry between the recent past and the threatening future. (Newton himself was an Anglican and believed that the beauty of the mathematical astronomy showed the Creator's genius.)

There are wider questions: is entropy bad because it is messy, or are there changes of *matter, mind,* and *energy* that are, in their own way, healthy? What constitutes a closed system? Can any system be entirely closed? Is the universe a closed or open system? What if time is reversible?

RANT 5: IT'S A SORRY...AND GUILTY...BIRD....

Of all the sections in this book, this one has been the saddest one for me to research, consider, and write. I recoil at the printed information, what the experts say, as well as the images from photos and cinema because they all show grave damage to our planet, specific impacts on plants, animals, and people, and a *continuing path toward harm's way,* especially when governments are unable or unwilling to take action.

It is frightening to read projections of increased daytime temperatures, more violent storms (including tornadoes and hurricanes), also losses of agriculture, corals, and fish, increases in wildfires, forced migrations, diseases, even armed conflict. Is it even possible to "weigh" such catastrophic threats?

In a Hollywood disaster movie, we can enjoy the drama because know that the imagery is part of the as-if world of fantasy. In the documentary "Before the flood," we see actual dead corals and dead animals, also today's flooded streets of Miami Beach. Watching this movie brings an array of difficult emotions: surprise that degradation is already happening, sorrow for the changes, shock at the enormity of some of the devastation shown in aerial views, grief at the current and projected losses, disgust that governments do little, and deep sadness for current young people who will deal with worse circumstances...and their children who will—in all probability—deal with far worse circumstances.

Sadness, despair

I feel betrayed by governments—and ours especially—that do little about Climate Change now and, worse, in the future. Donald Trump has called climate change a "hoax." He signs an executive order attempting to undo steps President Obama took to combat Climate Change. He withdraws from the Paris Agreement on Climate Change. "Before the Flood" depicts senators from carbon states Jim Imhofe (Oklahoma, oil) and Mitch McConnell (Kentucky, coal); they both deny that Climate Change is happening. How much of the Senate has sold out to carbon interests? Some of our country's leaders are unwilling to protect us from Climate Change, nor will they make the U.S. a leader in the world for renewable energy sources, and China and other nations will likely fill the gap.

While scientists and other commentators see the world headed into peril, today's captains of American industry are champions of carbon in coal, gas, and oil. They will not admit to the connection between their fossil fuels and the heating of the world, and they have sufficient control of government to maintain this dangerous status quo.

I can vote, sign petitions, and go on marches, but I have no direct power over governmental inaction. I feel helpless, a victim of those Carbon Curmudgeons, the plutocrats, the non-caring capitalists and the dark money that influences government. These are the today's robber barons, robbing from the earth, all living creatures, and generations to come.

"It's a sorry bird that soils its own nest," the proverb goes. I don't believe there's any literal truth to it because I've cleaned birdhouses and seen normal bird waste in the old nests. The metaphoric truth, however, applies to humans, similar to the judgmental "You've made your own bed, now lie in it." What will people a century later say about our failures today to protect the earth? How would we answer to our great-grand-children?

I don't accept blame for 1800 to, let's say, 1950, when carbon fuels were used less, but as scientific evidence mounted and our climate became warmer and warmer, governments and industries are now to blame for denial, inaction, and aiding and abetting continuing harm to the planet.

Nancy and I have changed our light bulbs, put in solar heating for water, bought an efficient car, and stopped eating beef—but we still travel and enjoy, in general, a carbon footprint larger than for many other citizens of the world. We too are complicit.

What about "progress"?

Continuous progress is a lie. There is not always a better tomorrow…what if the "best" may be behind us, not ahead.

And what about the continuous progress, Horatio Alger, and the "home on the range, where the deer and the antelope play"? These images point to the past, not a usable future.

Or will we muster new ideas about progress that deal with Climate Change, international cooperation, and wise use of resources?

Down deep: guilt

At some level, we know that we have messed up, screwed up, created a disaster. We have used up resources, polluted oceans, earth, and air. We have produced too many people and too many expectations for energy to run cars, air conditioners, heating, and so on.

WE BLEW IT.

If we realize this, we should ask, how can we atone, make up, make right, find solutions?

We need to change some of our deepest values.

How can God allow this? How? Why?

How can God watch as we destroy nature, even ourselves? Did God become the so-called *Deus absconditus*, who started the earth and then abandoned it? God let Adam and Eve choose to sin and were expelled them from Eden. God let Jesus die on the Cross. God has watched plagues, wars, and earthly disasters forever. Why would S/He intervene now?

Or: has S/He always allowed humans the freedom to make choices: to pick the forbidden fruit in every era, every land?

Christians and Jews acknowledge the Creator of the universe and all life described in Genesis; they share the praise of nature found in many Psalms (see 8, 29, 139, and 146-150). Psalm 104 praises not only God's creation but also God's continuing care of it. Many other religions believe in divine creation. How can believers understand our world under threat?

Theologians have constructed "theodicies," arguments to explain how God can allow evil and human calamity. Such arguments, it seems, are to help God out...how human!...and these are further proof of the limits of human knowledge, wisdom, imagination. God mocks Job's complaints with, "Where were you when I laid the foundations of the earth?" (Job 38.4), and the chapter continues to praise the intricacy and beauty of the creation. "Can you bind the chains of the Pleiades or loose the cords of Orion?" (38.31).

Another hope: God will miraculously save us. Should we pray for unimaginable forms of grace to emerge, to save us?

I've read Revelation's wild and whirling words. It suggests we'll all die and some will be taken to heaven, the New Jerusalem where peace, abundance, and glory reign forever (Rev, 21-22). Most ministers never preach from it. Most Bible readers, I'll wager, skip it. For me, it's a beautiful and inspiring account of what a Heaven might be.

Whatever the means, I imagine there will be an end to humans, and I prefer to believe that there will be a salvation well beyond our imaginations.

Let's be frank: we are talking about DEATH WRIT LARGE, death of species, death of ecological niches, death of food chains, death of coastal cities, deaths of lives of ease, and deaths of many other lives, easy or not, as well as worsening conditions for all future creatures—whether or not humans survive.

E. O. WILSON'S SUGGESTION TO SAVE THE EARTH; CAN WE CHOOSE BETWEEN TIMELINES OF DOOM, CHANCE, OR SALVATION?

Wilson proposed setting aside half of the earth so that it may recover. While I admire the bold imagination of this idea, I see no way it is possible. Who would do that? The UN? Not likely. What nations would give up large expanses of their territory? Even if half was set aside, it would probably be invaded by poachers, terrorists seeking safe havens, or other influences of rogue nations or failed states.

Today various towns and countries are studying how to "harden" bays, ports, river mouths, even coastlines into permanent forms. These may provide temporary mitigation...temporary. At some level, we know this isn't possible in the long run if (or when) seas rise 20 feet.

Nonetheless, bold imagination is needed to protect the earth and its inhabitants.

ESSAY 5: WEIGHING (?) CATASTROPHIC THREATS

How do we weigh these issues? Modern people love scales with numbers—even bathroom scales—and other instruments of measurement and analysis: these appear to give us control. Nowadays the term "metric" is in vogue; there may be an aura of European and scientific charm in the term because of the metric system as well as a general sense that numbers are the best measure of anything. No doubt numbers have many uses but they depend on information. Early users of computers referred, however, to GIGO as a trap: Garbage In, Garbage Out. GIGO is still a possibility, especially when there are many variables and some of them unknown. Further, science is always a work in progress; technologies of measurement change, and theories evolve. Unfortunately, the projections about Climate Change have been right in concept, although some have been too generous in predictions of time: *the world is heating up faster than expected.*

There are feedbacks that accelerate the process, such as the loss of arctic ice that previously reflected away the sun's rays. Melting permafrost, mentioned above, releases methane, a much worse gas than carbon dioxide for causing the greenhouse effect.

As much as we love information, especially when it suggests control, we are beginners in trying to understand impacts of Climate Change especially because this human-caused event has no precedent. *Humans are latecomers to the history of the earth.* We missed the dinosaurs by some 66 million years, even though some comic books and movies show an overlap of us and them.

The internal—that is to say mental—problems for us humans are also complex. We tend to perceive according to what we believe already— "confirmation bias"—therefore some people count changes in weather as "normal" and routine, even though recent years have been the hottest on record since 1900. Personal and social values are complicated, partially rational, partially irrational, and hard to measure in numbers. Poll results can be misleading or even wrong depending on the wording, the people responding, and hidden assumptions and values.

When nature is assumed to be "raw material," our risk increases.

Economic policy is to rip the good stuff out of the ground, the forests, even the seas. Pave over as much as possible. Frack the land, cause earthquakes, and pollute ground water. Dam up rivers until they silt up behind the dam. Let agribusiness pollute streams. Gather more carbon, burn it. To heck with the atmosphere also the oceans that uptake carbon dioxide and become more acidic. Rape the land, the sea, all of nature: she's just raw material for us to use and to use up.

These are perilous times with excesses of heat often symbolized by red, the color of stop signs and blood. Where's the green of plants and trees, of Hildegard's *viriditas*, that may symbolize prospects for a usable future?

As our teachers die, as our grandparents die, our parents, our teachers, our spouses and friends, and—even—some of our children, we struggle to come to some terms with these inevitable losses.

How can we come to terms with the possibility that humanity will also come to very hard times? Even perish? In the long term, it is normal for species to rise and fall. Estimates are that more species have already perished than are alive today.

But wait: extinctions aren't like turning a light switch to OFF. There are declines, with some pockets lasting longer than others. Wealthy people can move their homes. Wealthy nations can fence off the sea. There may be a rocky decline, as in the post-apocalyptic books and movies that show roving bands fighting with each other.

Some days, I'm afraid that we are doomed by any or all of the scenarios sketched above. I live in a nice home in the woods. Food and water and plentiful. My life appears a continuation of past normalcy, but there's news of migrations out of Africa and Latin America, record temperatures, cracks in Antarctic ice shelves, and disruptive and uninformed rhetoric from President Trump and others.

Are we doomed by Climate Change?

Maybe.

From a Christian perspective, I expect God to take us all home someday, but maybe not for centuries or millennia, and I hope that Climate Change could be a obstacle that we may surmount—one way or another—before then. As a person of faith, I believe this.

People of many sorts will work on local levels, and their cumulative minds may make a difference. The Chinese revived Qigong and other

old wisdom after the suppression by Chairman Mao. We may hope for similar rebounds after the disappearance of Donald Trump.

I don't want to be a Gloomy Gus, an elegiast for nature and all creatures, but the evidence is stark and, upon reflection, overwhelming.

I am open to fall-back positions: there are changes we can't predict, some technology will emerge to save us, nations may come together to act, a cooling effect may emerge.

Or are we entirely on our own, without progress, human solutions, even God?

At the end of Camus' novel *The Stranger*, Meursault (in prison for murder and sentenced to die) looks up at the stars and realizes, for the first time, the "tender indifference" of the world ("la tender indifférence du monde"). This phrase puzzled me the first time I read it in college; it puzzles me now. How can the universe be, simultaneously, indifferent and tender? Do we join Meursault in this assessment? The stars are distant and they seem cold. Nonetheless they are furnaces, like our sun. Do any heavenly bodies have intent? Are they *tender* in the sense of *vulnerable*? According to astronomy, they will, like our sun, expand and explode. Or do stars, our planet, and all the universe *care for us*, even as our lives are endangered? Or is talk about a comforting Universal Mind only talk…wishing thinking?

My reflections fail as a reasoned essay. There is too much I don't know. My emotions are a jumble. The topics defeat me.

Nonetheless, we will discuss strategies for renewal and hope, small in Sections VI and large in Section VII.

SOURCES 5

"Before the Flood," directed by Fisher Stevens, distributed by National Geographic Society, 2016.

Carter, III, Albert Howard. "Self-Help" in *Encyclopedia of Bioethics*, rev. ed., ed. Warren Thomas Reich. New York: Simon & Schuster Macmillan, 1995. Vol. 5, pp. 2338-44.

Facts about Climate Change and Global Warming. GOOGLE this topic for Websites by NASA, The National Wildlife Federation, Conserve Energy Future, Friends of Science, National Geographic, *The New York Times*, BBC

News, and others.

Francis, Pope, *Laudato Sí: On Care for Our Common Home.* Huntington IN: Our Sunday Visitor, Inc., 2015.

Gore, Al. *An Inconvenient Truth: The Planetary Emergency of Global Warming and What We Can Do About It* (New York: Rodale, 2006).

Klein, Naomi. *This Changes Everything: Capitalism vs. the Climate.* New York: Simon & Schuster, 2014.

Kolbert, Elizabeth. *The Sixth Extinction: An Unnatural History.* New York: Henry Holt, 2014.

Orr, David. *Dangerous Years: Climate Change, the Long Emergency, and the Way Forward.* New Haven: Yale Univ. Press, 2016.

Wilson, Edward O. *Biophilia.* Cambridge, London: Harvard Univ. Press, 1984.

_____. *Half-Earth: Our Planet's Fight for Life.* New York: Liveright Pub. Corp., 2016.

PART THREE

CHANGES WE CAN MAKE TO SAVE THE HEALTH OF THE EARTH AND ITS INHABITANTS, HUMANS INCLUDED

VI "SMALL" STRATEGIES: LOCAL, DOABLE WORLDS OF RENEWAL, HOPE, AGENCY, AND ADVOCACY

AFTER HARD NEWS...WHAT?

So...what's a person to do about Climate Change and other risks? A family? A neighborhood. A government? All of humanity—as if 7.5 billion of us (and growing) could ever be unified? Nonetheless...smaller numbers can make changes.

Yes, the news is hard, grim, and sometimes overwhelming, but people have faced terrible times of plague and war in the past and survived to make our lives possible today.

Even with hard news—and perhaps even because of such news—there are ways we can respond. The other part of Atul Gawande's notion of "hard conversations" (besides recognizing the risks) is to decide on courses of action. We now need hard conversations about Climate Change, even as we find it increasingly difficult in the Trump era of denial.

This chapter explores six small strategies that may bring about solace, hope, agency, and advocacy. Sometimes "small" means "large," as we saw in the cancer section. Even small strategies can expand outwards to communities and beyond.

This section will describe these smaller strategies with an emphasis on human nature. Section VII will discuss larger strategies for understanding physical reality, including "nature nature," so to speak.

The point is not to divide mind and nature, as did Descartes, but to show links between the two, for example our link of biophilia to external nature. Nature is, in many ways, our *home*—regardless of the many ways we have *polluted our home* or tried to *run away from home*. A tale of two rocks will illustrate these two natures and their overlap.

But first we need to consider spatial models.

IN WHAT "SPACE" DO WE LIVE: LINEAR VS. SPATIAL MODELS OF TIME

Both linear and spatial models are useful, but when linear modes dominate there are losses.

For a very long time humans have thought about ways of modeling time from the creation of the earth to its possible end. As we saw in Section V, many of these are linear. British physicist Arthur Eddington is credited with the concept "Arrow of Time." For him (and other physicists) the phrase had many meanings based on irreversibility (Eddington, 1928). In the popular mind, linear time often suggests necessity, fate, and lack of human choice. The basic metaphor of arrow = time suggests direction but also purpose. If we think about shooting an arrow, like Katniss Everdene in the movie (and books) "The Hunger Games," we think about aim, decision, and moral certainty.

Clearly, arrows can be weapons. In Homer's *Iliad*, archers were cowards who avoided hand-to-hand fighting and struck from afar. The English longbow has been called the "atom bomb" of its day. As graphic signs, arrows suggest direction, even accurate information and control. At least in the West, we seem to admire arrows for the clarity and power.

To the extent that the arrows of Climate Change have already been loosed, the future is grim. David Orr writes that even with great efforts to mitigate now, temperatures will still climb for a while, perhaps a long time (Orr, pp. 1-2).

Yes, linear models can be helpful, but they should not dominate.

Changing our concepts, however, we can see our experience through time surrounded by *spheres*, whether worlds, fields, or expansive images of our minds. These are spaces of potential, not prescribed linear directions. They are more systemic, more interactive, and less a single vector or set of physical forces.

Remember the medical resident discussing a patient's diagnosis? He said, "You just crank in the information and then ramify your way down." By contrast, a *spatial* model considers many surrounding factors, including particularities of any given patient, the multi-layered interaction between the patient and the care-giver, the state of medicine at the time, and so on.

Leaving Abstraction Ladders for Supportive Spheres.

Reviewing the Rants and Essays above, we see that reductive values narrow and distort our senses of matter, mind, and energy. Our perceptual lenses for *matter*, for example, include physical comfort, social status, monetary work, and the like. We can picture these as vertical abstractions rising above (and away!) from the material reality we experience. My argument calls for more horizontal dimensions, including nature, concepts of home, the Commons, and appreciation for materials as part of our home.

As for *mind*, our society favors reason, logic, strategy, and the like as superior and controlling uses of the mind, rising "above" the welter of our complex lives. A common philosophical model suggests that we create these concepts by moving up an "abstraction ladder" from specific data to big ideas, principles, and other formulas assumed to be universal. This is called "induction." When we move down the ladder, we apply these abstract ideas to local, concrete situations, in a process called "deduction." All these are quite useful, and they pervade much of our ordinary thinking, but they say little about emotion, collateral evidence and events, contexts, and more. What if there are spherical dimensions that may deepen the inquiry and allow for insights, complexities, even mystery?

The following figure illustrates a few of these contrasts. We should continue to use all that's useful from the Abstraction Ladders, but also use spheres as counterpoint, contexts, and nourishers of our minds. Note the vertical = signs (far left) between MATTER, MIND, and ENERGY; these suggest equivalence, conversions, and dynamism.

	ABSTRACTION LADDERS VERTICAL, LINEAR OMNIDIRECTIONAL	SPHERES OF ATTENTION HORIZONTAL, INCLUDING ALL CONNECTIONS
MATTER	Newtonian, raw material	Organic nature, the Commons,
	Physical comfort, raw material	Esthetic beauty, power, home
	Entropy, monetization	Permanence through changes
‖		
MIND	Reason, logic, strategy	Intuition, instinct, emotions such as love, celebration, joy
‖		
ENERGY	Linear, focused	Synergetic, cyclical, multiple foci
	Kinetic energy, carbon sources	Many forms: Qi, subtle energy, etc.
	Monetary worth, personal drive	Many resources of humans, all creatures

Figure 10: Values in Abstraction ladders vs. Spheres of Attention

If we shift our focus away from ladders and toward spheres, the next question is *how large a sphere*? The heating earth may be much too large to consider for many people, perhaps for any of us at any given moment. Thinking locally, however, may be not only possible, but also attractive and re-energizing. That we can and should do.

"Think globally, act locally"
This slogan has been used for decades now, on bumper stickers and elsewhere. According the Web, several people have been named as the originators,

another sign of its popularity. Why is it memorable and, for some, likeable? Because it suggests that the big picture is frightening and overwhelming; we can't solve it. Nonetheless, we can take on small, manageable goals. We take pleasure and modest pride in the phrase "my corner of the world" where we feel at home and capable.

Some readers will recall E. F. Schumacher's *Small Is Beautiful: Economics as if People Mattered* (1973; still in print).

Let's focus on two small rocks. The first is a stone axe that can fit in your hand. The second, is much smaller, but it came from far away: the moon. (See Section VII).

A TALE OF TWO ROCKS: (1) A HAND AXE CRAFTED BY HUMANS

A stone axe

When my family lived in D.C. some years ago, we visited the museums on the Mall. At the Smithsonian Museum of Natural History we went to a room that offered hands-on exhibits for children. My daughter Rebecca, in middle school at the time, and I were invited to hold a hand axe used by prehistoric people. It appeared to be a chunk of ordinary rock but carefully chipped to give it an oval shape with a long, sharp edge. We hefted it and agreed that it was heavy. We touched the knapped edge, and sensed that it could cut. I was puzzled, however, that the stone didn't fit my hand. Were the original users so different from me? Puzzled, I asked the attendant. She said, "Try it in your other hand." I shifted it to my left hand, where it fit *perfectly*.

I suddenly knew that I and the first users from many thousands of years ago were *alike*; not only early humans but prohominids made and used these. Further, the anatomy of our hands was the same: palm, fingers, muscles, bones, sensory nerves, and more. As I lifted it up and down, as if to strike and cut with a modern hatchet, we were alike as tool-using persons with intentions to change some material according to our wishes. As humans spread from Africa to Europe, Asia, and the Americas, they created and used hand axes. Because these axes don't physically decay, in modern times they are often dug up by archeologists. Anthropologist Ian Tattersall devotes several pages to them in *The Fossil Trail* (1995).

When my wife and I toured in France, we visited the prehistorical caves at Cougnac. The guide pointed out the hollows in the floor where the painters had mixed their paints, some 25,000 years ago. In the nearby Pech Merle cave someone created an image of a hand (his? hers?) by placing the hand on the wall and carefully putting red paint all around it. It is a hand much like yours or mine.

No other creatures on earth could create these stone axes or these paintings: only humans. What common faculties can we imagine that we share with the early tool users? Intelligence…strategy…shared learning of techniques…experiments with different rocks…satisfaction with a clean cut? And for the painters? Representation of animals and humans, uses of color, illuminating the cave, possible religious or mythic values, joy in an esthetic creation? Nature provided the matter of the rock; humans shaped it and used it according to their minds and energies.

Human nature has many qualities. We sense them in skyscrapers, gardens, music, and stories. For me, the hand axe was less important to me as a material rock than as a reflection of human mind, skill, and purpose. The hand axe was a small rock…smaller than my hand…but its implied meanings are large. Such axes were used for thousands of years, by untold numbers of beings like me, including the protohominids that later became humans.

The stone axe was a hand-tool that could skin and butcher an animal. It could cut small bits of wood or other substances. The axes had little impact on surrounding nature. They were an enormous distance from, say, oil derricks in the Gulf of Mexico or the explosives that destroy mountaintops for coal mining in West Virginia.

Unfortunately, single-minded urges for profit now dominate Western (and other) cultures. We use up natural resources with little care for the future impacts on all members of society and all inhabitants on earth, human and nonhuman.

SIX SMALL STRATEGIES TO CONSIDER FOR RENEWAL AND AGENCY

A journey of a thousand miles starts with a single step. – *Tao Te Ching*

While it is interesting to consider the many overlapping spheres, it is impossible for any of us to consider all of them at once. The 1990s fascination with "multi-tasking" went boom and bust as we learned that our minds need to *focus* on something specifically in order to do it well. Our attention can—and will—wander, but it can also be quickly recalled, as the Aikido master Morihei Ueshiba said.

It is often hard for Americans to focus because of their many interests and responsibilities. The demands are myriad: some of the external (family, friends, work, neighborhoods, various tribes, cohorts, and networks), some of them internal (self-imposed goals and deadlines, worries, fears, and doubts). Modern Western lives typically have little room for relaxation, meditation, evaluating new experiences or information, and the result is often pervasive stress. We are currently a very stressed society.

Given this complexity, what can we do?

May I suggest six approaches that can be sources of renewal, hope, and agency—affirmations that *a person can do something*?

1. Foregrounding an issue, a hope...anything

Foregrounding is a term from Thomas R. Kelly, a Quaker thinker, writer, and educator who died in 1941, the same year as his book *A Testament of Devotion* was first published. In the chapter "The Eternal Now and Social Concern," he puzzles about reconciling a life of religious faith with the injustices of modern society, or, in larger terms, cosmic responsibility and worldly suffering. He suggests that we should narrow our concerns to a selective foreground (a particular focus) within an immense and comforting background of universal concern "for all the multitude of good things that need doing." We cannot take up all good actions, but we can be aware of them, and leave them to others while we focus on the ones best suited for our specific talents (Kelly, pp. 65-85, p. 83).

2. Personal talents to apply

Thus, the second area: there is such variety among humans, and we don't often realize how our own gifts might be put to use. It makes sense to ask: what are my gifts, my experience, my heritage from family and culture?

Where am I now, and what is needed there? What can I supply to this instance? Something…nothing? Should I consider being somewhere else? Do my gifts allow me to advocate for any person, any threat, or any path for improvement?

3. Teamwork

When I was researching severe burns in the 1980s, I visited various units across the country. They used a *team concept* that allowed for comprehensive care: surgeons, nurses, nutritionists, social workers, case managers, psychologists, chaplains, as well as physical and occupational therapists. All these roles (and more) worked together on the severely injured patients in their care, often for many weeks. There was mutual respect between all of them and the shared goals of caring for the patients. Families can also work as teams, also neighborhoods, musical groups, and sports teams. When these work well, the members feel pride in their work and support from all the others. All can do more together than individuals working alone.

Teamwork is an important theme in Health Humanities and Integrative Medicine, discussed in Section IX.

4. Expansion of mental faculties

Our culture sets so much store by rational, linear thinking that it's easy to forget about other capacities that are also important: intuition, association, humor, emotions such as awe and wonder, and so on. If we just consider the range of our imagination, we see many resources: affirmation, play, creativity, appreciation of mystery, expectation, speculation, and more.

Many of these are nourished when we pay attention to nature. Other sources include hobbies, the arts, sports, dreams, and worship.

5. Play and rest

Modern culture emphasizes work, economic gain, duty, and the like.

Equally important are the renewals found in play. And there are many, many possibilities: sports, arts (museums, films, drama, music), games (playing or watching), conversation, cooking and eating, social events, making love, and more.

And rest.

6. Openness, imaginative intention toward luck, fortune, grace

If we can be quiet and open, maybe we can perceive a solution somehow coming toward us. It may be quite different from what we have imagined.

RANT 6: PRAISE FOR "SMALL" ACTIONS; THE PLEASURES OF RENEWALS

One of the pervasive and enduring qualities of nature is renewal in many forms. Day follows night; the seasons roll around. The moon follows its phases; birds migrate north and south. We enjoy seeing these rhythms; we recognize them as sources of order and renewal. Disruptions are followed by continuities and new disruptions. Dead trees and dead creatures make soil. The earth is always renewing itself.

Our bodies renew themselves as we sleep. Repairs to our skin, our muscles, our gut, and continuous divisions of cells in enormous numbers. When we recover from an illness or injury, we feel new energy and strength.

We enjoy repairing something at home. We like seeing a garden grow flowers or vegetables each year. We enjoy seeing old friends.

To what extent we can find renewals during or following Climate Change, however, remains to be seen. E.O. Wilson put the average recovery time from earlier extinctions at 10 million years.

What are some local, domestic renewals we take pleasure in?

In the sentences below, the opening line can indicate a woman or man, or grown-up or youth...maybe others. I can imagine this person smiling upon completing the action described.

Not only are they happy, they have often made others happy as well. Success in small actions is a source of hope because the people have affirmed their choices and their abilities to take action.

I realize most of these examples are from my experience; readers are invited to think of examples from around their world, from all levels of society, and during all human time.

_____ creates a dish, a dessert, a meal.

_____ cleans a car, inside and out...or a room.

_____ mows and trims a lawn.

_____ paints a room.

_____ repairs a child's bike.

_____ plants, tends, and harvests a crop, however small.

_____ builds a birdhouse or a doghouse.

_____ loses ten unneeded pounds.

_____ creates and follows a budget.

_____ stops smoking.

_____ apologizes and restores peace.

_____ patches a foundations crack.

_____ repairs a leaky faucet.

_____ cleans some windows.

_____ tightens a loose doorknob.

_____ cleans up a kitchen.

_____ does a large wash.

_____ listens attentively to someone's concern.

_____ volunteers for a community project.

_____ keeps in touch with a friend or relative by letter, email, phone, or visit.

_____ reads or sings with a child.

_____ advocates for another person, a living creature, the earth itself.

I am very grateful for all people who do small acts with care and kindness.

ESSAY 6: IN PRAISE OF HOMEBODIES

Let's hear it for homebodies! We don't hear this term much, and, if we do, there's often a disparaging suggestion that such people are limited, unambitious, or even unattractive, as in the term "homely."

Instead, let's *praise* these homebodies, brag on them, women and men alike who enjoy their homes, care for them, and take comfort from them and also share them with others.

One July 4th evening my wife and I watched the PBS celebration on the Washington mall, often called "America's lawn," even "America's front porch." In earlier years we had enjoyed picnics and fireworks there in person. Since then, we see this TV show often, enjoying that the participants are happy and very much at home as they celebrate the anniversary of our country's birth.

If we are lucky enough to be homebodies, we value such celebrations, but we know that this is not the case for many other people. In recent years, in the U.S. alone, many people have lost their homes because hurricanes, tornadoes, and floods. Worse still, are the millions worldwide who are exiled from their homes by warfare, agricultural or economic collapse, or failed governments in Syria, northern Africa, and Latin America, just to name a few. Many Syrians and South Sudanese (among others!) were homebodies earlier and are now exiles. They yearn to be homebodies once again. They will risk death to have a new home. In America there are street people, even settled people now threatened with deportation by the Trump government.

If Climate Change proceeds as projected, many, many more will lose their homes, including very wealthy people who have chosen to live near the oceans.

The early human (or human precursor) who made the hand axe choose that rock with care; he (or she) applied skill and patience to shaping the rock and testing and revising until it seemed just right. When we make or repair something—a lamp, a garment, a flower bed—we feel pleasure. We have agency, and this builds hope in our abilities and intentions.

"It takes a heap o' l'vin' i' a house t' m'ke it home," wrote the home-spun poet Edgar Guest. Within the physical structure, what do we think of? Surely the inhabitants, with their hopes and dreams. Their interactions, some loving, some competitive, some tragic. Also the decorations they have made, including objects that have large symbolic value: a painting, a piece of furniture, a rug. Their pets and plants. The groceries they buy and use, and many rituals that have meaning and order. The house gives them protection from weather, animals or other intruders…and privacy.

"A man's home is his castle," we read in British law, a relic of lines from Cicero, and, of course, today we speak of a man or a woman. Such a castle has stature, honor, security, even sanctity for Cicero, as if the gods approved.

If we were lucky enough to have a good home growing up, we remember it fondly, no matter where we are now. Or perhaps we imagine a home. American TV shows for children included *Mr. Rogers' Neighborhood* and *Sesame Street*, both idealizing home and neighborhood.

Homebodies in their homes sense energies and moods; they appreciate good meals, rituals and other celebrations, hugs from family members, making love, familiar objects, whether a room is messy or neat, and many other events.

Homes give context, meaning, and comfort. They are nests, dens, a sort of magnetic north. They can include the house where we live but even further: places where we work or play, places of study or worship. They can be places in nature, in memory, in dreams of the future. They can be as local as a hug, as cosmic as a view of the night sky.

Damage from a storm or a robbery can dramatically change the feel of their home…or a shouting match with harsh words…or, on a larger scale, war, plague, or Climate Change.

Homebodies pay their bills, get along with neighbors; they run errands and cook their meals. If there are others—children, grandparents, aunts, pets or plants—they take care of them. Homebodies clean house, cook meals, do upkeep to the interior and the exterior. They care for the yard, maybe a garden. In sum, they do a thousand small things, and there are billions of homebodies all across the globe and throughout human time, all contributing to the health and wellbeing of their communities and beyond.

Taking together, their small acts are large in value, and—as a whole— they demonstrate *intent that is caring*. If we think of all of homebodies' *minds* as a single, united force for order, care, and affirmation, we can speculate that this force of mind has influenced both the *energy* and the *matter* of this world, and perhaps even beyond. But will it be enough in the face of Climate Change, given the refusal by economic powers and the governments make changes that will mitigate the worst features of Climate Change?

We can emulate the happy Sisyphus and happy Atlas

Can we imagine Sisyphus or Atlas happy? Can we imagine ourselves saving the earth?

We are ordinary people, outside of the avenues of power and influence. What can homebodies do about climate change? The forces arrayed against us are huge, perhaps overwhelming. In the U.S. alone, dark money, Trump, and his ilk have gained extraordinary power. They care little for the "little man" (or "woman," or "child"), but, instead, for corporate interests that see unlimited horizons of power and wealth. Because of this, our democracy is in deep trouble: "dark money" can swamp a local election of homebodies with libelous and relentless advertising.

"If you see a scrap of paper pick it up": domestic and local agency

The exhortation from the environmental leader at my hospital suggests that little means big, both for the hospital and for all the employees. He set the mark for us to be homebodies—even in a facility of some 800 beds serving over 35,000 people a year. All of us can do the same for other public spaces, roadsides, parks, malls, where we see scraps of paper or other trash. Or maybe such is beyond our reach, so we report it to management to have it taken care of.

We can be active locally in politics, cleanups, and demonstrations.

We can educate children about nature and protection of it in schools, scouts, religious organizations, our neighborhoods, and beyond.

We can call/email/write/visit our politicians.

We can support local leaders by voice, dollars, even handshakes when they care for our environment.

What to do? Depends on who we are and what our gifts may be. Also, how we take care of our emotions in the face of tragedies unfolding and those projected.

In any case: *small can mean large.* By focusing on an aspect of a problem we can deal with, we commit our attention and intention, we use our wits (mind) to create a plan, and perhaps effect a solution. If the result is done, we are happy, proud of the moment's achievement. If not, we say, "so be it," and try again.

As contributors, we are agents, persons acting. We claim our agency, even if, in the long run, there are superior forces against us.

Will there be days of despair? Yes.

In addition to any strategies just mentioned, we will also need the wisdom about tragedy and loss.

Are we doomed by Climate Change?

Maybe. What we know so far is that the trends and predictions are bleak, and many of the harms have already been occurring faster than expected.

Nonetheless, what imaginative thoughts may we entertain?

In 2017, Michael Bloomfield, former mayor of New York City, and Carl Pope, previous chairman of the Sierra Club, published *Climate of Hope: How Cities, Businesses, and Citizens Can Save the Planet*. The subtitle ambitiously states their thesis, with modern cities at the center of new economies, revitalized neighborhoods, mass transit, and improved education. They write that we cannot depend on national government (pp. 234, 245), but must make progress at local levels. Good old Yankee can-do in a homebody approach.

In the future—some of it this year!—we will have news of further losses of corals, glaciers, species, habitats, food webs, arctic and Antarctic ice, as well as increases in greenhouse gases, storms, temperatures, as well as failures by governing bodies to mitigate these harms. We will do well to recognize these, mourn them, and return our intent to acting as we, even as homebodies, are able to do. No matter how small the action, the *intent* is there and multiplied many times as millions of other homebodies act similarly. I believe (and hope) that the power of mind is more than we currently understand. Are there fields of energy we can engage the concerted power of human minds? What about dark energy not yet understood but a large part of the "energy budget" of the universe, and might it somehow be of use?

From a Christian perspective, I expect God to take us all home someday, but maybe not for centuries or millennia, and I hope that Climate Change is an obstacle we may surmount—one way or another—before then.

Hope

We may consider hope as: the attitude toward the future that good things (however we define them) will happen to us and others, also that we will be able to make good things occur. I believe hope is a particularly human trait, especially because we can envision futures. We are gifted with this ability and should think about it carefully, rather than use it as an instinct with no reflection. The best hopes use information and consultation. They consider results and corollaries. Imagination should also play a role, as well as openness to solutions "coming toward us." I favor hopes that have a dimension of action.

We may feel small as individuals, but thousands and millions of us can have tremendous power. Homebodies often team up with national groups such as Sierra Club, Americans for Responsible Solutions, 350.org, Change. org, Earth Justice, Natural Resources Defense Council, Defenders of Wildlife, Food & Water Action fund, and/or Audubon Society. There are many political actions groups as well, and disaster relief groups such as the Red Cross.

We can celebrate the riches of our homebody minds with the intention and hope of reaffiliating with the material realities of our earth for the benefit all its inhabitants.

SOURCES 6

Bloomberg, Michael and Carl Pope. *Climate of Hope: How Cities, Businesses, and Citizens Can Save the Planet.* New York: St. Martin's Press, 2017.

Eddington, Arthur. *The Nature of the Physical World.* New York: Macmillan Co.: 1928.

Kelly, Thomas R. *A Testament of Devotion.* New York: HarperCollins, 1992.

"The Hunger Games." Directed by Francis Lawrence. 2012.

Orr, David. *Dangerous Years: Climate Change, the Long Emergency, and the Way Forward.* New Haven: Yale Univ. Press, 2016.

Schumacher, E. F. *Small Is Beautiful: Economics as if People Mattered.* 1973. Rpt. New York: Harper and Row, 2010

Tattersal, Ian, *The Fossil Trail: How We Know What We Think We Know about Human Evolution.* New York, Oxford: Oxford Univ. Press, 1995.

Tao Te Ching. Multiple editions available; also online; sometimes spelled Dao De Jing.

Wilson, Edward O. *Biophilia.* Cambridge, London: Harvard Univ. Press, 1984.

_____. *Half-Earth: Our Planet's Fight for Life.* New York: Liveright Pub. Corp., 2016.

VII STRATEGIES LOCAL AND BEYOND: WORLDS OF PERCEPTIONS IN SCIENCE AND NATURE

In this section we move from "small" to large, from daily life to the moon and beyond. How do we create and use our "big pictures" of reality for daily life as well as our place on earth and, even, the cosmos?

This section focuses on energy and mind. What are ways our minds perceive and understand energy? Which are helpful? Not helpful?

COSMIC ZOOMS, OUT AND IN, LINEAR AND SPHERICAL

"Cosmic Zoom" is a Canadian movie from 1968 just over eight minutes long. It starts with an aerial view of a boy in a boat, then zooms away from him and on out to show the river, the continent, the entire earth, then on past the moon, the Solar System, the Milky Way, to some huge, imagined, and far-distant perspective on, by implication, the entire universe. Next it zooms back to the boy's hand. The zoom continues into his body, through tissues and cells, to the atomic level, before pulling back out. I watched it with friends in the early 1970s. There is a similar film, "Cosmic Voyage," an IMAX film (1996). Such images can change our concept of reality large and small. In *Our Mathematical Universe*, Max Tegmark similarly organizes Part One: Zooming Out to discuss stars, the Big Bang, energies, and so on, while Part Two: Zooming In, presents nuclei, particles, Quantum theory, and the like. His approach is largely mathematical. As with cameras, the kind of zoom depends on the particular lenses that are chosen. Some people prefer words and graphic presentations; Tegmark uses all three, but he particularly loves math.

Such huge zooms are inspiring in both directions, and they suggest connections to, eventually, everything from the atomic and the universal. We can further imagine the zooms not just in a single long line, but also in every direction, so that they create a series of nested circles, spheres, or, as we've been saying, *worlds*. For Figure 11, a human is the starting point (and others can also be imagined: a plant, an animal, a rock, etc.)

SPHERICAL ZOOM OUT			SPHERICAL ZOOM IN
Family, friends			Area of the body
Home			Organs (or Organ System)
Neighborhood			Tissues (or Energy Paths)
Nature near and far			Cells
Country, Continent			Organelles
Earth			Molecules
Solar System			Atoms
Galaxy			Subatomic particles
The Universe	=	the Field =	Energy

Figure 11: Spherical Zooms, Out and In

Either zoom takes you to the Field, as large and small as we can imagine, and pervading everything. The phrase from the Beatnik '50s was "The way out is the way in." Or maybe it was the other way round…doesn't matter.

According to the concept of the Field, the universe is made of the same *matter* and *energy*, and so are all earthly objects including all living creatures, humans among them. There are large differences between a flea and a galaxy, but we can speculate that they made of the same matter and energy, even all three universals, thus including *mind*.

Less abstractly, we can say that each person has linkages, connections, equivalences to the earth, the moon, the stars, and beyond. When we experience and consider astronomic events, we may feel awe and wonder. We may feel gratitude. We may even feel at home. (The word "consider" has a Latin root *sidus*, or "constellation," as in studying the stars, or even, we might say, accepting their exemplary order and, therefore, influence.)

If there are aspects of mind that link matter and energy, how may we explore these?

IN WHAT HOMES MAY HOMEBODIES "LIVE"?

All of us homebodies need to live somewhere, physically and mentally. We especially like physical homes, domestic structures that surround and protect us. Sometimes we adopt other or secondary homes: a workshop, a place to eat, or a place for sports. Sometimes the natural world feels like home. Our minds receive sensory inputs from these places and our positive intent toward them grows. We like going to the gym, say, or a local park; such visits become rituals with both rational and affective meanings.

If we move into a new apartment or house, we sense many new things and slowly build up our concept of that space as our home. The upward movement of induction and the downward movement of deduction work together in our daily life usually without our specific attention. We move routinely from basement senses to attic principles, and the distance can be short or long, depending on our range of thought and awareness of it. In Qigong, the Rising Lotus exercise moves "from the muck to the stars," an exercise in our imagination both in the poetry of it and in our body's motions from the ground to above our head, as our imagined lotus blooms in splendor.

In the massage section, we discussed neuroscientist David J. Linden's explanation of how mind influences something we touch. The same is true of many other ways our senses perceive something. When we participate in large social constructions of meaning, we use principles that guide our perception. In Thomas S. Kuhn's *The Structure of Scientific Revolutions* (1996) there are examples of differing perceptions depending on prior knowledge. "Looking at a contour map, the student see lines on paper, the cartographer a picture of a terrain. Looking at a bubble-chamber photograph, the student sees confused and broken lines, the physicist a record of familiar subnuclear events" (p. 111). We perceive the world through our senses, the classic five of sight, hearing, taste and smell, and touch, and we interpret these perceptions by our mind's models of reality we already know—or think we know.

What we perceive is tied to our experience and our education, which may include the dominant paradigms of economics, social status, or the science of the day. If you were a carbon-loving businessman, you rejected the concept of Climate Change even as hurricanes Harvey and Irma, also Michael and Florence, intensified because of higher than normal seawater temperatures.

Regrettably, much of the West is now urban, rational, and economic in focus, so we have lost many connections to nature. People in other times and places have felt at home in the earth, and moderns can as well. If we consider that *we are part of nature and that it is our home*, it's a different way of living in the world, a stretching of our habits of abstraction and a deepening of our love of nature and, therefore, our likelihood to care for nature. Thoreau loved his Walden Pond. "Brother Sun, sister moon," St. Francis wrote.

Pope Francis entitled his encyclical *Laudato Sí: On Care for Our Common Home* (2015). "Laudato Sí" means "praise be to God," and "home" means our earth. Francis draws on current science and, of course, his Christian faith, two large realms that, for him, do not conflict. He describes and laments that our world is under attack from the sun's heat trapped in our changed atmosphere (pp. 7-23).

In learning skills for a hobby—needlework, sports, music, model trains—we build up our abilities, understanding, and imagination, all of which bring increasing pleasure and commitment. Thus, beyond linear, rational thought, there are expansions of, shall we say, horizontal emotions, intuitions, associations that bring us enjoyment, affirmation, and wishes to protect the materials and values of our hobby. The same is true of nature. As we enjoy nature, we are more likely to want to protect it.

This section focuses on the upper ends of the abstraction ladder, conceptual models for nature and the universe, with implications downward. Section VIII will focus on a more local worlds, the health of each person.

In this section we'll consider:

(1) views of nature, personal and scientific, that focus on energy,

(2) further scientific views, with focus on mind and intent, including Quantum physics and astrophysics with a surprising overlap to, and

(3) nature itself as a model for health

These models have very large ranges, and all of them place value on balance and harmony.

In what conceptual homes may homebodies choose to live?

All of us have many models choose from, depending on the culture, our personal leanings, and our interest in clarifying our thought and values.

Can we live in these homes all the time?

No, but we can visit various homes, depending on circumstances, like we shift gears of a car or, like Montaigne, use various gaits of a horse.

The world is complex, and our lives are always a mixed bag of experiences, ideas, and emotions. We return to our conceptual houses of choice through conversations with trusted friends, through rituals, through prayer, meditation, exercise, hobbies, and many other means.

WHAT IS UNIFIED? HUMANS KNOWLEDGE, NO, BUT NATURE AND THE UNIVERSE, YES

There is trouble at the top of our abstraction ladders. Nature is both more complicated and more unified than we typically assume, for example ecosystems where some 30 or many more species interrelate. While our instruments become more sophisticated in measurement, and our theories advance, there is always more to learn. Physicists seek a single grand theory that will explain astronomic and atomic activity in one unified theory. They push toward a Theory of Everything (TOE) or Grand Unified Theory (GUT) because humans are humans and want to *know*. But, because humans are humans, our knowledge is limited; my guess is that the universe will always have more to discover than we can explain. As Robert Browning wrote, "Ah, but a man's reach should exceed his grasp, / Or what's a heaven for?"

Large theories: temporary lenses for seeing nature

Large theories are often abstract, especially when expressed in numbers. Nonscientists, myself included, cannot understand the advanced mathematics. Worse, for some people "theory" is now a suspect word because it implicates scientists who, in their ignorant view, somehow perpetrated the "hoax" of Climate Change.

Some theories, like the notion of gravity, are fairly stable, but always open to refinement or modification. Other theories evolve over time as instrumentation improves, computing power increases, and new minds bring new insights. Neurologist Oliver Sacks discusses many developments in scientific thought in his essays collected as *The River of Consciousness* (2017).

Sometimes the theoretical math outruns the observed phenomena.

Einstein's calculations described gravitational waves, but the instrumentation of his time could not perceive them. It took a century for the Laser Interferometer Gravitational-Wave Observatory (LIGO) to catch up and verify them. Similarly, the Higgs Boson was proposed and eventually observed. I have read that the modern SQUID (superconducting quantum interference device) can measure very small magnetic energy in the human brain or as projected from an energy healer (Oschman, 2000), but I don't know whether any mathematicians are able to calculate so-called "subtle energies."

Clearly science has made wonderful progress with the study of energy in the last century or so, but more is yet to come. What is dark energy? What other energies might there be?

A TALE OF TWO ROCKS: (2) THE MOON ROCK TOUCHED BY MANY HUMANS

At the Air and Space Museum

Across the Mall from the Smithsonian Museum of Natural History, where we encountered the stone axes, is the Air and Space Museum. When my family visited there, we were surprised to learn that there was a rock from the moon installed so that visitors could put their fingers on it, stroke it, or rap it with their knuckles. We all touched this small, dark triangle with solemnity, even reverence, because it was, amazingly, *from the moon* and brought to earth by *astronauts, our fellow humans*. At the same time, it appeared to be an ordinary rock, much like many we had ever seen. A current website identifies it as basalt, the hardened lava welling out of the moon 3.8 billion year ago. (See Moon rock in Sources.) Later I came to enjoy basalts at Nehalem bay, Oregon, the Columbia gorge, and, mentioned earlier, the Isle of Staffa in the Hebrides.

By one theory, the moon was once part of the earth. By another, it was captured by the earth; by another, there was a collision of an external body so that matter from both objects created the moon. Regardless of the origin, *basalt is basalt*, From this perspective, not only are the 94 known elements the same throughout the universe, but basalt is a specific example of hardened lava *everywhere*. Similarly, the materials of our bodies are the same as all other living bodies, the matter of the earth, and the matter of all heavenly bodies. Similarly, energies are shared between us and the universe. And, I speculate, mind is similarly shared.

At the museum, our eyes perceived the dark stone and its setting. Next we wanted our fingers to touch it; they did so, and reported to our brains that the stone was smooth and hard. Next, we pondered the written material…how and when the stone came from the moon. At various times our emotions were involved: *a stone from our moon! brought by astronauts! displayed right here for us to touch!* We felt awe, wonder, admiration for the brilliant technology of space travel and the training of the astronauts. We were connected, by touch, to the heavenly body we have seen many times in our skies. Perhaps we felt gratitude for the museum authorities who imaginatively created this display for thousands, even millions of people, even risking theft or damage. If we looked at information on the web, we learned that, while the travel to the moon in 1972 is well documented (except for people who thought it was faked), the ancient origins of the rock and the moon itself are not currently clear: there are several theories; perhaps one (or another) will become considered the most likely. Despite such mysteries, we affirm a closer relationship to the moon when we touch this piece of rock. If we learn that it is basalt, we can say, yes, the moon and earth are continuous in matter, and we may assume and celebrate that other heavenly bodies share the same elements and materials.

With the now famous "Earthrise" photos of Apollo 8 and 11, we see that our earth may also be considered a heavenly body.

Whatever our concepts about reality, our senses and our minds can feel gratitude for our various homes: as large as the universe and as close as nature in our front yard. Perhaps we can extend Wilson's biophilia to the moon, planets, and myriads of stars as *geophilia*, or just plain *love*.

We turn now to a first group of explanations, views of nature, personal and scientific, that focus on *energy*.

PERCEIVING EARTH THROUGH ENERGIES CURRENTLY RECOGNIZED, SUCH AS SIGHT AND SOUND

We perceive the world through our senses, the classic five of sight, hearing, taste and smell, and touch. These seem to do well in our daily lives, even though what we hear is a limited to a part of the range of sound waves, and what we see is a very limited part of the electromagnetic spectrum. Further, there is variation among species. Wilson remarks, "Human beings live in a world of sight and sound, but social insects exist primarily by smell and taste. In a word, we are audiovisual where they are chemical" (*Biophilia*, p. 31).

We live with a many familiar rhythms of energies: day and night, phases of the moon, high and low tides, change of seasons, fair weather and so-called foul, planting and harvesting; these are some of the most obvious large rhythms. Some functions are circadian, organized around the 24-hour day. We work, we rest. Jetlag is a good example of a human out of sync with geographical time. Our bodies live with heartbeats, breaths in and out, sweating or shivering as needed, effort and rest. We hunger, we eat. Our brains function in measurable waves (Delta, Theta, Alpha, Beta, etc.). We are born, we mature, we die.

There are other rhythms we are less aware of, and these can influence sleep-wake cycles, hormone release, body temperature, and other important bodily functions.

WE DO NOT SENSE THE SCHUMANN WAVES AROUND THE EARTH

Numerous websites, including NASA's, discuss Schumann waves that circle the globe, day and night, year in, year out, but we are not aware of them. This is a standing wave of electromagnetic energy and resonates between the earth's surface and the lower border of the ionosphere. It was predicted mathematically by Winfried Otto Schumann in the 1950s and verified experimentally in the 1960s. The basic energizer of the atmospheric cavity is the lightning occurring regularly in the tropics that circle the earth. The fundamental wave at around 7.8 Hz, is considered extreme low frequency on the electromagnetic spectrum, but multiple partials (like overtones for sound) exist in higher frequencies.

While this energy is clearly recognized, its possible relationships to humans are not. Comments on the Web are pro and con for any possible links. Time and science will tell, but it seems likely that there will be links found, given that humans evolved on the earth living continuously with Schumann (and other) waves. It would make sense that we are attuned to them, including our brain waves. Indeed the basic Schumann wave is near the border of frequencies for Theta (relaxation state) and Alpha (meditation state) waves.

ENERGIES ROUTINELY IGNORED BY MODERNS

Given our Newtonian world-view, we typically ignore perceptions of ley lines, feng shui, dousing for water, and other earth energies sensed by ancient people, some of today's indigenous people, and various individuals. Such energies have disappeared from Westerners' awareness because they don't match modern paradigms. (See Furlong, 2003.)

Also, given our Newtonian world-view, we (for the most part) pay no attention to energies of Quantum physics. Fritjof Capra's *The Tao of Physics: An Explanation of the Parallels Between modern Physics and Eastern Mysticism* appeared in 1976 and Gary Zukav's *The Dancing Wu Li Masters: An Overview of the New Physics* appeared in 1979. Over the last 40 years serious readers have enjoyed these and similar books, but the dominant modern worldview ignores the Quantum world, which appears non-Newtonian, even downright loony.

ENERGIES NOT UNDERSTOOD (OR YET-TO-BE UNDERSTOOD): "DARK" AND "SUBTLE"

There are two ranges of energy currently not well understood.

Dark energy

In *Our Mathematical Universe*, Max Tegmark discusses the interactions of ordinary matter, dark matter, and dark energy. He writes the ordinary matter—the stuff we see and use every day—plus dark matter comprises only 30% of the "cosmic matter budget." This is calculated from gravitational effects on cosmic clustering. (See Chapter 4, "Our Universes by Numbers.") The remaining 70% is considered to be dark energy. Both dark energy and dark matter are invisible as "light or other electromagnetic phenomena" and poorly understood. While measurements of the total "budget" agree—even with different investigators and different methods, dark matter and dark energy somehow interact to attract and repel, cluster or not cluster, build galaxies or tear them down (p. 94). Even the term "dark" is a placeholder until better concepts solve some of the mysteries. Astrophysicists assume it exists but don't know, yet, how to describe it.

As speculation is free, I'll bet that dark energy and dark matter will have ways of transforming back and forth.

Further, I'll bet that they relate to Qi and other energies that indigenous people have sensed and used for millennia.

Subtle energy

The second range is so-called "subtle energy," also a wastebasket phrase to describe the many sorts of energy of indigenous cultures, non-Western cultures, and the worlds of Psi—Qi, prana, psychokinesis, remote viewing, remote healing and so on—discussed in Section IV. The Latin roots of "subtle" have the meaning "below the weaving," that is, below the material, sensate world. Such energies may be considered esoteric, not measurable, analyzable or definable, and therefore dismissible as anecdotal, woo-woo, Twilight Zone, etc. Some, however, have been measured, and some await (we might say) developments in instrumentation and theory, say a paper like Tegmark's on consciousness and matter that show relationship of mind and matter.

Nowadays we see headlines about the discovery of the Higgs Boson and gravitation waves, perhaps our world picture of matter and scientific inquiry is evolving, but our layman's concepts of energy are still Newtonian mechanics. For now, it appears that Quantum physics is the best explanation for subtle energy and related phenomena with as psychokinesis, remote viewing, clairvoyance, and other so-called "psychic" phenomena.

TEGMARK'S MATHEMATICS OF THE UNIVERSE

Mark Tegmark's *Our Mathematical Universe* is a fascinating book. While his technical paper "Consciousness as a State of Matter" has math I cannot comprehend, this book does not. Indeed, it explains complicated matters clearly (and often with wit), and provides helpful illustrations. Tegmark discusses multiverses, fractals, the Big Bang, and the inflation of the universe among other complex topics.

He stops short of putting matter, energy, and mind as universals that interchange. He writes, "My guess is that we'll one day understand consciousness as yet another phase of matter" (p. 295). As we saw before, if consciousness (or mind) equals matter and $E = mc2$, then consciousness must equal—at least at times or in special conditions—energy…or so it seems to me.

In a fascinating section "What Are You?" Tegmark begins, "we don't fully understand what we are," including the mysteries of consciousness" (p. 281). He goes on to describe aspects of the human body (blood, brain) that are made of particles operating in patterns that have analogues throughout the universe with the same mathematical structure (pp. 281-84). In that sense our body and everything else share the same mathematic principles, a home, of sorts, for Tegmark. Numbers, words, drawings, music of the spheres—many modeling systems have suggested the overlaps of matter, mind, and energy.

And still there are mysteries, like the five models he discusses for how the universe might end. And there's one more, "none of the above." He writes, "I love mysteries, and find paradoxes to be nature's best gifts to us physicists, often providing clues to future breakthroughs" (pp. 367-70). He discusses human-caused risks to earth (Climate Change, ecosystem collapse, etc.), but he is more worried about nuclear war and artificial intelligence taking over in a "Singularity" (pp. 378-86).

Even with risks and grave threats, Tegmark believes that science, improved education, and—even—marketing and fund-raising to publicize "the facts" can be helpful to humans and the earth. Indeed, humans (and perhaps other life) give the universe meaning (pp. 386-91). Even with enormous uncertainties in our future, humans have, from his perspective, important work to do.

I VISIT A PHYSICIST

Yearning for further clarification, I make an appointment with Jonathan Engel, professor of physics and associate chair of the department at the University of North Carolina-Chapel Hill.

It's a gorgeous autumn day on the UNC campus. Cool dry air strips the leaves from trees and blows them across lawns and sidewalks as determined students walk to class. As I approach Phillips Hall, it seems strangely familiar, with battlements on top and white borders around large windows. I've seen campus buildings like this before.

The hallways are, indeed, functional and the ceilings are high. Posters describe technical work by professors and bulletin boards show active groups for physics students, many of whom, now, are women. An editorial in *The Daily Tar Heel* posted inside says the building, built in 1919, is "Collegiate Gothic" and should be torn down. The writer dislikes its strange layout, crowded hallways, and black chalkboards. None of these bother me; I like to imagine the wealth of knowledge implied by years of teaching and learning in this space. Looking into classrooms I see blackboards but also alert students, all with their laptops.

Prof. Jon Engel and I approach his door at the same time. I recognize him from years ago when we both sang in the Choral Society of Durham. He asks about the group and what we are doing this year.

His office is large enough for a small group to meet and, sure enough, there's a large blackboard covered with equations. On his desk (very neat, by my standards) are two large monitors. When he turns them on, they show a lovely autumn scene from this campus.

I explain the flow of my book and my need for some physics help. He studies nuclear theory and its applications to particle physics (neutrino physics, CP violation, dark matter) and astrophysics (nucleosynthesis in supernovae and neutron-star mergers). I'm very aware of my amateur, lay, humanist status.

Over the next 40 minutes he explains that physics recognizes strange phenomena such as entanglement, but that these are subtle and must be looked for in very controlled conditions. There are, he says, some "correlations that could not be caused by an influence traveling at the speed of light or less," but they are hard to study. He strides to his black board and draws two particles, each with a different spin, that are far apart. "When the spin of one particle is measured, something about the other entangled one changes instantaneously, but anyone looking at that particle would not notice anything unusual and the phenomenon can't be used to send instantaneous signals," he says. A "subtle" correlation means, for him, hard to measure, with no influence traveling from one particle to the other.

I'm sneaking up on my grand triangle of universals, so I ask first about the 94 chemical elements and the assumption that these are the same throughout the universe. "As far as we know," he agrees. I ask about dark energy and dark matter. He says these are placeholder terms for things that are known to exist, but haven't been identified. Dark energy is particularly poorly understood. As for dark matter: "We have a better idea." He also agrees that we generally assume that physics works in the same ways everywhere, by "mathematical laws" that physicists consider universal. Matter and energy seem to be solid universals.

What about *mind*? I describe my experience—even as a skeptic—at the Rhine Center with spoon bending. Jon is doubtful and asks about *trickery*. I describe what I saw and, perhaps callously, suggest that he go to a similar open house, see spoons bent, and "blow his mind." We both laugh.

We talk about problems in defining mind, not just in humans alone, but also in animals and plants. Jon describes artificial neural networks that can perform deep learning, even facial recognition, but says these are too complex to be understood intuitively at present.

I ask, does the universe, in any fashion, "know" what it is doing? "We don't know," he says. I mention the lunar eclipse of the previous summer, the excitement the viewers felt, and the pin-point accuracy of scientific prediction for the shadow's path. We agree that this example suggests order in the universe, some of it understood by humans and some not. (See Essay 7 below).

We take a few minutes to discuss the double arrows between the three "universals." Matter and energy make sense, from $E = mc^2$, but trouble arises, he says with matter and mind also energy and mind, at least from current, standard science. When I show some quotations from physicists, including David Bohm, who contemplate versions of panpsychism, Jon says, "Some physicists were interested in that at the dawn of Quantum theory, but today they are less so."

As I leave, classes are changing. The students look focused, earnest, and intent. Perhaps some of them will solve some of the problems not yet understood.

In another setting I mention spoon bending to another physicist. She asks about *trickery*, as did Jon.

In another setting, I ask another physicist about physicists who think spoon bending involves trickery. He laughs and says, "They're just not paying attention!"

I recall David Hoekema's phrase regarding panpsychism: "It depends who you ask."

I think of an old joke: "If you present a problem to three economists, you'll get four answers."

It appears that people with different perspectives (training, research, standards for validation) have different perceptions and draw different conclusions.

We turn now, to our second group, further scientific views, with focus on mind and intent, including Quantum physics and astrophysics with a surprising overlap to mysticism.

TWO VIEWS OF CHOPPING WOOD: NEWTONIAN AND NEW PHYSICS

1. A Western (Newtonian and results-minded) chopper

If we think of a human chopping wood, westerners typically understand this activity as an application of kinetic, *mechanical energy* of the *material* ax cutting the *material* fibers of the wood. This formulation is useful for our daily lives. There are, of course, other energies in the person cutting: the neurological energies that fire the muscles and make vision work in eye and brain, plus the digestion of the last meal, plus temperature regulation through sweat, plus faster breathing, and much more—all wonderful energy systems within body and mind. Scientists and other specialists study all these in detail but tend not to see connections between and among all of them as a totality.

Most people take *mind* for granted, because it guides us subconsciously and routinely, but the mind of the chopper is active on several levels. There is the assessment that the wood must chopped, the decision to do it now, the intent to chop, the strategies of placing the wood and aiming each blow, the proprioception of the swung arms, feeling the impact in hands and arms, and much more. There is the hearing the sounds, the smelling of opened wood, and the feeling pleasure in performing a task well (if that is indeed the case).

2. A new-physics (fields, Quantum physics, matter = energy = mind) chopper

In the new physics, Quantum and beyond, there are other dimensions that we don't ordinarily consider. Although we perceive and understand the wood and ax as solid matter, both are largely space at an atomic level, and that "space" is not empty or neutral. Further, the matter of the wood may be understood as a particular concentration of energy and mind, resonating with the entire universe as subatomic particles come in and out of existence continuously. The same is true of the chopping human who is also made up of atoms with plenty of space. All objects, be they axes, people, or stars, resonate with all other objects, and *mind* unconsciously and/or consciously travels through the medium that is called the "Field," to be discussed below. While we most commonly understand that the ax will cut the wood, in a Quantum view, there are endless other possibilities including that the ax will disappear, become flame, or be cut in two by the wood, possibilities that are statistically remote but possible nonetheless.

Today these are uncommon notions. Even if we give them credence and mental space, we routinely revert back to standard views of causation, matter, and other Newtonian concepts that pervade our culture and seem to make daily life possible.

If, however, mind is quieted and open, we can consider other formulations. If we practice such awareness, some humans (all potentially?) perceive other minds, whether in persons, or places, pets, plants, even celestial bodies. Indeed, in one theory, our brains are already connected (or distributed) with a universal mind (Figure 6 above) something like the contemporary iCloud for computers.

MODELS FROM PHYSICS ALLOWING RENEWAL AND HOPE

Our second group of models includes three models from physics that go well beyond Newtonian physics and, it seems to me, have promises for hope: the physics of fields, "new" kinds of energy, and Quantum physics.

A. PHYSICS OF FIELDS...WITH SOCIAL INTENTS

In Section IV we met Larry Merculieff, the Inuit elder who spoke about his upbringing; when he went hunting, he saw native men all sensing the energy of a seal. This sounds odd to Westerners, but a theory of the Field casts a new light on it.

Lynne McTaggart was an investigative journalist who has interviewed many of the leading figures in modern physics. Her three books emphasize the development of Quantum physics through the last century and applications, especially in communication through the "Field." While these books are somewhat repetitive of concepts, they vary in coverage of scientific evidence and the applications possible today. As the books evolve, McTaggart sees more and more social and practical applications, including active groups joining together to join their *intent* to improve society.

In 2010, I attended a conference of the International Society for the Study of Subtle Energy and Energy Medicine (ISSSEEM) in Westminster, Colorado. Travel-delayed, I walked into a large hall where Lynne McTaggart was speaking. She was talking about the "Field" and how information traveled through it instantaneously. Field? Like a meadow? A crop? I thought of the "Elixer field" in Qigong, the small area just below the navel, but clearly this was not her subject...rather something much larger, even across the universe. I slowly understood from her (and later from her three books), that she was talking about *the entire universe that is uniformly energized and can, therefore, relay information between all of its parts.*

Founded in 1991, ISSSEEM ran conferences and produced publications for some 20 years. I attended the last two major conferences. I heard not only Lynne McTaggart, but William A. Tiller, Mae Ho Wan, and other luminaries such as Eric Pearl, Beverly Rubik, Charles Tart, Rollin McCraty, and Larry Dossey. These conferences provided a bookstore and an "Expo" with displays of products (including Tesla Energy Lights) and practitioners, including a Qigong therapist who used electrical readings to test meridian flow. The two conferences were exciting in their variety of presenters and persons attending, a true hub for learning about energy practices and their applications. The economic downturn apparently ended the large conferences, although, after a break, smaller versions started up again, now sponsored by Holos University.

The Field

In physics, there are various kinds of fields (gravity, electromagnetic) but what they all share is a volume of space and energy (see Field in Sources).

For McTaggart, "The Field" is short for the "Zero-Point Field," meaning the lowest possible energy of a field, for example absolute zero for an environment of heat. Even in that case, where atomic motion would cease, *subatomic particles are zipping in and out of existence and can be carriers of information.*

The Field pervades the entire universe, including every object and every living creature. It functions as an information medium as energy and matter keep changing back and forth in quantum-level fluctuations. Humans are especially involved in the Field because of their minds. Two humans meeting exchange subconscious energy through the field. Aware or not, we "read" the energy of others, sensing friendliness, threats, or other qualities.

In *The Field: The Question for the Secret Force of the Universe* (2001, rev. 2008), McTaggart discusses the scientific work of Harold S. Burr, Fritz-Albert Popp, Hal Puthoff, Herbert Fröhlick, Karl Lashley, Karl Pribram, Sir John Eccles, Stuart Hameroff, and others. In exploring implications for the human mind, she reports on Helmut Schmidt, J. B. Rhine, Robert Jahn, Brenda Dunne, and others. Can our minds work by energy that connects to other minds without a mechanical link such as a telephone?

In brief: because the Field exists everywhere, it functions as a transmitter of *energies*, including thoughts and feelings of *minds* in the *material minds* of humans and other creatures, even RNGs. RNGs (random number generators) showed that these mechanical objects were influenced by the minds of people through entanglement and the transmission of coherence; for example, RNGs detected New Year's celebrations, when social energies change. Also TV coverage of the Academy awards, the Olympics, and the Super Bowl, when thousands of viewers created a coherence (or "superradiance"). In an experiment in Washington D.C., a superradiance group actually lowered crime (p. 211). When the group stopped their collaboration, crime rose again.

This theory helps us understand awareness at being stared at, telepathy, hypnosis, healthy communities (Roseto, Pennsylvania), remote viewing, precognition, psychokinesis, distant healing, and more. McTaggart reviews the work of many scientists, including William Braud, Elizabeth Targ, Ingo Swann, Dean Radin, Rupert Sheldrake, and Gary Schwartz. Typically these scientists came to these topics with skepticism and methodological care to limit variables in their explanations. In the realm of health, therapeutic touch, remote prayer, were all shown by scientific studies to have positive effects. Measurement of Qigong masters found measurable photon emissions and electromagnetic fields.

Vibrating molecules, atoms, and subparticles, are basically mechanical. We need to add mind's *intent* for enhanced healing capacity. I recall a Florida OB/GYN who said she always prayed for her patients before doing surgery on them.

In *The Intention Experiment: Using Your Thoughts to Change Your Life and the World* (2007), McTaggart goes deeper into ways mind can influence energy and matter, still using scientific studies as evidence. Again, resonance, entanglement, and entrainment are all connections of mind using the Quantum particles of the universe. Multiple studies, including "the Love Study," show that loving couples have EEG synchonizations of the brain, and this can have healing effects. "Compassionate Intention" can be sent between people (p. 57-61). There are further reports about the efficacy of Qigong and ways meditating monks can change their brain rhythms, also the qualities of prayer, sacred sites, the relation of geomagnetic energy and the sun. Gary Schwartz, drawing on Popp and Stuart Hameroff, did experiments with healers that suggest that healing energy by intent is actually pure light (p. 32).

The last third of the book explores intention and various impacts of intention on plants, cancer cells, and water. Groups of people can "power up" and send energy to "targets." McTaggart describes a world-wide experiment; she used scientific advisors and invited anyone who could access the website to join in group intention. She found success with some targets, but not with others.

In *The Bond: How to Fix Your Falling-Down World* (2011), several of these themes continue. She posits that the most helpful view of the world (and the universe) is as a whole, not a realm for individuals to compete and seek domination. Quantum physics allow us to contemplate an inherent bond between all people that might be called a "superorganism." Again, entanglement leads to coherence, even in the case of natural rhythms such as day and night and the sun's 11-year cycle of solar flares, which appear linked to heart attacks and cardiovascular deaths (pp. 39-42). Brains can synchronize in healing or in martial arts as practiced by Qigong masters (p. 63). According to scientist Fritz-Albert Popp, all living organisms exchange biophotons, or "light emissions" (p. 64), another source of unity, and "our constant impulse to merge" (p. 65). In another publication, Popp and

Wolfgang Klimek provide photos of humans whose stomach and bladder meridians are lit up in their bodies by photons, an effect known as "light piping," or "photon sucking" (2007, pp. 29, 30).

McTaggart argues that we are "hard-wired" to help our fellow humans. She cites a book by R. Wilkinson and Kate Pickett *The Spirit Level: Why More Equal Societies Almost Always Do Better* (2011; pp. 134-35) and other studies, examples, and traditions that support her thesis that *humans do better when they cooperate instead of compete.*

In her three books, she progressively moves toward social implications of worldwide, energetic communication and the shared intents that may be helpful to humanity.

Speculation: could the intent of billions of people can have a favorable impact on Climate Change?

B. SCIENCE EXPLORES "NEW" KINDS OF ENERGY

I doubt that there are new forms of energy, only forms that are new to human consciousness. We may be aware of some of them already, for example awareness of being looked at (see Sheldrake below). As for specific healing methods like Qigong, humans have known, studied, and practiced these for millennia. Some scientists are studying these forms today and some write interesting—even entertaining—books for general audiences.

Lynne McTaggart referred to three scientists at work on unusual kinds of energy.

Dean Radin, Ph.D.

Radin is Laboratory Director of the Institute of Noetic Sciences in Petaluma, California. His *Entangled Minds: Extrasensory Experiences in A Quantum Reality* (2006) discusses a wide range of psychic phenomena, such as telepathy, clairvoyance, and psychokinesis. He describes many laboratory studies proving their existence and discusses field theories and Quantum theories that suggest how they work. While there are competing theories with much complexity, it seems clear that the intersections of physics, neuroscience, and psychology support a theory that "our bodies, minds, and brains are entangled in a holistic universe" (p. 264) and that Quantum theory explains psychic connections beyond our typical conscious awareness that is (regrettably) "heavily driven by sensory inputs," such as sight and sound (p. 265). In other words, if we relax, rest, and open our minds, we can reaffiliate with the universe.

Gary E. Schwartz

Schwartz is a Harvard Ph.D. now at University of Arizona, where he does research on healing, psychology, and biophotons. He wrote *The Energy Healing Experiments: Science Reveals Our Natural Power to Heal* (New York: Atria Books, 2007).

In brief, he says that all matter vibrates, even at absolute zero, thereby sending and receiving energy, and especially living matter, and ever more especially living matter that can think, since mind and matter overlap, and mind can influence matter. Matter is exchanging information constantly and across the universe.

He writes, "When we integrate concepts of memory and information with network and systems models, we realize that energy and information are constantly circulating within and between material systems, and in the process of circulation—created by what are termed 'feedback loops'—storage, memory, and evolution can occur" (p. 186).

ELF (Extra-Low Frequency) fields help tissues heal, even bone. ELF fields and biophotons appear to be the two most powerful healing influences (pp. 104-114).

Schwartz believes we are all energy healers, although persons who study and develop this resource can do so more powerfully. Reiki, Johrei, Qigong, therapeutic touch, healing touch, acupressure, acupuncture, quantum touch, vortex healing, Bio-Touch are all examples of disciplines that develop energy healing. Also, "caring energy and loving intentions are the key to healing and health" (p. 147). Trained healers can increase or decrease biophoton emission in plants as much as five- or ten-fold (p. 127; 137), and "all electromagnetic signals are made up of photons" (p. 138).

We could be called "human cell phones" or "living antennas" always in touch, subconsciously, with the universe and all our fellow creatures, especially humans.

Rupert Sheldrake, Ph.D.

Two books by biologist Sheldrake are informative scientifically and entertaining with many examples of everyday events. In *The Sense of Being Stared At and Other Aspects of the Extended Mind*, he describes many instances and studies that show humans are aware of eyes upon them and that they have abilities for remote viewing. He argues that materialist theories of the mind dismissively refer these as "paranormal" phenomena when, in fact, they are normal. Many examples and studies attest to the reality of telepathy. One example is the group behavior of large groups of birds or fish that turn exactly together much faster that classical neurology of sensing, processing, and acting can explain. He describes these as a sort of group mind of "a single large organism" functioning within a morphic field (pp. 113-117). Like Schwartz and Radin, he finds promise in Quantum concepts such as non-locality or entanglement to explain this behavior (pp. 271-272).

In *Dogs That Know When Their Owners Are Coming Home and Other Unexplained Powers of Animals* (2011), Sheldrake shows how similar telepathic behaviors are common throughout the animal kingdom. More than 1,000 reports describe dogs knowing the instant an owner has decided to return to home. Chapter 2 describes many charming examples, including Queen Elizabeth's gundogs at Sandringham that start to bark in their kennel when she reaches a gate a half a mile from the House (p. 43). The revised edition reports that he has 4,500 cases histories of animal navigation (homing pigeons, turtle and salmon migrations), premonitions of disaster (earthquakes and avalanches), and social behaviors of insects, birds, and fish. Abandoned dogs have followed owners over enormous distances to new homes.

Again, Quantum theories may offer some explanation, also other energy fields such as magnetics, electromagnetic fields, and, in general, an understanding of "the universe as a vast network of interacting particles" that create a "single Quantum system" (p. 403); Sheldrake is quoting Paul Davies and John Gribbin (1991).

Animal intelligence

While Westerners speak of "dumb beasts" and "bird brains," it appears that many (if not all) living creatures have a wide variety of awareness and intelligence, even if mind in these cases is instinctual. We have created this cultural divide and maintain today, despite much evidence to the contrary.

A recent *New York Times Book Review* (May 1, 2016) was titled "Wildly

Intelligent: Two books turn to recent revelations in the study of animal cognition" (p.16): *The Genius of Birds* by Jennifer Ackerman, and *Are we Smart Enough to Know How Smart Animals Are?* by Frans de Waal. I should mention as well *Beyond Words: What Animals Think and Feel* (New York: Picador, 2015) by Chris Safina. I haven't read these books and don't know whether energy was discussed, but I'll wager it is a basic factor in the minds and actions of animals.

Clearly humans are not the only tool-users or interpreters of their environment. The best-seller *Soul of an Octopus: A Surprising Exploration in the Wonder of Consciousness* by science writer Sy Montgomery (2015) plumbs these ideas with a lesser known, reclusive animal, the octopus. Oliver Sacks apparently concurs: "if one allows that a dog may have consciousness of a significant and individual sort, one has to allow it for an octopus too" (2017, p. 76).

Holy and beloved places

Holy places receive the energies of visitors, pilgrims, and tourists, and have done so for millennia; how are their energies imprinted, and can this be measured by any scientific device?

Natural places, valleys, mountains, and caves, similarly have been visited by people (and animals?) with energetic intent. European tourists have enjoyed a large cave on Isle Island in eastern Scotland for two centuries. At sea level, the cave receives waves and amplifies their sounds in wonderful, powerful sounds. Mendelssohn wrote his "Hebrides Overture" (aka "Fingal's Cave Overture") after visiting there. Folklore suggests that Staffa is the end of Finn's causeway arching over from Ireland. Among other notable visitors was the artist J. M. W. Turner, who painted a dramatic version of the island, with the sun sinking in the West. In 2015, Nancy and I enjoyed the sounds and sights of this strange, remote place. There is a primal feeling about the island, including its geology: it consists largely of crystallized tubes of basalt that were volcanically extruded, covered with ice, and much later revealed— all millennia ago.

Web searches for "energy sacred places" turn up many sites that discuss ley lines, vortices, and the suggestion that the earth itself has a set of chakras (energy centers) that parallel the human body. Perhaps it's electromagnetic energy, perhaps tectonic plates play a role, perhaps…? These websites are largely non-scientific but interesting; just the large number of them is suggestive. Perhaps before long there will be scientific explanations.

In summary

The evidence for all these energetic phenomena is enormous. Scientists have carefully observed animal behaviors and created laboratory tests. Cultural blinders, however, still prevail and dominant models can't explain them. The theories from physics that give some insight are complex and new to many people, nor is there one convincing summary that can be readily taught in schools or put into educational programs on TV.

Nonetheless, such energies can be used for maintaining health and healing. (See Section IX.)

C. QUANTUM PHYSICS AND, ONCE AGAIN, QIGONG

A short book by Imke Bock-Möbius goes a long way to help us further. The title is *Qigong Meets Quantum Physics: Experience Cosmic Oneness* (2010), and the author is a German scientist with a Ph.D. in nuclear physics and particle physics. Following her studies in Germany, she spent six month in Beijing studying and working with other physicists. When she had a serious sinus infection, she received a single Qi treatment that cured her, much to her surprise. Thus she joined Bernard Lown and Lisa O'Shea in a cohort of Western professionals who experienced Eastern healing and therefore knew first-hand about its efficacy. How can Bock-Möbius reconcile Western science with an Eastern Worldview?

Her book presents Quantum physics' explanation for how Qigong works (and, by extension, other healing approaches using energy). Further, she suggests how science and mysticism can complement each other. I can't follow much of the mathematics she provides, but I understand her prose.

Key Points

In classical mechanics, she writes, we expect a coffee cup falling to the floor to shatter because of our experience with physical reality; we are oriented to matter and understanding it by our culturally formed minds. In our cultural views of Newtonian causality, A leads to B which leads to C. In Quantum mechanics, she explains, causality is no longer linear and expected but rather sets of statistical probabilities, some more likely than others. In Quantum mechanics, the cup will probably shatter, but it may, possibly, go right through the floor, since anything is possible, and material objects are made up largely of space (p. 64).

Cartesian thought separated (1) God from (2) mind (*res cogitans*) and from (3) corporeal substance (*res extensa*, or "stuff," objects). Quantum physics reunites all of them: human observation (mind) can influence reality, causing an indeterminate state to "collapse" into a particular state of *matter* or an *energy wave*. Light can exist simultaneously as waves or photons, but in the classic dual slit experiment measurement, observation (mind) causes a "collapse" into one form or the other. Schrodinger's famous cat in a black box is both alive and dead until the box is opened and perceived by a human.

For her, Qigong, mysticism, and Quantum physics, although "three completely different areas...all have the same basic theme, which they call by different names: Dao, oneness, and entanglement" (p. 85). For her, deep meaning comes from nature, the Dao, the outer reality. She quotes Einstein, "It is wrong in principle to try to reduce a theory solely to observable factors, because in reality the opposite is true: the theory decides what we can observe" (p. 93). For her, *stillness* is how we allow understanding into "the inherent process of nature, as much as the connections between humanity and nature" (p. 93).

In summary: a universal reality

Quantum physics assumes one reality: integrated wholeness, as all parts of the universe influence each other. The physicist David Bohm, a student of Einstein, titled his book *Wholeness and the Implicate Order* (1980). The implicate (or "enfolded" order) universe is the strange world of atomic particles—as opposed to the explicate (or "unfolded") order we know in our daily lives—and these atomic particles influence each other. They are "entangled" or linked energetically because they all originated in the Big Bang that began the universe. A related and/or overlapping term is "resonance." If you feel you are in "sync" with another person, you share "coherence," even in the inelegant phrase of "photon sucking," as you mutually share energy. **From this point of view, we share energy all the time with all people, alive or dead, and even more so if we have intent to relate to them. Entrainment increases with supportive familiarity: colleagues, families, lovers,** people and their pets, sports teams, musical groups, people who worship together etc., etc.

Connections with the absolute

We have mentioned the possible entanglement arising from the Big Bang. According to Chinese cosmology, the original unity of the Dao (the Way), split into the Qi energy of Yin and the Qi energy of Yang. Particles of matter may be considered Yang, and waves are Yin. The split was not absolute, and Yin and Yang continued to work in dynamic harmony (Tai Chi), producing the 10,000 things or the daily reality we experience.

Human observation (mind), in science for example, can influence reality, causing an indeterminate state to "collapse" into a particular state of matter or an energy wave.

Bock-Möbius sees the Dao as a mystical point that includes reality and nonreality; she discusses mysticism in Chapter 4. Indeed, her book's subtitle is "Experiencing Cosmic Oneness." Chapter 6, "Synthesis: Pulling It All Together" also emphasizes this notion.

As we saw earlier, Daoism (a basis for Qigong) is a steady-state philosophy: the earth always changes within its dynamic harmony. Any and all such changes are normal in the long term, and that earth will be fine however it evolves, with or without humans. The earth has always been in danger; this is normal. There were the five extinctions before. Plagues and famines come and go. Whatever our philosophies or senses of history writ large, it is helpful to understand that change always occurs. This is a lesson of one of the earliest pieces of literature we have, the Babylonian *Epic of Gilgamesh*: change is the law of life. It appears that mystery surrounds the absolute, the ultimate, no matter how much science, philosophy, or religions may press for answers.

In summary

McTaggart emphasizes the Field as a medium for human mind to operate for the good of society. Bock-Möbius is more personal, seeking balance and calm through meditation, but both writers believe in the usefulness of Quantum physics to explain continuities that Newtonian physics cannot, phenomena that we push into the realm of the so-called "paranormal." Neither author touches on Climate Change. Tegmark admires how mathematics reveals the complexity and order of the universe. Although our contemporary world is in danger from threats that include Climate Change, he believes our abilities to think may be able to save us and the world. All three authors see science as incomplete in explaining reality but always evolving toward better understanding of the magnificently complex universe and our place in it.

SCIENCE AND MYSTICISM AS COMPLEMENTARY

"Mystery," "mysticism," "mystic"

In modern usage we use "mystery" to mean something we don't understand or cannot solve. Sometimes we assume a solution is impossible, so we relegate mysteries to the Twilight Zone, part of the "woo-woo factor." Sometimes, however, the word suggests a challenge to investigate, as Tegmark wrote. In popular literature there's another usage: a mystery story describes a crime that is, typically, solved.

In the mystery plays of the middle ages, there was a pun on "mystery" as revelation of religious truth and "mystery" as in handicraft or trade, as in the guilds that sponsored the plays; there may be a further meaning of "mastery," as in learned persons or adepts who know the mysteries of a craft, a sport, or an academic field.

Today we take the related words "mysticism" and "mystic" to indicate people and thoughts beyond the pale of normal thought or experience. Earlier meanings included special knowledge that was revealed from worlds beyond and/or gained by meditation, study, or openness to new ideas, but these meanings have faded. For many today, a mystic is a weirdo outside of normal life.

Science and Mysticism as Complementary

With the positive meanings of "mysticism" in mind, we can consider the following figure; I've expanded Bock-Möbius' drawing (p. 82) considerably. She emphasizes that science and mysticism are polarities that profit from combination. (When science, by itself, is insular, we may call it "scientism," a belief system owing to 19th century Positivism.)

For a true, deep, and imaginative understanding of the worlds around us and within ourselves, we need both sides of the figure below. (Once again, there are force-field aspects to this figure with reductionism in both columns.)

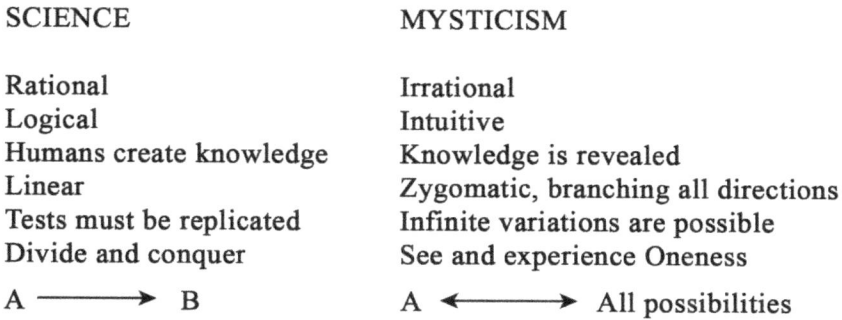

SCIENCE	MYSTICISM
Rational	Irrational
Logical	Intuitive
Humans create knowledge	Knowledge is revealed
Linear	Zygomatic, branching all directions
Tests must be replicated	Infinite variations are possible
Divide and conquer	See and experience Oneness
A ⟶ B	A ⟷ All possibilities

Figure 12: Science vs. Mysticism

I recall a urologist at the University of Iowa who enjoyed making a fountain for his back patio. He made jokes about that fountain and his professional field. He—and many others—have found ways to mix the rational and the irrational.

Now, we turn to the third set of concepts: nature itself as a model for health

NATURE AS HOME, ACCORDING TO LEOPOLD, BERRY, AND POPE FRANCIS

Three powerful voices, from different perspectives and in different formats, speak to the nature as unified and healthy model for humanity: Aldo Leopold, Wendell Berry, and Pope Francis. Each one speaks of matter, mind, and energy as well as health, wholeness, or unity. Each discusses the values of understanding the natural world as home.

Aldo Leopold (1887-1948): humans as members, not conquerors, of nature

While Thoreau's *Walden* (1854) is often taken as a forerunner of ecology, it is Aldo Leopold's *A Sand County Almanac* (1949, 1987) that is usually cited as a cornerstone of modern ecology, conservation, and environmental care.

After a career in forestry, Leopold was appointed Professor of Game Management in the Agricultural Economics Department at the University of Wisconsin, the first such professorship of wildlife management. Two years later, he and his family started to restore a worn-out farm, planting many pine trees and restoring prairies. His *Almanac* describes the beauties of nature month by month with brief observations. For example: "Every farm is a textbook on animal ecology; woodsmanship is the translation of the book" (p. 81). Part III of the book, "The Upshot," discusses the "Land Pyramid," his image for how land (the bottom of the pyramid) supports plants and animals with food chains going upwards. "Land, then, is...a fountain of energy flowing through a circuit of soils, plans, and animals. Food chains are living channels which conduct energy upward; death and decay return it to the soil" (p. 216). He argues that humans should understand this cooperative model and honor it. In his triangular pyramid, we may see matter, energy, and mind all interrelating.

Further, Leopold presents a "land ethic," a philosophy of how humans should fit into the symbiotic world of nature. Here is an oft-quoted paragraph: "In short, a land ethic changes the role of Homo sapiens from conqueror of the land-community to plain member and citizen of it. It implies respect for his fellow-members, and also respect for the community as such" (p. 204). Regrettably, modern man is separated from the land, "something he has outgrown" (p. 224).

A scientist, Leopold studies matter and energy. A manager of his own property, he draws on his rational mind and also his love for nature. He sees nature as wonderfully rich: "The only conclusion I have ever reached is that I love all trees, but I am in love with pines" (p. 70).

In an essay "Conservation: In Whole or In Part?" (1944), there is a section titled "Land-Health." He writes, "The land consists of soil, water, plants, and animals, but health is more than a sufficiency of these components. It is a state of vigorous self-renewal in each of them and in all collectively." He continues, "Such collective functioning of interdependent parts for the maintenance of the whole is characteristic of an organism. In this sense land is an organism, and conservation deals with its functional integrity, or health" (*The River of the Mother of God and Other Essays*, 1991, p. 310).

Although Leopold sometimes uses terms and phrases from the Bible, his writings are more scientific and philosophical rather than religious.

Wendell Berry (1934–   ): creation is one continuous fabric of matter and spirit

Wendell Berry taught creative writing at the University of Kentucky for many years, and, at the same time, lived on and worked a farm. Much of his writing reflects a rootedness in nature and emphasizes respect and care for the natural world.

In October of 1994, he delivered a speech "Health Is Membership" at a conference "Spirituality and Healing" at Louisville, Kentucky. A holistic and synthetic thinker, Berry writes that health has many dimensions beyond mere absence of disease: wholeness, belonging to others and to place, community, and holiness because the earth is God's creation. He quotes the English agriculturist Sir Albert Howard: "the whole problem of health in soil, plant, and man [is] one great subject." In Berry's words, "Creation is one continuous fabric comprehending simultaneously what we mean by 'spirit' and what we mean by 'matter.'" He criticizes modern reliance on machines, information, and industrialization, especially when these are applied to medicine.

Instead, we should focus on how the body, with food and rest, restores itself. "The patient is restored to family and friends, home and community and work." Further, "rest and food and ecological health" should be "basic principles of our art and science of healing." These words fit well with our term *reaffiliation*.

The second half of the speech describes his brother John's heart attack and medical treatment. He criticizes the mechanistic care that ignores the family's needs and deals poorly with the notion of death. He observes: "The world of love includes death, suffers it, and triumphs over it. The world of efficiency is defeated by death; at death all its instruments and procedures stop. The world of love continues, and of this grief is the proof."

Strongly Christian, Berry interrelates the matter of nature and human bodies. He sees mind as multi-dimensional, including "sensation, emotion, memory, tradition, communal life, known landscapes, and so on." He believes that God's love pervades creation and "summons the world always toward wholeness," citing the Gospel of John. We might say that for him: *love and spirituality are the energy of the universe, our home writ large.*

Pope Francis (1936–   ): all creatures are connected; the world is on loan to us

Aldo Leopold died a decade before a "global warming" was theorized, and Wendell Berry's text precedes today's fact of Climate Change. By contrast, Pope Francis (and his team) write specifically about global warming and Climate Change. *Laudato Sí* is an encyclical letter sent, in theory, to all humans. The subtitle "On Care for Our Common Home" emphasizes that the earth is home for all humans (and more), and humans must take care of it.

The phrase *Laudato Sí* is short for, in English, "Praise be to you, my Lord," repeated words in the "Canticle of the Sun" of St. Francis of Assisi. St. Francis praises God through the sun, the moon, wind, water, fire, and the sister/mother earth: all wonders created by God.

Pope Francis follows his namesake in considering the earth as God's creation of a home for all nature, including humans. Regrettably, we have not done well in caring for the earth. Francis decries the abuse of earth: "The earth, our home, is beginning to look more and more like an immense pile of filth" (p. 19). A strong advocate, he calls for an "integral ecology" that considers biology, culture, our shared earth and its climate, quality of life, and social justice, especially for the current poor and for all generations that will follow us. Dialogue and education will be important. Aware that not all religious persons are Christians, Pope Francis concludes this book with two prayers, one for Christians and one for the earth and including an "all-powerful God" (pp. 158-59).

Francis discusses scarcity of loss of biodiversity and glaciers, scarcity of safe water and rise of water-borne diseases, loss of corals, urbanization, and loss of nature in general. He emphasizes interrelationships: "Because all creatures are connected, each must be cherished with love and respect, for all of us as living creature are dependent on one another" (p. 31). He decries income disparities, neglect of the poor, war, and lack of social justice.

More than Leopold and Berry, Francis speaks of the mystery of creation and a universal communion of all created things. Leader of the worldwide Catholic Church, his vision is, of course, strongly Christian. Humans should honor and take care of God's creation, the earth and all its creatures. Chapter Four calls for the integral ecology that is both social and environmental and allows for "intergenerational solidarity." He powerfully writes that the environment "is on loan to each generation, which must then hand it on to the next" (pp. 106, 158-160).

Note: the word "ecology" has the Greek roots of *oikos* or "home" plus the suffix *logy* or "word," in the sense of "study."

These three voices, Leopold, Berry, and Francis—from different times and places, with different language and concepts—all agree on the wondrous complexity of nature, the harm that modern humans have caused nature, and the need to restore and reconnect in order to care for the earth, this *one home* that is—or should be—healthy, sustainable, and common to all.

CHOICES FOR UNIFYING VISIONS AND IMPLIED VALUES

Choices of unifying vision and implied values: from Aldo Leopold to Max Tegmark

Humans have created a wide variety of models for the earth and our place in it. I imagine that every culture from ancient time to the present has one or more accounts—whether in scripture, in creation myths, or in mathematics—of the how the universe came into being and how it functions in the present.

For the modern West, there are several models: Newtonian physics, religions, Quantum physics, and various renewals of ancient "whole world" systems, including Qigong and Ayurveda. These give explanations and values that unite and guide a culture; we might consider them as conceptual homes.

Reaffiliations in our minds

When R. Buckminster Fuller visited my college in the early 1970s, he said, "Knowledge is anti-entropic. The universe is not running down!" I was very glad to hear this, because the notion of entropy was much in vogue, suggesting that we were doomed by a failing universe. According to Fuller, thinking, educating, and learning all keep the world (and more) going. In his *Synergetics*, he wrote, "The physical is inherently entropic, giving off energy in ever more disorderly ways. The metaphysical is antientropic, methodically marshaling energy. Life is antientropic. It is spontaneously inquisitive. It sorts out and endeavors to understand" (1975, p. xxx). I take this to suggest that human thought, among other forms of mind, is a nourisher and maintainer of the universe.

Beside speculative thought, we need our emotions, instincts, and other resources from our long history with the energies of nature. With these too we can support the universe—that is, if we chose to use them.

RANT 7: LOSING EARTH...ON OUR WATCH! HOW GUILTY ARE WE?

Let's imagine that the Climate Change continues with heavy, even disastrous, costs to humans and other life.

Let's further imagine a colossal Court of Justice. The plaintiffs are humans from two generations in the future as well as all the animal and plant species that, by then, have gone extinct as a result of Climate Change, loss of habitat, breakdown of food webs, poaching, and so on.

The defendants are today's humans, especially wealthy people who used much energy, businessman who would not give up carbon fuels, and the governmental leaders who denied Climate Change, loosened regulations protecting nature, did not promote renewable energy sources, and the like. Or caused nuclear war.

Others are charged because they, as Neodarwinists, valued their own success far more important than the success of all humans and all creatures. As they gained wealth, they shared little or none, and they cared not for future life.

There are no lawyers in this court. It's a meeting of minds.

The minds of the plaintiffs form these thoughts:

YOU STOLE OUR HERITAGE
YOU ARE GUILTY OF DOMESTIC VIOLENCE
YOU WERE HOMEWRECKERS
YOU MADE A MOCKERY OF JUSTICE AND HUMAN
   SCIENTIFIC KNOWLEDGE
YOU ARE VANDALS, SABOTEURS, CRIMINALS
YOU HAVE CAUSED MASSIVE DISTRESS AND SUFFERING

WHY DID YOU CONTINUE TO CANNIBALIZE NATURE?
WHY DID YOU NOT PROTECT THE EARTH, THE ONLY
ONE WE HAVE, OUR HOME?

we are bereft
we are hungry, thirsty, sick, and afraid
you, our ancestors, betrayed us
you stole our nature, our sustenance
you were greedy or stupid or both

The minds of the defendants form these thoughts:

[Another thought experiment: the reader is invited to imagine the responses.]

ESSAY 7: HOMEBODIES AT HOME AND VARIETIES OF HEALING: REAFFILIATIONS, INCLUDING THE SOLAR ECLIPSE OF AUGUST 2017

Today many humans are estranged not only from nature but also the richness of their own minds. Unfortunately some of them are in positions of power through business, government, and military, mass media, or other avenues of power. Some of them are stuck in the mental habits discussed earlier of capitalism, Newtonian reductions, limited versions of rationality (calculation, urge to gain wealth and control), competition, and so on. Their homes are not only fancy buildings but also symbols of the many things they own or can buy, their travel, and their sense of themselves as economic successes and, therefore, superior to other people. For them nature, if noticed at all, is a resource to be *used*, even *used up*.

In Greek myth, the wrestler Antaeus had power only when touching the earth. Hercules, locked in struggle with him, made no progress until he lifted him off the earth and could then crush him. Many modern humans are similarly disconnected from the earth and disempowered.

Will Climate Change crush us or can we re-unite, reaffiliate with nature, with our own minds, with our fellow humans, and all other living creatures?

Today we ponder this frightening notion: *humans may have wrecked their home, the earth* (and, worse, will continue to do so) through the use of carbon fuels, creating an increasing estrangement between all living creatures, humans included, and the earth. Because they came to understand energy as economic and based on carbon fuels, they lost their sense of other kinds of energy…between people…with nature…with mystery.

How can we bring about healing reaffiliations?

During my early years, I worked summers for the *St. Petersburg Times*. I recall a tough reporter asking the city desk editor how to write about a man who had caused, unintentionally, his own tragic death. The reporter appeared to enjoy the irony of the accident, even to the point of laughter. The editor paused then said thoughtfully, "Just write it as if he was a member of your own family." I interpret this as a gesture toward reaffiliation with a fellow human through caring intent, regardless of the unfortunate circumstances.

If we consider the earth as a family member, we will behave differently.

The eclipse of August 21, 2017: a dramatic instance of reaffiliation

Our minds and actions are routinely involved with chores, jobs, relationships with people, money, and all the stuff of daily life. Many of these concerns disappeared during the eclipse of the sun in August of 2017. Whether at home or in a public space, people came together without status, wealth, or competitive urge to watch this unusual astronomic event, the moon blocking the sun's rays from the earth. Viewers had fun with their special glasses and pinhole versions. At home, Nancy and I experimented with two colanders, even the holes in the hat she was wearing and took photos to send to friends and relatives.

Instead of daily ideas and emotions, people across the country felt wonder, awe, amazement. Some professional TV newscasters were overwhelmed and actually had trouble talking on camera. People gathered together in squares, stadiums, or libraries were suddenly friendly and speaking their shared joy to each other. The next day there was a flurry of photos sent around by email and conversations in person—even at work—about the eclipse.

The eclipse allowed for reaffiliation on several levels. As for mind, people felt deep emotions of joy and wonder, leaving their routine thoughts and strategies. In modern life, we often lose, forget, or deny deep joy, dreams, euphoria, and awe, as well as sharing them with other people. Nationwide, there was an enormous *eclipse party*, thousands of people sharing as they could, some traveling to see totality, many staying home with pinhole cameras, many watching on TV. Some groups I saw on TV clapped or cheered. Some had no words, but cried out in primal sounds of wonder and pleasure. In a time of political turmoil, people were glad to reaffiliate with each other and some of the natural signs on earth. In places where the birds started to nest for the night, people felt connections with nature's creatures. Even in places of partial eclipse, people felt temporary twilight and coolness.

Further (clearly *further*), people felt wonder and joy as they reaffiliated with our moon and sun. From our earthly point of view, these two celestial bodies improbably joined exactly together in time and, as we viewed them, the exact same size. Because of the enormous distance between them, sun and moon perfectly matched up their spheres. We also felt awe that the sun—our source for all our heat and light directly and, indirectly, through potential energy of fuels, and kinetic energy of wind—could be blotted out with no choice on our part. Newspapers spoke of rituals in earlier cultures to save the sun from the dragon, and so on. Our sense of a primal connection to the sun is usually unconscious, but it was reawakened by viewing this eclipse.

Many people were impressed by the accuracy of the scientific predictions of time and place, and the clarity of explanations on how eclipses occurred. These were widely available in print, on TV, on the Web, for example the excellent *Sky and Telescope* website. This knowledge was heartening in a time of dis-affiliation for some from intellectual inquiry, including science. Some reports discussed experiments during the eclipse, showing that science is an on-going investigation.

In Summary

Our nation's experience of the eclipse, allowed us to reaffiliate (1) with our minds, their many faculties, including wonder, awe, joy, and love, (2) with our fellow humans as equals in celebration, (3), with birds and pets that reacted in tune with the loss of sun, (4) with our senses of the moon and sun, their orderly astronomic activity, and the sun's power that makes life on earth possible, and (5) with the importance of scientific observation, theory, and prediction.

Homebodies expanding senses of home

With the eclipse, we suddenly had new sense of home. If we watched it at home, the eclipse came into every part of our yard, also our entire neighborhood. It traversed all of our nation from coast to coast with extraordinary speed. We may have even felt that this was a private viewing, an American eclipse just for us and a rare sign these days of national unity and celebration.

If we traveled to the path of totality with our preparations (glasses, maps, cell phones, and so on), we stretched our concept of homes to that location, and joined other homebodies there. If we went to a public square, a stadium, or a bar, we joined a social community in celebration. If we went to a library where large screens showed TV coverage, we joined people there and electronically around the world. Across the country there was a moveable feast, a festivity, a solar party. While most of North Carolina had good access, a small storm blocked the view of people in Greensboro. A TV reporter mentioned the disappointment but showed nonetheless cheerful people enjoying the event on cell phones and TV.

As the time of the eclipse approached, we left behind daily chores, worries, and obligations. Our consciousness—our home—expanded horizontally and vertically…93 million miles, give or take, to the sun and even beyond: those in totality saw planets and winter stars vast distances beyond the sun. We had an insight into the how the earth, moon, and sun all work in orderly and predictable patterns; we could consider that our heavenly bodies are emblems for the entire, orderly universe.

Homebodies expanding our consciousness

Experiencing the eclipse, we felt new—or renewed—connections with nature and with the stars, local and large senses of home. We could reaffiliate with birds, our pets, and our fellow humans who *were just like us* in enjoying the eclipse. The experience of awe, wonder, even gratitude brought relief from our rational minds and a renewal in a fuller range of consciousness.

Perhaps we felt an order of the cosmos we had forgotten: sun, moon, earth, and all other bodies *know what they are doing*, and it was thrilling to see a clear instance of this order.

More examples of our minds appreciating nature

Another dimension of our minds is a sense of awe. Richard Louv entitles Part VII of his book "To Be Amazed" (*Last Child in the Woods: Saving Our Children from Nature-Deficit Disorder*, 2008). In his next two books he moves on to adults, families—that is, all of us who have, one way and another, become estranged from nature. (See Sources.) Louv refers to Thomas Berry, Aldo Leopold, and many others in his well-researched books. See also the just published *The Stories of Science: Integrating Reading, Speaking, and Listening into Science Instruction, 6-12* (Janet MacNeil et al., 2017).

For extraordinary discoveries into forests, see *The Hidden Life of Trees: What They Feel, How They Communicate* by forester Peter Wohlleben. For example, they communicate by their roots: "When trees grow together, nutrients and water can be optimally divided among them all so that each tree can grow into the best tree it can be" (p. 16). *If only humans could do the same!*

For many years, the Ohio Department of Natural Resources, Division of Wildlife maintained a peregrine falcon project, including cameras at a nest box on the 41st floor of the Rhodes State Office Tower in downtown Columbus. According to administrator Donna Schwab, tens of thousands of visitors from 131 countries watched the parents tending the eggs, the babies emerging from the eggs, and the nestlings taking their first flights; those numbers suggest the eager enthusiasm of the viewers.

Reassessing deaths, of people, of the earth

If we consider that matter, mind, and energy all feed each other, we can imagine that none of us completely die: we transform our matter into other forms of matter and energy and mind. I like to think that our minds—in whatever form—are eternal. I'm not a true Daoist; I'm a student of Qigong that draws on Daoist concepts, and I find Qigong compatible with my Christian beliefs and helpful for considering the possible end of human civilization.

My nuclear family fell away from me, my parents first, then my beloved sister, in her early 60s. Avise was brilliant, funny, irreverent; we exchanged letters, emails, and phone calls to say anything that we wanted to say. After her early—and, to my mind, absurd and unfair death—I missed her terribly. From the outlook of transformations, however, I have reaffiliated with her, my parents, and many other persons who are dead in the usual sense, but still alive to me as souls made up of energies and minds. I talk to them, wish them well, ask for their guidance and blessing.

As for another transformation, over a few years I asked three women if I could adopt them (informally, that is) as a sister: would they consider me an adopted brother? Each said yes, and these have been rich friendships with sharing and mutual support, and, also with the safety of incest taboo: like massage students, we are sibs, not romantic possibilities.

As for humans, I am sad that our future appears grim because of Climate Change. There may be positive changes that no one can foresee now, but current trends and predictions suggest a world where either we cannot live or a world where life is very difficult.

Can we imagine further reaffiliations that will help?

If the human race must come to an end, we should celebrate the time that we had on earth, mourn the mistakes we made, and enjoy whatever existence we have in whatever worlds beyond this on.

If the materials of our houses, our cities, and our bodies all disappear, all these still exist as mind and energy and live forever perhaps reconfiguring into matter in new and wonderful ways, because that's how our world and the universe have always operated and will always operate.

As for the earth, I believe that it will continue with or without humans.

And Dante wrote...

Dante concluded his *Divine Comedy* with lines that affirm that there are mysteries beyond our calculation. At the end of the *Paradiso*, the pilgrim Dante in the epic story has a vision of God and tries to make sense of it, but he finds that he cannot. Nonetheless, in a flash of insight he understands that his own will and desire are moved by "the Love that moves the sun, and the other stars" ("l'amor che move il sole e l'altre stelle" (*Paradiso*, xxxiii, 145)—the last line of entire poem.

Written centuries ago, these lines unite matter, mind, and energy and invite us to do the same.

SOURCES 7

Berry, Wendell. "Health is Membership." https://home2.btconnect. com/tipiglen/berryhealth.html, accessed August 21, 2017.

Bock-Möbius, Imke. *Qigong Meets Quantum Physics: Experiencing Cosmic Oneness.* St. Petersburg, Fla., Three Pines Press, 2010. Also published in German in 2010.

Bohm, David. *Wholeness and the Implicate Order.* London: Routledge & Kegan Paul, 1980.

Capra, Fritjof. *The Tao of Physics: An Explanation of the Parallels Between modern Physics and Eastern Mysticism.* New York: Bantam Books, 1976.

Carson, Rachel. *Silent Spring.* Boston: Houghton Mifflin, 1962.

"Cosmic Voyage." Directed by Eva Szasz. 1968.

Dante Alighieri, *La Divina Commedia*, Vol. III Paradiso, ed. Natalino Sapegno. Firenze: La Nuova Italia Editrice, 1958, 1968.

_____. *The Divine Comedy: Paradise* in *The Portable Dante*, ed. and trans. Mark Musa. New York: Penguin, 1995.

Field. See https://en.wikipedia.org/wiki/Field_(physics), accessed Sept. 14, 2016.

Furlong, David. *Working with Earth Energies: How to Tap into the Healing Powers of the Natural World.* London: Piatkus, 2003.

Pope Francis. *Laudato Sí: On Care for Our Common Home.* Huntington, IN: Our Sunday Visitor, Inc., 2015.

Fuller, R. Buckminster. *Synergetics: Explorations in the Geometry of Thinking.* New York: Macmillan Publishing Co. Inc., 1975.

Kuhn, Thomas S. *The Structure of Scientific Revolutions* 3rd ed. Chicago: Univ. of Chicago Press, 1966.

Leopold, Aldo. *A Sand County Almanac.* New York: Oxford Univ. Press, 1949.

_____. *The River of the Mother of God and Other Essays*, eds. Susan L. Flader and J. Baird Callicott. Madison: Univ. of Wisconsin Press, 1991.

Linden, David J. *Touch: The Science of Hand, Heart, and Mind.* New York: Viking, 2015.

Louv, Richard. *Last Child in the Woods: Saving Our Children from Nature-Deficit Disorder*, rev. ed. Chapel Hill: Algonquin Books, 2008.

_____. *The Nature Principle: Human Restoration and the End of Nature-Deficit Disorder*. Chapel Hill: Algonquin Books, 2011.

_____. *Vitamin N: 500 Ways to Enrich the Health & Happiness of Your Family & Community*. Chapel Hill: Algonquin Books, 2016.

MacNeil, Janet, Mark Goldner, and Melissa London. *The Stories of Science: Integrating Reading, Speaking, and Listening into Science Instruction, 6-12*. Portsmouth, NH: Heinemann, 2017.

McTaggart, Lynne. *The Intention Experiment: Using Your Thoughts to Change Your Life and the World*. New York: Free Press, 2007.

_____. *The Field: The Quest for the Secret Force of the Universe*. New York: HarperCollins, 2008.

_____. *The Bond: How to Fix Your Falling-Down World*. New York: Free Press, 2011.

Montgomery, Sy. *The Soul of an Octopus: A Surprising Exploration in the Wonder of Consciousness*. New York: Atria Paperback, 2015.

Moon rock. See https://airandspace.si.edu/stories/editorial/long-journey-our-lunar-touchrock.

Oschman, James L. *Energy Medicine: The Scientific Basis*. Edinburgh: Churchill Livingston, 2000.

_____. *Energy Medicine in Therapeutics and Human Performance*. Edinburgh: Butterworth Heinemann, 2003.

Popp, Fritz-Albert, and Wolfgang Klimek. "Photon Sucking as an Essential of Biological Regulation," Chapter 2 of *Biophotonics and Coherent Systems in Biology*. Beloussov, V. L., V. L. Voeikov, V. S. Myrtynyuk, eds. New York: Springer, 2007, pages 17-32.

Radin, Dean. *Entangled Minds: Extrasensory Experiences in a Quantum Reality*. New York: Pocket Books, 2006.

Sacks, Oliver. *The River of Consciousness*. New York: Alfred A. Knopf, 2017.

Sheldrake, Rupert. *The Sense of Being Stared At and Other Aspects of the Extended Mind*. New York: Arrow Books, 2003.

_____. *Dogs That Know When Their Owners Are Coming Home and Other Unexplained Powers of Animals*, rev. ed. New York: Three Rivers Press, 2011.

Schwartz, Gary. *The Energy Healing Experiments: Science Reveals Our Natural Power to Heal*. New York: Atria Books, 2007.

Tegmark, Max. "Consciousness as a State of Matter" (March 18, 2015), https://arxiv.org/pdf/1401.1219v3.pdf, accessed May 8th, 2015.

_____. *Our Mathematical Universe: My Quest for the Ultimate Nature of Reality.* New York: Vintage, 2014.

Wilkinson, R., and Kate Pickett. *The Spirit Level: Why More Equal Societies Almost Always Do Better.* London: Allen Lane/Penguin, 2011.

Wilson, Edward O. *Biophilia.* Cambridge, London: Harvard Univ. Press, 1984.

_____. *Half-Earth: Our Planet's Fight for Life.* New York: Liveright Pub. Corp., 2016.

Wohlleben, Peter. *The Hidden Life of Trees: What They Feel, How They Communicate,* trans. Jane Billinghurst. Vancouver: Greystone Books Ltd., 2016.

Zukav, Gary. *The Dancing Wu Li Masters: An Overview of the New Physics.* New York: Bantam Books, 1979.

VIII PERSONAL, LOCAL APPLICATIONS FOR HEALTHCARE AND MEDICINE

In this section, we move from large theories back down the abstraction ladder to practical applications for the health of all humans, sick people, well people, also healthcare workers. We discuss improvement we can make now, even if the long-term prospects for the earth are poor. Our discussion understands the universals of matter, mind, and energy as resources for health and medicine.

Health as "assets-based"
Because of its main focus on illness or injury, the world of modern Western medicine is typically "deficits-based," meaning, basically, *what is wrong*. To be sure, society needs these services, but they should not, eclipse the worlds of health maintenance or health promotion that is "assets-based," in Edgar Rivera Colón's term (Charon et al., 258). By "assets" we mean the health resources of a person when well but also even when sick, injured, very young or aged. Athletes, dancers, and practitioners of Asian and Indian arts (Qigong, Yoga, etc.) tend to think this way: they enjoy and do the best they can to maintain their marvelous bodies. They know both mind and energy can play a role in performance and health. They understand balance of effort and recovery, the harmony of coordination, the importance of good nutrition, rest, and reduction of stress. Sooner or later, all of us become injured or ill and must draw on our innate healing abilities.

"Let's get Mr. Jones back home as soon as we can"
Nurses and doctors often remark that patients heal better and faster at home than in the hospital, so they try to discharge them as soon as possible. The reasons include risk of infection from the many germs in the hospital, especially for patients with weakened immune systems, but also the many disruptive routines (for example, blood draws in the night), noise, and, in general, the foreign nature of rooms, hallways, food, and more. All these raise stress for patients. By strong contrast, at home—a good one, that is—a sick

person relaxes with her or his familiar surroundings, companions, food, pets, music, plants, and so on: in short, all the usual energetic inputs of sights, sounds, smells, tastes, and touches. All these are healing resources.

"VINEGAR SOAK WORKS WONDERS ON INJURED FOOT!" CONSCIOUSNESS AND ENERGY HELP EXPLAIN PLACEBO EFFECTS AND SPONTANEOUS HEALING

A newspaper headline reads "Vinegar Soak Works Wonders on Injured Foot." Who knows what the absolute truth was? Perhaps the foot healed on its own, regardless of the vinegar soak, or resting the foot to soak it and sending it good energy allowed the healing, or other possibilities. For all of them, however, there was the belief of the person that the soak would be helpful, and this belief may have been a placebo factor. In terms of our three universals, the patient's mind sends energy to the body's matter and renews or reaffiliates the coherence and health in the foot.

There is an extensive literature on placebos that I will not review here. Many believe that patients' beliefs are important in their healing, and they can boost medical treatment or even promote healing on their own. "Placebo," we recall from Section III, is Latin for "I will please." (See Lemoine, "The Placebo Effect: History, Biology, and Ethics," 2015). In the early days when medicine was quite limited, Ambroise Paré (1510–1590) suggested that the physician's duty was to "cure occasionally, relieve often, console always" ("*Guérir quelquefois, soulager souvent, consoler toujours*"). Accordingly, placebos were widespread in medicine until the 20th century, and were even endorsed as necessary deceptions.

Andrew Weil writes, "*Even when treatments are applied with successful outcomes, those outcomes represent activation of intrinsic healing mechanisms, which, under other circumstances, might operate without any outside stimulus. The main theme of this book is very simple: The body can heal itself.*" (*Spontaneous Healing*, p. 5, his emphasis). Further, "The evidence is incontrovertible that the body is capable of healing itself. By ignoring that, many doctors cut themselves off from a tremendous source of optimism about health and healing" (p. 65). The opposite "nocebo" has been proposed, translated as "I shall hurt." Years ago the phrase "sophomore syndrome" was coined for second-year medical students who, upon reading about a disease, immediately imagined that they suffered from it. Even today, lists of side effects (as in my cancer treatment) are either kept from patients or downplayed because patients might "give themselves" a side effect because they knew it was possible.

Regrettably, mainstream medicine today is suspicious of placebo effects and typically ignores it as a potential asset in a patient's healing.

CARE OF SELF AND WELLNESS

Besides healing from injury or illness, there is the important and pervasive area of *health maintenance*.

Selfcare on a par with a valuable car

During my years at Eckerd College, students were required to take "Values Sequence" courses in order to graduate, and they could choose from a large variety of courses. Biologist John E. Reynolds and I offered "The Human Body as an Environment." It was well attended, probably because students were aware of their own bodies and, certainly, the bodies of their peers. For one class, John and I invited a drug educator (name long forgotten) to talk about drugs.

She asked the students if they liked cars.

"Of course," they said.

"If you could have any kind you wanted, what would you prefer?"

Answers abounded.

"And what if we said that was the *only car you were ever going to have*, what kind of care would you take of it?"

Answers abounded.

"And what kind of gas would you use?"

"Regular servicing?"

And so on.

Pretty soon the students caught on—and she finally said—that the illustration had the parallel topic of *their bodies, the one vehicle they were ever going to have*. Beyond this *material* reality she cleverly evoked the values and emotions of their minds, even the possible *energies* involved in car care and body care. Their involvement in the discussion was intense, because she elicited their emotions of care and protection, even celebration and honor transferred from cars to bodies, bodies they often took for granted. (It's commonly observed that the young feel *invulnerable* and take for granted that *they will live forever*.) At the end of the course, student presentations reflected a wide range of topics: improved diets and exercise, research on a familial disease, smoking cessation, and so on. One man held up a sealed plastic bag containing his last pack of cigarettes and his lighter. One woman held up a radio she bought with the savings from no longer buying cigarettes.

My college students had young healthy bodies. They had smart, imaginative, inquiring minds. They had plenty of energy, but they seemed unaware of the depth of any of these resources or how to recognize and manage them. Some had issues of self-doubt, anxiety, and depression. College youth (and many other people, some with less privilege) need avenues for affirming their assets to maintain and extend their health.

Whether a person is young or old, caring for health has immediate and daily benefits. The proverb runs, "An ounce of prevention is worth a pound of cure." According the CDC, "Chronic diseases and conditions—such as heart disease, type 2 diabetes, obesity, and arthritis are among the most costly, and preventable of all health problems" (https://www.cdc.gov/chronicdisease/about/prevention.htm, accessed Aug. 12, 2017).

The wellness boom

The 1970s and 1980s saw the rise of many avenues for wellness, witness Yoga, healthy diets, running and other aerobic sports, gyms, sports clubs, and recreational leagues, as well as magazines, newspaper columns, and books on health and medicine. I still have my copy of Donald B. Ardell's *High Level Wellness* (1979) and Mike Samuels' and Hal Zina Bennett's *Well Body, Well Earth* (1983), a Sierra Club Environmental Health Sourcebook.

Ardell stresses self-responsibility, nutritional awareness, stress management, physical fitness, and environmental sensitivity. A 67-page "honor roll of books" provides an annotated (with strengths and weaknesses) bibliography. Samuels' and Bennett's book describes pollutions of radiation, chemical, and poisons, as well as the importance of clean air and water. Their earlier *The Well Body Book* (1973) sang the praises of the human body as a "three-million-year-old healer" that knows *what to do to stay healthy*.

Access to such books and other resources was easier for wealthy and urban people, while rural or inner city people often had fewer opportunities. Years ago I lived in a ghetto where none of this was available. Today my local YMCA has outreach programs for inner-city kids, but disparities still abound in the U.S. and elsewhere.

Care of others, advocacy for others

We all know this: when we help a child, another family member, a pet, a neighbor, we feel better ourselves.

Perhaps we cannot directly care for the poor, the immigrant, the sick, the aged, but we can advocate for them.

Some businesses have a wellness or health promotion officer, a nurse, even a doctor on staff. Besides the practicality of quick access, there symbolic value in such staffing; it shows caring intent on the part of management. A doctor in such a job told me it was, over all of his career, his favorite setting for providing healthcare.

How about, as well: working for and advocating for clean air and water, gun control, healthy food, civil and informed discourse?

Care of the earth

Isn't this obvious? Each person lives on this earth. The earth is, of necessity, our home. A healthy earth supports each person's health. A sick earth impacts each person's health. We should care for this earth, and we will feel better as we do that.

ALL HEALTHCARE, INCLUDING STANDARD MEDICINE, IS ALREADY ENERGY MEDICINE

We can often sense healers' intentions, empathy, and healing energy. When crying children come to a parent who kisses a child's "boo-boo" to "make it better," and the crying stops.

According to cell biologist James L. Oschman, all persons have energy fields that interact if they are within, say, six feet of each other, and even more if they touch. (Qigong assumes much wider distances, as does Quantum physics.) In his *Energy Medicine in Therapeutics and Human Performance* (2003), he describes how concepts from Quantum physics explain that the energies of the body are subtle, instantaneous, and highly efficient in regulation and healing at the cellular level. For him, the body is best understood as a *living crystal* which can semiconduct energy, translate physical energy (such as touch) to piezoelectricity, all for the purpose of maintaining coherence and continuity in our bodies. We do this automatically, day and night. Mammals have been maintaining and healing themselves for eons, also early humans, well before there were shamans, doctors, hospitals, insurance companies, or HMOs. When there are minor injuries, such as muscle tears or damage to skin, our bodies automatically carry out repair. Families and caring people support each other; groups of animals, even trees, have ways for supporting each other.

Energy healing, in one form or another, is worldwide and has been with us as long as we have written accounts and surely before that. Nurses, sensitive doctors and techs all know that patients are happier when greeted pleasantly, but Standard Western medicine does not recognize energy per se as a healing entity, nor does much of the Western public. This is unfortunate because there are many benefits that could improve medicine and health in general. Medicine already uses many forms of energy: X-rays, CT scans, MRIs, sonograms, various nuclear tags, laser scalpels, thermograms, gamma knives, and other energies applied to the *matter* of patients' bodies. Besides technical uses of energy, caring doctors and nurses boost our spirits by their intent and coherent energy as they attend to us.

Energy healing in hospitals

Energy healing could deepen and expand integrative medicine by providing another source for healing. Today we have "hospitalists," commonly internal medicine doctors, who watch over patients (and lab results, etc.) in a given hospital. Can we imagine an "integrationist," a doctor, a P.A. or a Nurse Practitioner, who explains a wide range of supportive care and helps a patient choose? Just to mention two, massage and energy work are forms of witnessing and companioning that can relieve loneliness, sadness, and despair as well as making possible nourishment, enhanced healing, and various reaffiliations. In cases beyond physical healing, energy work could assist palliative care, even when a patient is active dying.

How long standard medicine will be resistant to such resources cannot be known, but I imagine it will happen, much like the shift to the Temporal Artery Scanner we saw earlier. Indeed medicine can be a change agent for the culture at large in using so-called subtle energy, often at lower cost and with enhanced participation of patients. I imagine that there will be scientific studies that show the efficacy of energy work and also that there will be growing patient demand for energy healing as part of medical treatment.

THE FIELD FITS A COMMUNICATIONS MODEL

Taking a classic model from communications studies, we consider that the Field connects all objects, energies, and minds. The impacts may be very small or miraculously pivotal, like the famous butterfly in chaos theory that causes a hurricane thousands of miles away.

SENDER		CHANNEL MEDIUM		RECEIVER
ANY OBJECT, IDEA	←→	THE FIELD	←→	ANY OBJECT
ANY IDEA, EMOTION	←→	THE FIELD	←→	ANY IDEA, EMOTION
ANY ENERGY	←→	THE FIELD	←→	ANY ENERGY
ANY LIFE FORM	←→	THE FIELD	←→	ANY LIFE FORM
ANY HUMAN	←→	THE FIELD	←→	ANY HUMAN
THERAPIST	←→	THE FIELD	←→	CLIENT/PATIENT

Figure 13: The Field in a Communications Model

The notion of "channel" is commonly linear: a letter, a phone line, or a cable network. With email, we have the more diffuse concept of the Internet, a global system of interconnected computer networks, a parallel with the Quantum world where everything is always linked. By this theory, every-thing—all forms of mind, matter, and energy—are interlinked throughout the entire universe.

A channel is most efficient when noise or static is quieted or eliminated, for example, the monkey mind, Cartesian dualism, materialism, positivism, individualism, Neodarwinian competition, hatred, anxiety, anger, loneli-ness, despair, and other stresses, etc. The most efficient receivers would be those more entrained with the sender as a dyad: friends, partners, spouses, parent and child, also a dog or horse and owner, healer and patient, therapist and client.

A channel is most efficient when there is coherence and positive emotions such as love, care for others, and a sense of home. Healers know how to focus into a channel, to sense the energy of the receiver, and to transmit appropriate healing energy.

Humans as antennas

By this theory, human beings are antennas to each other but their sensitivity and awareness varies. We are potentially related to all matter, mind, and energy but our focus is typically local, governed by our conscious attention, cultural beliefs, and so on.

The coherent and intended energy of a pet dog, another person, or a loving group of people influences the energy of another person. Such persons can even be doctors, especially those who interact with a patient as a fellow human being. I've observed a gastrointestinal doctor treat a patient with technical excellence then stay a few more minutes to ask about the person's hometown, work, and the like. It seemed to me that both aspects of the encounter were healing, and the patient was not only less nervous but also more happy and confident. I've known doctors who routinely pray for their patients.

We are antennas, sending and receiving day and night, from all other senders, but especially fellow humans with whom we are entrained/entangled. Long-term partners sometimes dream together. Our communication with humans (or others) can be deepened by the mind's intent to care.

RANT 8: WESTERNERS SHOULD EXPAND THE USE OF ENERGY IN MEDICINE

Can we imagine that:

Doctors, nurses, and all health personnel have training in use of energy, both in their education and in the later careers.

Doctors recommend use of energy healers.

Energy healers work in hospitals, along with massage therapists.

Education allows children and adults to use energy for themselves and others.

Public ceremonies use energy to heal a community, a nation, the entire earth.

Why can't we grasp, use, and enjoy the scientific theories?

The scientific work described by McTaggart describes comes largely from the 1980s and 1990s with some later applications, but, as I write in 2019, these concepts and applications are not in mainstream culture, our media, education, or consciousness. Nor are they in standard medicine, where they could be effective at low cost.

Here are some factors:

Our world picture is still Newtonian, materialistic, bias against the "woo-woo factor."

Science uses mathematics, strange symbols and formulae, that laypersons can't read. Many concepts are abstract and/or counter-intuitive or, like the coffee cup going through the floor.

Things nuclear, atomic, subatomic have a bad reputation: bombs, reactors going haywire.

There hasn't been a Neil DeGrasse Tyson explaining Quantum theory. Max Tegmark says scientists should do a better job of reaching the public.

Experiments are hidden away in labs.

Some people—including political leaders!—have biases against science, theories in general, and specifically against Climate Change.

In a high-demand, high-stress society, we have trouble being calm, receptive. We get high on Yang and have no time for Yin. We can't empty our minds, allowing nature, inspiration, muses, the universe, or new ideas come to us.

We have lost our awareness of all persons' healing resources of body and mind. Perhaps they seem too good to be true. Instead, we accept the dominant models that medicine is provided only by professionals and, often, through high-tech, high-expense methods. Energy healing can combine with any healing modality and does so especially if there is awareness in the healer, the recipient, or ideally, both.

Energy medicine has many advantages: it is low tech, low expense, available in various spaces, and sometimes without travel, waiting, or lengthy time for treatment. It can operate remotely, as we saw with Qigong. Many churches routinely offer prayer for congregants who are ill or facing challenges, also for governmental officials, even for the earth itself. My Presbyterian church—mainline, with a traditional liturgy—does this every Sunday, and also through an email alias (a "prayer chain") available all weeklong.

People who were prayed for when sick later say that prayer was helpful to them. They felt support, care, and love. Such feelings lower stress and support mind and body; these responses aid in healing.

Does energy healing seem to good to be true? Can we accept that it is free, endless, and already part of our basic human nature of cooperating matter, mind, and energy?

And the usual obstacle: for patients and caregiver alike: *we've always done it this way.*

ESSAY 8: LOCAL, ASSETS-BASED HEALTH…EVEN HEALTHY DEATH

Musing on the solar eclipse of 2017, I recall of a poem by Walt Whitman:

> When I heard the learn'd astronomer,
> When the proofs, the figures, were ranged in columns before me,
> When I was shown the charts and diagrams, to add, divide, and measure them,
> When I sitting heard the astronomer where he lectured with much applause in the lecture-room,
> How soon unaccountable I became tired and sick,
> Till rising and gliding out I wander'd off by myself,
> In the mystical moist night-air, and from time to time,
> Look'd up in perfect silence at the stars.
>
> – *Leaves of Grass* (1865)

The speaker leaves the rational explanation, the talk, the applause, and, instead, reaffiliates with silence, "the mystical moist night-air," and—with no rush or plan—looks at the stars. The speaker has, at that moment, *all that is needed.* We may imagine her or him relaxed and renewed, perhaps even made whole, as were the viewers of the 2017 solar eclipse. Indeed, the word "health" relates to similar words: *hale, holy, healing, whole, wholesome.*

Humans as lotus plants and other assets

There are many roads to reaffiliation. I mentioned earlier the Qigong standing exercise "The Rising Lotus," that links the Yin of the earth to the Yang of the skies. You imagine that you are that plant. First you bend over and scoop up the mud with both hands from the bottom of the pond where the lotus is rooted. Next you slowly straighten up and raise your hands together in front of your body; your hands make the long, green stem that rises through the water. As they enter the air, about chest high, your hands are back to back to represent the plant floating on the water. Higher still, your hands become a bud, then they open outward above your head as a lovely, multi-petaled lotus flower—pink or white or any color your chose—blossoming to the sky, the sun, the universe. In this exercise out bodies and minds participate in recognizing and honoring the harmonies of nature. In imagining ourselves as a lotus plant, we reaffiliate with nature, the earth, and our own health—both physical and mental.

Like many other approaches, Qigong understands that all our bodies—at any age—are wonderful, even miraculous. Further, they are microcosms of the beauty, majesty, and complexity of the universe.

Whether we study Qigong per se or not, we can take pride in our bodies and take care of them. The basics are well known: exercise, healthy food, stress management, medical care, basic hygiene, and limited alcohol, drugs, and risky behaviors. We have less control over pollution of air, water, and soil, noise and light pollution, and other threats to the health of our bodies, especially those now posed by Climate Change. We need to work through government and other avenues to lessen these.

Homes and other assets

In our discussion of nested spheres, our sense of home expanded out in conscious and subconscious ways. We had insights in the coherence of local and cosmic worlds.

Our first home, we might say, was our birth mother. Our second home, as the Eckerd College students realized, is our own bodies, from our infancy to our death. Besides these wonderful bodies, what are some of the assets we can count on and reaffiliate with?

People

Who are the potential assets? Our closest family and dearest friends also neighbors. Someone who helped us, a good Samaritan, or otherwise touched our lives. How about fellow travellers on this earth? In 1955 an exhibition called "The Family of Man" opened at the New York MOMA. (We'd say

"Man or Woman" today.) It toured for eight years. A book representing all 503 photos sold over four million copies and is still in print. The images show people from around the earth…and in a variety of activities, moods, and cultures, and we see resonances between them and us. We grasp that all humans are much alike, similarly amazing creatures in body, mind, and energy: an extraordinary reaffiliation of a world-wide family.

Nature

It's hard to miss the web of life, even in a large city. A tree, a bird, a plant. Maybe a memory from a trip or growing up, perhaps a powerful thunderstorm or a spectacular sunset. An animal, a pet. Maybe a place that seemed awesome, holy. All of these touch our biophilia, as E. O. Wilson called our innate love for nature.

Earth

So many ways. Feel it beneath your feet, in your hands. Plant something. Look at hills, mountains, valleys. Rock, clay, sand, and humus have all provided food for us. Their elements become ours. When we die, our elements become, sooner or later, part of the earth.

At night our planet turns away from the sun to cool, to rest, to renew; similarly, we need sources of rest and renewal.

Local and current applications

"Pray for the dead and fight like hell for the living," urged social activist Mary Harris "Mother" Jones in 1902. The motto is applicable today, both for those who have died and for those alive today, and even for changes in the Carbon Curmudgeons who abet Climate Change, fight efforts to save the earth, and have no care for future generations. In the last section, we cited *Climate of Hope*, the inspiring book by Michael Bloomberg and Carl Pope.

Even a healthy death…of each of us…or even all of us

Poet and cardiologist John Stone wrote "health is whatever works / and for as long."

In medicine, curing is not always possible, but healing is, even at the end of life. As Atul Gawande wrote, the best possible dying has many features: pain control, life review, making peace with others, and enjoying final weeks and days. We honor our bodies, even as they inevitably fail, and understand that they will return their materials—which have been on loan to us—back to the earth.

For those who believe in a world beyond this one, there is reaffiliation there, probably with other minds/souls.

And if there is to be an end of the human race, can we imagine it also to be a healthy event? *We had our run, did what we could, and now recycle our bodies, minds, and energies into the earth, the universe.*

Our one earth

Like my students' bodies and the allegory of the one car they might ever own, humanity is given just one earth. How will we appreciate, use, and nurture this planet? Yes, there are scenarios that some fraction of us may escape to another better (?) planet, but don't hold your breath.

SOURCES 8

Ardell, Donald B. *High Level Wellness*. Rev. ed. New York: Random: 1979.

Bloomberg, Michael and Carl Pope. *Climate of Hope: How Cities, Businesses, and Citizens Can Save the Planet*. New York: St. Martin's Press, 2017.

Colón, Edgar Rivera, "From Fire Escapes to Qualitative Data: Pedagogical Urging, Embodied Research, and Narrative Medicine's Ear of the Heart," in Rita Charon, et al., *The Principles and Practice of Narrative Medicine*. New York: Oxford Univ. Press, 2017. Chapter 11, pp. 251-267.

Lemoine, Patrick. "The Placebo Effect: History, Biology, and Ethics," http://www.medscape.com/viewarticle/852144_5, accessed Sept. 16, 2017.

Oschman, James L. *Energy Medicine: The Scientific Basis*. Edinburgh: Churchill Livingston, 2000.

_____. *Energy Medicine in Therapeutics and Human Performance*. Edinburgh: Butterworth Heinemann, 2003.

Samuels, Mike, and Hal Zina Bennett. *The Well Body Book*. New York: Random House/Bookworks, 1973.

_____. *Well Body, Well Earth*. San Francisco: Sierra Club Books: 1983.

Steichen, Edward. *The Family of Man*. New York: Ridge Press, 1955.

Stone, John. "He Makes a House Call," *In All This Rain*. Baton Rouge: Louisiana State Univ. Press, 1980, p. 4.

Whitman, Walt. "When I Heard the Learn'd Astronomer."   https://www.poetryfoundation.org/poems/45479/when-i-heard-the-learnd-astronomer, accessed July 10, 2017.

Wilson, Edward O. *Biophilia*. Cambridge, London: Harvard Univ. Press, 1984.

Weil, Andrew. *Spontaneous Healing: How to Discover and Enhance Your Body's Natural Ability to Maintain and Heal Itself*. New York: Alfred A. Knopf, 1996.

IX PRACTICAL APPLICATIONS FROM INTEGRATIVE MEDICINE AND HEALTH HUMANITIES FOR HEALTHCARE PROVIDERS AND INSTITUTIONS

In this section we circle out to clinics, hospitals, and all means of supporting health for humans and the earth itself. We also circle back in, because institutional values shape all caregiver-patient interactions. As in the previous section, these are improvements we can make now (or soon) regardless of how Climate Change impacts our earth.

MUCH MODERN MEDICINE IS "DATA DRIVEN"

I've heard "data driven" from doctors who accept the concept as a crucial pivot for what counts in healthcare. The phrase usually implies "Evidence Based Medicine," which means consulting current scientific studies. Treatment may based on Randomized Clinical Trials (RCTs) that are, as the common phrase has it—the so-called "gold standard" for medicine. RCTs have problems (among others) of which patients were included in the trial, how the trial was conducted, the particularity of the drug being tested, reliance on the middle of a bell-shaped curve, and exclusion of other factors. Data-driven medicine may ignore patient variability, even a physician's own experience and judgment.

I don't dispute that these philosophies produce many helpful techniques, but much of the research depends on the funding of drug industries and the underlying philosophies are often reductionist (based on chemistry, molecules, and genes) and, later, limited in their applications by what insurance plans will fund. Further, these are, as we said earlier, disease- and injury-based, not assets-based. Still further, research dollars for treatment of disease far outweighs research for health maintenance.

The data-driven pathways typically exclude Eastern and other holistic approaches. They have, as far I've seen, no strategies to turn into numbers values of love, caring, respect to the dying patient, or supportive bonds between caregiver and patient.

American medicine—especially at tertiary-care centers—focuses on the very ill, part of a complicated and heavily funded industry. Such centers have high status in the eyes of young doctors who are considering their futures and who typically carry considerable debt from medical training. They also know that some specialties, like Family Practice, Emergency Medicine, and Integrative Medicine, are less well paid.

ROUTINE HEALTHCARE FOR ALL

While the U.S. has an excellent trauma system and wonderful high-tech medicine that can perform, for example, organ transplants, its routine health care is good only for certain clienteles. People in rural areas or inner cities often do not have good routine care. Regrettably, the U.S., despite its high expenditures, does not compare well with other countries, especially in childbirth statistics. We do poorly with underserved populations and with health-maintenance in general, favoring trauma and other rescue strategies. We are going backwards on clean air and water, we are unable to control guns, and we have an epidemic of drugs, both opioids (many prescribed by doctors) and street drugs. This is a shame because preventive medicine and supportive health care in general is low-cost, low-tech, and can use, besides physicians, nurse practitioners and other allied healthcare workers to reach people who may not have access to care otherwise.

I'm lucky to have excellent support. My doctor has a good sense of humor; he speaks of my "career with him" as he reviews my blood lipids, inoculations, and body systems. His nurses and clerical persons are friendly and supportive. Sometimes they ask me screening questions: smoking…alcohol…depression…any falls in the past year? A fortunate man, I do well—currently—on these questions but, more important, I feel that the office cares about my health and that I should as well. The physical surroundings are pleasant. There's an urgent care option outside of office hours for emerging problems, and I have used this on a Saturday. The nurse practitioners that staff this service are smart, caring, and effective. I wrote a thank-you to one of them, and she was surprised and pleased. Evidently she rarely receives feedback. I would like everyone to have the care I receive, but I know that they do not. I also know that I will have a fall someday, various body systems will be compromised, and—one way or another—I will die.

Why should I get excellent care when many in my country and elsewhere do not? Across the globe there are people who have never seen a doctor of any kind. Even in North Carolina (my state) some people have never seen a dentist; when clinics travel to poor areas, the dentist must extract many teeth, even in young people. In my county 60 per cent of people are "doing OK or better" economically, but 40 per cent are at the poverty level or below. Rachel Pearson's recent book *No Apparent Distress* (2017) lambastes the disparities of care in Texas.

This section looks at practical applications for health care with an emphasis on assets-based, preventive care. Keeping in mind the body, mind, and energy of all patients and healthcare providers (at times patients themselves), we discuss (A) integrative health care, (B) patient- and family-centered care, and (C) a new subfield, Health Humanities.

ALL PEOPLE ARE HEALERS

Recalling Oschman's point about humans as senders and receivers, we may understand that all persons are potential healers. Using the assets of mind (including intent), energy, and the matter of their bodies, they can help others heal. Family members routinely do this for each other, also close friends, even pets. Caring communities can be extended. Groups using prayer and meditation often pray for the entire earth, including all life.

Especially in a hospital, human-to-human support should be a pervasive value promoted by administration. In a hospital, *hospitality* should be a main value.

Values of human-to-human care vs. merely technical care

I've had phlebotomists take my blood as if they were checking a car tire. I've had doctors, arriving late, rattle through questions as if my house needed pressure washing. Only a few seconds of human-to-human communication could have changed the mood for me and for them.

Even in standard medicine: all employees can share a smile or other clues of social equality, a comment about the weather, or just "How's it going today?" Medical employees (nurses, docs), ask diagnostic questions, of course, but they can ask also how the illness or injury *feels* for the patient.

For cases of serious illness, patients need emotional sympathy from caregivers and, when appropriate, support from hospital chaplains and other religious professionals, ministers, imams, priests, rabbis, whom a patient wishes to see.

Besides social support, patients deserve accurate and timely information about their health. A welcomed and informed patient has less stress and more likely to bring her or his healing assets into play.

INTEGRATIVE MEDICINE REVISITED

Integrative medicine has been around since Andrew's book *Spontaneous Healing* (1996). We reviewed the "Defining Principles of Integrative Medicine" in Section III. These fit well with traditional science-based medicine, and they emphasize health for caregiver and patient, including body, mind, and community.

In some university medical centers, integrative medicine is, oddly, a department that is separate from the clinics and hospitals that deliver medical care. If integrative care is to be the norm for patients—and for all employees—a hospital, clinic, or health center needs strong leadership and a shared vision that becomes the norm throughout the institution. Hospital architects often do a fine job with materials, creating a welcoming and pleasant space, but the employees also need to embody healing mind and energy for the best treatment of patients.

Some hospitals and retirement communities use the word "health" more, for example, a health center, emphasizing wholeness, even when the body is injured or sick.

The concepts of energy fields and Quantum energy deepen the possibilities of integrative medicine.

Massage therapists and other therapists

Let's imagine: energy healers working in hospitals and clinics, practitioners of Qigong, Johrei, Reiki, Reconnection, hypnosis, and similar modalities so that patients feel less pain and isolation, they are happier and heal faster, or, when necessary, they may die better supported and cared for. A patient might choose not only the meals, but also movies, music, hypnosis, or clown visits offered by the hospital. We can imagine primary doctors who refer to energy healers. We can imagine the arts, in many settings, nourishing the imaginations of patients and caregivers alike (see next section). For one extraordinary example of the power of music, see Andrew Schulman, *Waking the Spirit: A Musician's Journey Healing Body, Mind, and Soul* (2016). Schulman himself was awakened from a coma with recorded music and later, when healed, he provided guitar music for surgical patients.

HEALTH HUMANITIES: AN ALLY AND ASSET FOR STANDARD MEDICINE

Another area of interdisciplinary study that can strengthen and expand medical care is called "Health Humanities." This is a new, interdisciplinary academic field that is an extension and expansion of its forerunner Medical Humanities. Arising in the 1970s, Medical Humanities made many contributions to broader approaches to medicine through such disciplines as bioethics, literature, history, and religion. These and other humanistic approaches, helped widen our understanding of illness and medical care. Health Humanities stresses *health*—personal, social, national—more than *disease*. It fits well with Edgar Colón's words, "the push toward an assets-based approach to public health challenges, as opposed to a deficits-based and pathology-replicating paradigm" (Charon, et al., 2017, p. 259). Further, health humanities broadens our notions of caregivers well beyond doctors and nurses to include all hospital personnel (including cleaners and diet personnel), also allied healthcare workers (some 40 kinds), and nonpaid caregivers at home—probably the largest group of all. Still further, Health Humanities includes the creative arts: drama, film, dance, music, and more.

There are two pivotal publications: *The Health Humanities Reader* (2014) and *Health Humanities* (2016).

The Health Humanities Reader, edited by Therese Jones, Delese Wear, and Lester D. Friedman is a collection of 46 original contributions not easily summarized. Drawing on (among others) disability studies, race and gender studies, social studies, and religion, the authors expand the Medical Humanities to a larger meanings of health and recognize the many people (beyond doctors and nurses) in the health field and even family members and patients themselves. Many disciplines can contribute, from the arts to public policy, from economics to global studies. There is increased attention to the complexities of patients and their needs for hope and benefits of placebos. The notion of home is cogently explored in Chapter 13. Problems in medicine such as errors, paternalism, and patient safety are also explored. Medical education should widen the medical gaze, emphasize teamwork and democratic values in a global, medical culture that promotes health. Each of the 12 sections ends with "an imaginative or reflective piece by an artist, writer, teacher, or scholar" (p. 9), illustrating how creativity can take many forms and provide different perspectives on topics such as disease, aging, science, and technology.

Health Humanities, a monograph by Paul Crawford, Brian Brown, Charley Baker, Victoria Tischler, and Brian Abrams, is more philosophical and theoretical. A pervasive theme is that medicine per se has been too science-based and too disease-oriented, but critical theory and the arts can be "enabler[s] of health and well-being" with many applications to hospitals, clinics, homes, and neighborhoods. This book promotes a balance between giver and receiver, neither paternalism nor patient "autonomy." It suggests that healthcare became narrowed in the 20th century by positivism, reductive science, Evidence Based Medicine, economics, and public policy. We should keep all the benefits of modern care but also expand our understanding to include other resources, such as the arts, and other intellectual traditions, such as feminist and postmodern inquiry.

The authors argue that the dominance of the pharmaceutical industry in research and delivery of medicine has meant less emphasis on preventive healthcare, often a more efficient and economic path. The book is critical of paternalistic medicine and a sub rosa ideology of elitism and power relations in, at times, a straw-man rhetoric. (Health Humanities itself also has various ideologies inherent in it. Caregivers, scholars, and thinkers will need to sort out which ideologies serve healthcare best.)

Drawing on both books, we can entertain a force-field model for Health Humanities and standard medicine. (Such a model, it may be recalled, ignores overlaps and middle ground between the polarities.)

HEALTH HUMANITIES	STANDARD MEDICINE

FOCAL CONCEPTS, IDEOLOGIES

Health, wellness, wellbeing	Disease, injury, suffering
Care	Cure, rescue
Health promotion, maintenance	Diagnosis and treatment of disease, injury
Patient as person, partner, resource	Patient as puzzle for diagnosis
Many branches of knowledge	Science
Team concept, including the patient	Doctor-patient dyad
Death: natural, inevitable, can be managed	Death: a failure to be feared, taboo, to be avoided at all costs

ACTIVITIES, CONTEXTS

Prevention, screenings, health maintenance, also, good care for injury, illness	Intervention, rescue
Patient in context: family, neighborhood, heritage, core beliefs	Patient as cells, systems, a "case"
Locales: urban, rural, nature, the world, home	Clinics, hospitals, treatment rooms

PERSONNEL WITH THE MOST POWER

Multiple: doctors, nurses, patients, unpaid family, clergy, 40+ areas of allied health workers	Limited: CEOs, lawyers, governing bodies, doctors

IDEOLOGIES, THEORIES, TACIT OR EXPLICIT ASSUMPTIONS

Many kinds of knowledge	Science working with matter
Mind-body syntheses, feminism, postmodern approaches, Quantum physics, energy, etc.	Descartes, Positivism, Big Pharma
Research standards: multiple	"Gold Standard" of RCTs, metastudies
Experience Based Medicine/Practice	Evidence Based Medicine

Reason, intuition, emotion, imagination right brain, multiple links with left brain	Logic, left brain, decision trees
Inductive: from experience to concepts	Deductive: from rules, principles, protocols to application to specific cases
Caregivers as fellow humans, colleagues	Caregivers in superior status

Figure 14: Health Humanities and Standard Medicine in a Force-Field Comparison

We turn now next to other fields, many older but still valuable and important. Some of these overlap, and we should consider that all of these are helpful and welcome.

PATIENT-CENTERED AND FAMILY-CENTERED CARE
Based in Bethesda, Maryland, the Institute for Patient- and Family-Centered Care was founded in 1992. According its website:

> Patient- and family-centered care is an approach to the planning, delivery, and evaluation of health care that is grounded in mutually beneficial partnerships among health care providers, patients, and families. It redefines the relationships in health care by placing an emphasis on collaborating with people of all ages, at all levels of care, and in all health care settings. In patient- and family-centered care, patients and families define their 'family' and determine how they will participate in care and decision-making. A key goal is to promote the health and well-being of individuals and families and to maintain their control (http://www.ipfcc.org/about/pfcc. htmlk, accessed June 12, 2018).

Specific applications of this approach are presented in the book *Privileged Presence: Personal Stories of Connections in Health Care* (Croker and Johnson, 2014; see also Johnson, B. H. and Abraham, M. R. *Partnering with Patients, Residents, and Families: A Resource for Leaders of Hospitals, Ambulatory Care Settings, and Long-Term Care*, 2012). The former volume presents some 58 brief stories of patients, family members, nurses, nurse practitioners, and some administrators. While many are favorable about a specific instance of medical care, some are critical of impersonal treatment, several by physicians. In some stories, complaints by patients or families directly to hospital administrators led to changes to improve care, for example not treating family as visitors, with limited *visiting hours*, but as members of the healthcare team, welcome and considered helpful at any time.

The titles for divisions in the book are, in themselves, inspiring: "As Unique as Snowflakes: Responding to Individuals," "When Life Is Threatened: The Importance of Support," "Natural Allies: Partnerships in Care," and "More than Words: Feeling Heard and Being Valued."

OTHER ESTABLISHED FIELDS

With no pretense of full coverage, I mention established fields that are multidisciplinary and values oriented. They too can benefit from the integrative and energy-related resources for health and medicine discussed here. One example of evolution is the CDC, formerly the Centers for Disease Control and changing its name in 1992 to the Centers for Disease Control and Prevention with programs in disease and injury prevention and health promotion.

Health promotion, public health, and epidemiology

We have well-baby visits, some screenings, and many good publications (newsletters, journalism, books), but nonetheless many Americans are overweight, overstressed, and under-exercised. More is needed!

In the larger spheres of neighborhoods, communities, and nations, we think of public health that provides the structures of water, sanitation, vaccinations and, of course, absence of war, violence, pollution of air, water, and land. Public health improves health and quality of life through treating disease and other physical and mental health conditions. Tools range from education about hand washing and condoms to biostatistics, environmental health officers, and community development workers.

A cornerstone of public health is the field of epidemiology; it studies diseases among a population, including newsmakers like HIV/AIDS, Ebola, and Zika. I've put my hand on the pump in London that John Snow figured must be source of deadly cholera in Soho in 1854. With the removal of the handle and the outbreak ceased.

What if we, also, removed many guns from our society, especially automatic weapons? Australia has had two buy-back programs and an amnesty. Rates of murder and suicide have decreased.

World Health Organization

Health promotion is the process of enabling people to increase control over, and to improve, their health. It moves beyond a focus on individual behavior towards a wide range of social and environmental interventions. According to the WHO website: "Our goal is to build a better, healthier future for people all over the world. Working through offices in more than 150 countries, WHO staff work side by side with governments and other partners to ensure the highest attainable level of health for all people" (http://www.who.int/about/en, accessed December 5, 2017).

But it can't just be health care for humans

The earth is where we all live; it provides the food, the water, the air, but also pollutants, harmful bacteria and viruses, and a myriad of stresses brought about by pollution, poverty, disenfranchisement, and violence on many levels: domestic, neighborhood, inter-tribal, or all-out warfare.

Unfortunately, Climate Change is an immense threat to all living creatures, including humans. Like humans, the earth needs both routine healthcare and, now endangered, focused interventions.

RANT 9: "ALL FOR ONE AND ONE FOR ALL"

"All for one and one for all."

We've all heard versions of the Musketeers' motto, but there were only three musketeers.

We now have 7.4 billion (or more) humans plus trillions more of other creatures on earth. The days of Neodarwinism ("eat the other guy's lunch"… or "eat *him* as lunch") should be over if we prefer that all of us to survive. U.S. coins have "*E pluribus unum*" written on them: "From many, one." While the phrase originally celebrated our new country made from 13 colonies. it's a good motto for any community seeking unity, success, and health, even the entire world.

Such mottoes suggest values of kindness, care, and treating all persons and creatures as family, affiliated (or reaffiliated) through shared beliefs.

Such values call for social justice, which might mean:

- Better care for children, aging adults, persons with health or mental deficits
- Health maintenance for all persons, all living creatures, and nature
- Healthy neighborhoods, towns, cities, and open, accessible lands
- Efforts to ameliorate poverty, pollutions, and violence
- Meaningful work and volunteer opportunities
- Economic policies that promote jobs and fair taxation
- Saving the climate that surrounds the earth

As a goal, *All for one* is different from humans as apex predators or ravishers of land, sea, and air. Humans who are *all for one* would be cooperative beings who support the lives of others and, even, advocate for them.

HEALTH AND MEDICAL EDUCATION: ALL LEVELS

Various levels of medical education can benefit from Health Humanities and Integrative Medicine.

Pre-med

A recent publication *Health Humanities Baccalaureate Programs in the United States* (2016) by Sarah L. Berry, Erin Gentry Lamb, and Therese Jones discusses "Longitudinal Benefits of Health Humanities Education." The authors review studies published in *Academic Medicine* and describe four areas of achievement. Undergraduate students with Health Humanities courses do as well on MCAT as premedical science students; they do as well or better in medical school or residency; they are more likely to earn academic honors and do clinical research. They are well suited to fields of primary care, pediatrics, and psychiatry, fields involving whole-person care. They have enhanced interpersonal skills of empathy and communication. They are likely to assume roles of service and leadership. Such benefits, we may assume, will continue throughout their professional lives. A chart shows rapid growth in these programs, some 60 now, in 26 states. (An update the next year listed 70 programs.)

Luckily for me, two such programs are new in North Carolina, one at UNC-Chapel, the other at Duke. Both use the acronym "HHIVE," standing for Health and Humanities: an Interdisciplinary Venue for Exploration, as if the participants are vital and productive bees. At UNC there's a student run *Health Humanities Journal* on line at hhj.web.unc.edu.

Med schools

Graduate schools for professional training routinely have challenges covering every area and all new knowledge. As a result, curriculum space is always tight. Nevertheless, Integrative Medicine and Health Humanities can be a pervasive influence in many if not all courses.

At some schools there are already changes as standard medicine becomes more flexible. Young doctors have more training in complementary medicine, more awareness of stress as a factor in illness, and, typically, more good care of themselves. There is more research now about efficacy of massage, Qigong, and other modalities. And there is "market demand": people are more open with their doctors about other modalities and how they can complement standard care.

SOCIAL VALUES AND CHANGE: "THE CUSTOMER IS ALWAYS RIGHT" AND "THE PATIENT IS ALWAYS WELL"

"The customer is always right"; values from the hospitality industry

Good waiters have heard the motto, "The customer is always right," suggesting that any customer complaint or question should be respectfully heard and appropriately addressed. I was trained this way when I served at a faculty club. We didn't work for tips because there were none, but the explicit values were enough to help us give good service. Similarly, all healthcare personnel should serve every single patient or family member with good care, perhaps guided by hospital values that suggest "the patient is always well"…and also a fellow human being. When patients are treated as fellow humans, they feel connections to other humans, not isolation or rejection.

Whatever the level of illness of injury, patients always have some level of wellness.

Even the actively dying patient has body systems that know what to do.

For many years hospital food was second rate, but there have been changes for some hospitals. The UNC Hospitals provide patients with menus for full service 13 hours of the day and a late night menu for the rest of the time. Clinical Dietitians are available, doctor guidelines are followed for special needs, and guest trays at a reasonable price can be ordered so that visitors can eat with patients in their rooms.

"If you see a piece of paper, pick it up, just like you would at home"

We recall these words from the housekeeping leader at UNC Hospitals. Similarly, "If you see a stranger, welcome him or her—whether patient, family member, a new nurse or doctor or other staff member." Every medical encounter should start with a human-to-human interchange, a greeting, a "Can I help you find something?" Whether a parking attendant, a receptionist, a tech, a nurse, or a doctor—anyone can suggest a social relationship of equality. "Where's your home?" "How was your drive?" "Traffic and parking OK?" Such phrases symbolize equality and caring intent. When my group went for hospital massage training at the Oregon Health & Science University, we were required to watch several videos beforehand on the Web. One was about dealing with patients coming to a facility; it included greeting them in a friendly way and offering help with directions. They, and their families, are in an alien place and need welcoming acceptance and, often, directions.

Also, advocating for patients, especially if they are unable to do so.

ESSAY 9: REVIEWING AND TEMPERING ALL RANTS; LIVING TOGETHER IN HARMONY

Let us review the Rants of this book, including 10 (in Section 10 below). There are three groupings: our losses, outworn social values, and our possible gains.

Losses

Tragic stories, plays, and poems all center on separation and loss. Antigone,

Lear, and Anna Karenina come to mind. Losses can be personal, social, or across the world.

Rant 2 expressed the difficult emotions a patient often faces with a serious illness.

Rant 5 expressed sorrow and guilt that we are killing the earth and, it appears, that God allows it.

Rant 7 imagined a trial in the future; the plaintiffs testify about their losses due to this generation's failures to protect the earth.

On a personal level, We need to continue evaluating our attitudes about death and futile medical care. Each one of us will die. How may that happen supported by the best of care?

On a social and global level, I think we need to look hard at the evidence of rising temperatures and sea levels, increases in hurricanes and wildfires, losses of habitats, fish stocks, and agriculture, migrations of people who have lost food production, and so on.

The Carbon Curmudgeons and other deniers of Climate Change want to stick to economics and technologies of the 19th century that favored the Robber Barons and assumed endless resources.

Damage to our climate has already occurred. How far it goes, will depend on what responses we make.

In any case, we need to accept and manage both our personal mortality and the vulnerability of the human race to many losses, including the possible end of humans' time on this earth.

Outworn cultural values

Our behaviors and activities are grounded in our beliefs, conscious or unconscious.

In large and complex societies there are many strands of values, but some of them are dominant.

Rant 1 complained about "Yankee malaise," the result of a society stressed by individualism, Neodarwinism, and loneliness. It saw competition and tribalism as causes for separation and stratification by wealth. It urged cooperation, shared goals, and massage—particularly for a touch-aversive society.

Rant 4, more dramatic, spoke of disharmonies and prisons of modern life, strictures we live with and act out daily, with little awareness. We get high on high stress; we strive for perfection and exceptionalism; we admire the wealthy and focus on building personal net worth. Not only are we losing contact with nature, but also we are using up resources of nature, which we largely see as "raw material." Our mental lives are often dominated by calculation, survival, competing, and strategizing, with losses of joy, mystery, awe, or the holy. Many we live unhealthy lives and favor rescue medicine over health promotion. Qigong—and many other practices and traditions—allows for rest, renewal, and dynamic harmony.

Sources of hope

Even if our earth is compromised, there are avenues for evolving our values and acting on them.

Rant 3 called for massage in all hospitals.

Rant 6 praised small actions, often local and domestic, by homebodies.

Rant 8 called for wider uses of energy healing in hospitals and elsewhere.

Rant 9 urged that we think more often of common ground, social welfare, on the scales of neighborhoods, countries, and the entire world. A suitable motto is "All for one, one for all."

Rant 10 will suggest ways we can, personally and together, use language to define and affirm values that are clarifying and supportive.

Is One-ness problematic?

Yes. While we are alike physically in many, many ways, we are different in many wonderful ways. Strict unity should not (and cannot) be an absolute goal. There are cultural differences, subcultural differences, and a magnificent range of personal differences. Nonetheless, we can share goals of health for all of us and our earth.

What are the ranges and of the universals matter, mind, and energy?

The range of each is, of course, universal, and the combinations myriad as they convert back and forth in dynamic harmony. Because our bodies and minds share in all three, we can consider the earth, the solar system, our galaxy, and the universe as our home of many levels. Further, these realms operate by known laws, have beauty, and provide mysteries for us to continue to study.

There is a special challenge for us to free and expand our minds. The rich resources of reason, analysis, and strategy do a lot for us, but they can contribute to the monkey mind. We need, as well, times of no thought and the rest that provides an open field for new insights and ideas. We can use fantasy, imagination, awe and wonder, instinct and intuition to seek the common good. If millions—even billions—of people join in caring intent, good things may be possible for us and for the earth.

Harmonious living, Yin and Yang

Montaigne, following his horse collision, focused on life, not death (Bakewell, pp. 12-22).

For all humans: peace, not war and shared resources, not modern versions of colonialization, economic devastation, and other threats.

For all living creatures and all resources of earth: respect and careful use with future generations in mind.

The "two-fish in a circle" emblem of Yin and Yang illustrates the dynamic harmony of major forces in our lives and across the earth. A fallen tree nourishes the base of the food chain. Floods bring nutrients across a landscape. There are mysteries of death and rebirth throughout the universe.

Resonances from ancient India, Paul's letter to the Romans, and Gregorian chant applicable to healthcare and all human life

The ancient Hindu greeting, still used today, is "Namaste," translated as "I bow to the divinity in you," and often understood as "The divinity in me acknowledges the divinity in you." Hands are joined as prayer in front of the heart, and the greeter makes a slight bow.

Paul advises the early church in Rome: "Rejoice with those who rejoice, weep with those who weep. Live in harmony with one another.... If possible, so far as it depends on you, live peaceably with all" (Rom. 12.15, 16, 18).

"Ubi Caritas" was sung as a Gregorian chant from between the 4th and the 12th centuries. It is still popular today and set by many composers. There are three short stanzas and the first line for each is "Ubi caritas et amor, Deus ibi est," usually translated, "Where charity and love are, God is present."

All these celebrate the harmonies of people supporting people in larger contexts.

SOURCES 9

Bakewell, Sarah. *How to Live, or: A Life of Montaigne*. New York: Other Press, 2010.

Berry, Sarah L., Erin Gentry Lamb, and Therese Jones. *Health Humanities Baccalaureate Programs in the United States*. Hiram, OH: Center for Literature and Medicine, Hiram College. December 2016. See http://www.hiram.edu/wpcontent/uploads/2016/11/HHBP_12_2_16.pdf,   accessed May 20, 2017.

Charon, Rita, Sayantani DasGupta, Nellie Hermann, Craig Irvine, Eric R. Marcus, Edgar Rivera Colón, Danielle Spencer, and Maura Spiegel. *The Principles and Practice of Narrative Medicine*. New York: Oxford Univ. Press, 2017.

Crawford, Paul, Brian Brown, Charley Baker, Victoria Tischler, and Brian Abrams. *Health Humanities*. New York: Palgrave Macmillan, 2015.

Croker, Liz, and Bev Johnson. *Privileged Presence: Personal Stories of Connections in Health Care*, 2nd ed. Boulder, Colorado: Bull Publishing Co., 2014.

Johnson, B. H., and Abraham, M. R. *Partnering with Patients, Residents, and Families: A Resource for Leaders of Hospitals, Ambulatory Care Settings, and Long-Term Care Communities*. Bethesda, Maryland: Institute for Patient- and Family-Centered Care. 2012.

Jones, Therese, Delese Wear, and Lester D. Friedman, eds. *Health Humanities Reader*. New Brunswick, NJ: Rutgers Univ. Press, 2014.

Oschman, James L. *Energy Medicine: The Scientific Basis*. Edinburgh: Churchill Livingston, 2000.

_____. *Energy Medicine in Therapeutics and Human Performance*. Edinburgh: Butterworth Heinemann, 2003.

Pearson, Rachel. *No Apparent Distress: A Doctor's Coming-of-Age on the Front Lines of American Medicine*. New York: W. W. Norton & Company, 2017.

Weil, Andrew, M.D. *Spontaneous Healing: How to Discover and Embrace Your Body's Natural Ability to Maintain and Heal Itself*. New York: Ballantine Books, 1996.

X   CONCLUDING ESSAY AND RANT: SINGING THE BODY ELECTRIC, THE MIND ELECTRIC, AND ALL ENERGIES

MORTALITY: BOTH A CURSE AND A BLESSING

Late in life, Dorothy (my mother-in-law) had lost her husband to Parkinson's. Her own health was declining. Nonetheless, she would remark, "Every day is a good day." Nancy and I took this as valuable wisdom about gratitude. Similarly, my mother Marjorie was a widow; at 92 she had many health limitations, including the unraveling of her mind. One day at her health center, we had a lunch for her with close relatives gathered around. The only word she was able to speak was "rainbow." She said this repeatedly and with a smile. We took this as her symbol for connection and gratitude. Both women, despite their difficulties, expressed their intent to share gratitude, well-being, and love with family and friends. While medicine could not cure them, medical and family caregivers could provide loving care.

The length of a given person's life—and all the conditions—is not known. The psalmist wondered centuries ago, what might be *the number of his [her] days* (Ps. 39.4). The length of the entirety of humanity's life—and of our imperiled earth—is also not known.

If we consider (1) that humans are continuous with the matter, energy, and mind of the universe, and (2) that nature, both external and internal, has dynamic balance, there is cause for not only gratitude and hope, but also celebration. We have this earth, this galaxy, this universe as home. It is orderly and can take care of us. Our current knowledge and behaviors are not, however, orderly or congruent with our earthly home, and we have, unfortunately, caused considerable damage to the earth and to many creatures (some now extinct) and people.

On the other hand, we can contribute our minds, wills, and actions to the earth's maintenance; we are aware of its deep history and the current threats to its health. We know some steps we may take to mitigate Climate

Change. We need to seek other steps as well, especially in dealing with energy sources and usages as well as wise governmental policies.

We know that each of our individual lives will end (gradually or suddenly). This knowledge can lead to depression, despair and inaction, or it can motivate us to focus on issues important to us and to commit ourselves to action. Actions can be directly physical: local...national, political, etc., but they can also be mental by intent. One of my ministers spoke of an elderly church member who had formerly been very active in the church and the wider world. Later in life, she (or he—I never knew the gender) could no longer get out or do much of anything. Nonetheless in her apartment she went through the church directory page by page and prayed for each person and each family.

WALT WHITMAN'S "I SING THE BODY ELECTRIC"

"I sing the body electric," wrote Whitman, the first line of his long poem with that title and theme. Joanne Trautmann Banks used this line for the title of a course at the Pennsylvania State College of Medicine. In 1972 she was the first appointment of a literature specialist at an American university medical school. Banks was one of the founders of literature and medicine as an academic field and a mentor to me.

In this poem, Whitman praises many of the material aspects of men and women. He praises their power, beauty, and generativity. There is erotic energy between them, and the bodies, he concludes, are not just material parts, but they embody, demonstrate, and give witness to "the soul." The souls in each one of us are also reflected in Nature (with a capital N). In section 7 and 8, he describes and mourns the bodies of men and women at slave auctions, a corruption of the sacred. We can see a parallel with the current servitude of impoverished people and despoilment of our earth.

The poem concludes in section 9 with a catalogue naming and celebrating parts of the body. The final lines run:

> The beauty of the waist, and thence of the hips, and thence downwards toward the knees,
> The thin red jellies within you or within me, the bones and the marrow in our bones,

The exquisite realization of health;

O I say these are not the parts and poems of the body only, but of the soul,

O I say now these are the soul.

In his vision, body parts, a person's soul, and the universal Oversoul (from Emerson) all merge.

The adjective "electric" in English was new in Whitman's day, and "the body electric" an unusual, perhaps startling, phrase. The term "electricity" dates from the 18th century and draws on the Greek and Roman words for "amber"; the ancients knew that rubbing amber could create static electricity.

We tend to see the *body material*; can we also see and celebrate the body electric?

WHATEVER OUR BELIEFS, WE CAN ECHO WHITMAN

We may consider Whitman's affirmation in two ways.

First, we can celebrate the magnificent bodies we live in, its physical materials, the electricity (and other energies), and the mind that allows us to sing in praise using our lungs and vocal apparatus.

Second, we can imagine a sentence similar to Whitman's that represents our hopes for health. This experiment explores ways humans can situate themselves within the nested spheres discussed earlier and also cultivate positive and comforting emotions of belonging to our world, our galaxy, even the universe.

A healthy motto (an experiment in tagmemics)

Readers may recall "sentence patterns" in language arts classes; students were given a structure and asked to fill in words. My peers enjoyed nonsensical (but technical correct) assemblages, such as "Chocolate dogs smile sternly the soccer ball." Darn right: noun phrase, verb phrase, direct object phrase—*so there, teacher!*

Tagmemics is a branch of linguistics that studies such places in a sentence or "slots" for nouns or verbs (or phrases thereof), modifiers, etc.; the possible words or phrases are called the "filler class."

	1	2	3
Whitman's words:	I	sing	the body electric

	1	2	3
With new fillers:	We	affirm	the ultimate's love

Figure 15: A Healthy Motto (Whitman Reformulated)

For each slot we can create a filler class of related words for any person to choose from.

1. We: all humans, creatures, Leopold's "biotic community," as well as inanimate objects; obviously matter, but with conversions to matter, energy, and mind.

2. affirm: live within and celebrate vital being. If we convert entirely to energy and mind at death (body parts left behind before later conversions), we continue to dwell, live, exist.

3. the ultimate's: pertaining to God/Allah/Jehovah/Creator/any and all deities such as Mother Nature, Gaia, etc., whatever the comforting, spiritual "surround," even rational or humanistic beliefs and values. The universal sense of matter, energy, and mind may be considered an ultimate order.

love: care, concern, coherence, dynamic harmony, great balance (Tai Chi), ultimate worth.

The ultimate's love: The universe protects and shelters us; we are at home here, embraced. The universe is permeated with positive mind and energy to match the matter that makes up all heavenly bodies and everything on earth including humans. Christianity often uses the phrase "underneath are the everlasting arms" at funerals (from Deuteronomy 33.27).

Perhaps many people already subscribe to something similar to Whitman's affirmation, consciously or unconsciously.

Or perhaps they believe (and live out) the opposite:

1	2	3
I	suffer	from random chaos

1. I: poor loner me, separated from others by individualism, pride, fear of connection.

2. suffer: feel pain, loneliness, guilt, insufficiency, neglect, separation.

3. from random chaos: there is no ultimate order in nature or in myself; anything can happen next: I feel uncertainty, dread. The universe is running down. The earth is doomed.

We have wide choice for the affirmations we choose and live by.

CHOSING WORDS TO GUIDE OUR MIND, ENERGIES, AND BODIES

Affirmations: a uniquely human gift

Humans enjoy verbal affirmations in many forms, slogans, mottoes, mantras, catch-phrases, prayers, proverbs, and more. In repeating them, our minds affirm their values and join with countless other people who have said or written them. In speaking them, we embody their meaning with our breathing and with our material vocal apparatus vibrating with energy. Some are well known, as in "Two wrongs don't make a right" or "Better late than never." Some are personal, as in "I can make it through one more day," but probably also universal because everyone has felt that way at one time or another. Some are used in therapy. My massage therapist asked me to talk to my tight muscles. In this book I suggest to all readers to talk to their own bodies and to the health of the world, its leaders, and all of humanity because I believe in the power of mind working with matter, other (and all minds), and energy.

Doing it yourself...or together

People often post affirmations they like by the bathroom mirror, in a day-book, in a cell phone, or on a computer—anywhere for easy access. Some people use them as a morning prayer, at meals, or before sleep...or upon waking in the dark of night, when worries are readily available.

I think these have more power than Pascal's wager that there might be a God. Our affirmations can affirm our values and links to all people, all creatures, all matter and energy, the universe as our home.

Not claiming words?

If we don't claim our words, other words will operate on subconscious levels because we know them—and live them—from dominant cultures. It's better to choose words that nourish us.

Affirmations based on universality of matter, mind, and energy.

Here are some affirmations based on the proposed universality of matter, mind and energy. In them we affirm the body local and the body universal.

My matter and universe's matter are one, the same, and healthy.
My energy and the universe's energy are one, the same, and healthy.
My mind and the universe's mind are one, the same, and healthy.

I am at home in my home
I am at home in my neighborhood
I am at home with friends
I am at home in my country
I am at home in the world
I am at home in the universe

Matter becomes energy
Energy becomes mind
Mind become matter
Matter becomes mind
Mind become energy
Energy becomes matter

Matter becomes energy becomes mind becomes matter becomes mind becomes energy....

Let us affirm that the health of humans, all of nature, and the earth is crucially important and worthy of our best efforts to protect and maintain it.

SINGING ONE WAY OR ANOTHER

We sing many ways: humming, crooning a lullaby, chanting, singing a spiritual, an aria, a popular song, or a sports cheer. We sing in groups, in our cars, or in the shower where tile walls reflect and magnify the voice. One way or another we sing our beliefs out loud or silently. I think it is good to claim them in words, even sung words.

"But I can't carry a tune in a bucket!"

No matter…a single tone, or even a wandering one has its own music.

Further, song can be in the mind only: "Heard melodies are sweet, but those unheard / Are sweeter," wrote Keats ("Ode to a Grecian Urn").

Maybe the "song" is tactile (a cross, a mezuzah, a rosary) or visual (an icon, an altar, a shrine).

What's important is to find links to the values that nourish and bodies, minds, and spirits.

Where to sing? When? What?

As you choose: anywhere, any time, anything. We make our song our own according to our needs, gifts, whims, and choices.

Even in the face of calamity, we can sing

Whitman saw horrors in the Civil War.

The band on the sinking Titanic continued to play.

Our loved ones will become injured, ill; they may become estranged to us, a kind of social death; they will lie on their deathbeds.

We may sing to them. We may sing dirges after they die. We may sing praises for their lives, past, present, and future.

If necessary, we can sing dirges for the earth.

We can also sing—in whatever voice, ritual, or medium—praise of life, surrounding nature, this earth, the universe, the universal continuities of matter, mind, and energies, all of which make up our various homes.

FAREWELL: GO, LITEL BOOK

Chaucer's envoi (send-off)

Given that Chaucer's epic of *Troilus and Criseyde* is over 8,000 lines, there is charming authorial modesty in his phrase "litel book" in his envoi at the end of it. I borrow some of his words for my farewell to my/our "litel book":

Go, litel book	Go, little book
…	…
And red wher-so thou be or elles songe	And wherever thou art read or else sung,
That thow be understonde I god beseche!	I beseech God that thou be understood!

SOURCES 10

Chaucer, Geoffrey. *Troilus and Criseyde*. Multiple editions available.

Keats, John. "Ode to a Grecian Urn." Numerous editions available, even online.

Whitman, Walt. *Leaves of Grass* (1855). Many editions, also online.

ACKNOWLEDGMENTS

I acknowledge the help and inspiration provided by scholars past and present in the worlds of literature and medicine, Medical Humanities, and Health Humanities: Felice Auel, Joanne Trautmann Banks, Jay Baruch, Rita Charon, James F. Childress, Jack Coulehan, Carol Donley, Rebecca Garden, Anne Hudson Jones, Therese Jones, Martin Kohn, Laurence B. McCullough, Kathryn Montgomery, Lois Nixon, Warren Thomas Reich, John Stone, and Mahala Yates Stripling, and Delese Wear. I also thank Lucy Bruell, editor-in-chief, New York University Database for Literature, Art, and Medicine, a very helpful resource.

I thank the Cornucopia Cancer Center of Durham for my time with them, providing massage and Qigong for patients and their family members, part-time, for ten years. Directors during my time included Becky Carver, Maxine Turner-Fitts, and Mary Lawrence. Other faithful people there were Mindy Gellin, Rosie Smith, Will Pulley, and Olivia Stancil.

I thank my colleagues in the Patient and Family Resource Center at the North Carolina Hospital, Tina Shaban, the coordinator, Stephanie Nussbaum, my fellow massage therapist, also Pam Baker and Jennifer Hanspal, as well as Don Rosenstein, director of the UNC Comprehensive Cancer Program that includes the Resource Center. I provided massage for some 4,000 patients or family members during my time there; I am grateful to all of them.

I thank the successive chairs of the Social Medicine Department, School of Medicine, University of North Carolina-Chapel Hill, where I am adjunct professor, Gail E. Henderson and Jonathan Oberlander. Also Barry F. Saunders, Jane F. Thrailkill, and others involved with the Health Humanities program there.

I also thank Jeffrey P. Baker, executive director, Raymond C. Barfield, and Nikki Vangsnes of the Trent Institute for Bioethics, Humanities, & History of Medicine, Duke University & School of Medicine.

I thank my teachers at The Humanities Center Institute of Allied Health/School of Massage, Pinellas Park, Fla., where I studied massage, and Gayle MacDonald, who taught cancer massage at the Oregon Health & Science University.

I thank my teacher Lisa B. O'Shea, founder and director, the Qigong

Institute of Rochester (N.Y.) with whom I trained. While members of my Qigong study group in Chapel Hill changed over the years, I thank Myra Emerson, Pamela Hayward, Deb Hlavaty, Cindy Johnson, Kate Vosecky, and Richard Wilson.

I think medical librarians at UNC and Duke, also librarians at the Chapel Hill Public Library. I thank neighbor Christopher Carr for retrieving an online article for me.

I thank Ian MacNeil for the cover art and help with figures.

For answering technical questions, I thank Jonathan Engel, Laurie McNeil, Ken Turner, and Laura Hansen.

I thank various readers of portions of the book, Stephanie Nussbaum, Tina Shaban, Andy Peterson, and especially Kate Vosecky, and Nancy Corson Carter.

For friendship, inspiration, and support I thank Janet Boudreau, Carol Ann Carter, Courtney and Jerry Chavez, Laura Herbst and Joe Baysdon, Ken Jens and Sandy Milroy Jens, Michael Jokich and Elizabeth Welton, Charlie and Nancy Zimmerli, Carl and Janet Edwards, Al and Janet Rabil, Jerry and Betty Eidenier, Tom and Nancy Trueblood, Peter and Jeanne Meinke, Rich and Mimi Rice, Tom and Marian Price, Carolyn and Lloyd Horton, Bill and Diane Savage, Bob and Mary Hetzel, Steve Stambaugh, George Meese, Colleen Adomaitis, Nikole Weir, Cathy A. Wagner and staff, and the Bozo Baritones of the Choral Society of Durham as well as the director Rodney A. Wynkoop. Also members of the Church of Reconciliation, Chapel Hill, and Pastor Mark and Allison Davidson.

Also physicians who have cared for me: Richard Oldenski, Andy Peterson, Eugene Orringer (now deceased), and Arthur Axelbank.

I thank my closest family, Rebecca Alice-Carter Rincon and her husband Gustavo Rincon, also my wife, Nancy Corson Carter, all helpful and supportive in every step of a long, long adventure.

BIBLIOGRAPHY

Books, articles, films

Ardell, Donald B. *High Level Wellness*. Rev. ed. New York: Random: 1979.

Arsan, Emmanuelle. *Emmanuelle*, trans. Lowell Bair. New York: Grove Press, 1971.

Bakewell, Sarah. *How to Live, or: A Life of Montaigne*. New York: Other Press, 2010.

Barea, Christina J. *Qigong Illustrated*. Champaign, Ill.: Human Kinetics, 2011.

"Before the Flood." Directed by Fisher Stevens, distributed by National Geographic Society, 2016.

Beinfield, Harriet and Efrem Korngold, *Between Heaven and Earth: A Guide to Chinese Medicine*. New York: Ballantine Books, 1991.

Bender, Tom. "The Physics of Qi." tombender.org/energeticsarticles/ qi_physics.pdf. accessed March 9, 2016. This is a "brief summary" of his DVD "The Physics of Qi" (2006).

Bernstein Jerome S. *Living in the Borderland: The Evolution of Consciousness and the Challenge of Healing Trauma*. London and New York: Routledge, 2005.

Biel, Andrew. *Trail Guide to the Body*, 2nd ed. Boulder: Books of Discovery, 1997.

Blakney, R. B., trans. See Lao-Tse, *The Way of Life* [*The Tao Te Ching*] below.

Bloomberg, Michael and Carl Pope. *Climate of Hope: How Cities, Businesses, and Citizens Can Save the Planet*. New York: St. Martin's Press, 2017.

Bock-Möbius, Imke. *Qigong Meets Quantum Physics: Experience Cosmic Oneness*. St. Petersburg FL: Three Pines Press, 2010. Published in German also in 2010.

Book of Changes (The I Ching). Many editions available; also online.

Bohm, David. *Wholeness and the Implicate Order*. London: Routledge & Kegan Paul, 1980.

Brennan, Barbara Ann. *Hands of Light: A Guide to Healing Through the Human Energy Field*. New York: Bantam Books, 1987.

Braun, Mary Beth, and Stephanie J. Simonson. *Introduction to Massage Therapy*, 3rd ed. Philadelphia: Lippincott Williams & Wilkins, 2013.

Campbell, Joseph. *The Hero with a Thousand Faces*. Princeton: Princeton Univ. Press, 1968.

Camus, Albert. "The Myth of Sisyphus" (1942). Available in various editions and on the Web.

Cannon, Walter. *The Wisdom of the Body*, 1932. Various editions available.

Capra, Fritjof. *The Tao of Physics: An Explanation of the Parallels Between modern Physics and Eastern Mysticism*. New York: Bantam Books, 1976.

Carter, III, Albert Howard. *Clowns and Jokers Can Heal Us: Comedy and Medicine*. San Francisco: Univ. of California Medical Humanities Press, 2011.

_____. *First Cut: A Season in the Human Anatomy Lab*. New York: Picador, 1997.

_____. *Our Human Hearts: A Medical and Cultural Journey*. Kent, Ohio: Kent State Univ. Press, 2006.

_____ and Jane Arbuckle Petro. *Rising from the Flames: The Experience of the Severely Burned*. Philadelphia: Univ. of Pennsylvania Press, 1998.

_____. "Self-Help" in *Encyclopedia of Bioethics*, rev. ed., ed. Warren Thomas Reich. New York: Simon & Schuster Macmillan, 1995. Vol. 5, pp. 2338-44.

Cassell, Eric J. *The Healer's Art*. Cambridge, Massachusetts: MIT Press.

Charon, Rita, Sayantani DasGupta, Nellie Hermann, Craig Irvine, Eric R. Marcus, Edgar Rivera Colón, Danielle Spencer, and Maura Spiegel. *The Principles and Practice of Narrative Medicine*. New York: Oxford Univ. Press, 2017.

Chödrön, Pema, *The Places That Scare You: A Guide to Fearlessness in Difficult Times*. Boston and London: Shambhala Publications, 2001.

Cohen, Kenneth S. *The Way of Qigong: The Art and Science of Chinese Energy Healing*. New York: Ballantine, 1997.

Cohen, S., Janicki-Deverts, D., Turner, R. B., & Doyle, W. J. (2015). "Does hugging provide stress-buffering social support? A study of susceptibility to upper respiratory infection and illness" *Psychological Science*, 26(2), 135-147.

Colón, Edgar Rivera, "From Fire Escapes to Qualitative Data: Pedagogical Urging, Embodied Research, and Narrative Medicine's Ear of the Heart," in Rita Charon, et al., *The Principles and Practice of Narrative Medicine*. New York: Oxford Univ. Press, 2017. Chapter 11, pp. 251-267.

Connor, Melinda H. *See Auras*. Marana, AR: EarthSongs Holistic Health, LLC, 2009.

"Cosmic Voyage." Directed by Eva Szasz. 1968.

Crawford, Paul, Brian Brown, Charley Baker, Victoria Tischler, and Brian Abrams. *Health Humanities*. New York: Palgrave Macmillan, 2015.

Croker, Liz, and Bev Johnson. *Privileged Presence: Personal Stories of Connections in Health Care*, 2nd ed. Boulder, Colorado: Bull Publishing Co., 2014.

Dante Alighieri, *La Divina Commedia*, Vol. III Paradiso, ed. Natalino Sapegno. Firenze: La Nuova Italia Editrice, 1958, 1968.

_____. *The Divine Comedy: Paradise in The Portable Dante*, ed. and trans. Mark Musa. New York: Penguin, 1995.

Downing, George. *The Massage Book*. New York: Random House/The Bookworks, 1972.

Dunn, Tedi, and Marian Williams. *Massage Therapy Guidelines for Hospital & Home Care*, 4th ed. Olympia: Information for People, 2001.

Eden, Barbara. *Energy Medicine: Balancing Your Body's Energies for Optimal Health, Joy and Vitality*, 2nd ed. New York: Jeremy Tarcher/Penguin, 2008.

Eddington, Arthur. *The Nature of the Physical World*. New York: Macmillan Co.: 1928.

Eisenberg, David. *Encounters with Qi: Exploring Chinese Medicine*, rev. ed. New York: W.W. Norton, 1995.

_____. Ronald C. Kessler, Cindy Foster, Frances E. Norlock, David R. Calkins, and Thomas L. Delbanco. "Unconventional Medicine in the United States—Prevalence, Costs, and Patterns of Use." *New England Journal of Medicine* 28 January 1993. 328:246-252. http://www.nejm.org/toc/nejm/328/4/DOI: 10.1056/NEJM199301283280406

Field, Tiffany. *Complementary and Alternative Therapies Research*. Washington D.C.: American Psychological Association, 2009.

_____. *Massage Therapy Research*. Edinburgh: Churchill Livingstone, 2006.

_____. *Touch*. Cambridge, Mass.: MIT Press, 2003.

_____. *Touch Therapy*. Edinburgh: Churchill Livingstone: 2000.

Frame, Donald. See Montaigne below.

Frampton, Susan B., Laura Gilpin, and Patrick A. Charmel, eds. *Putting Patients First: Designing and Practicing Patient-Centered Care*. San Francisco: Jossey-Bass, 2003.

Francis, Pope, *Laudato Sí: On Care for Our Common Home*. Huntington IN: Our

Sunday Visitor, Inc., 2015.

Frank, Arthur W. *The Wounded Storyteller: Body, Illness, and Ethics.* Chicago: Univ. of Chicago Press, 1995.

Freeman, Lyn. *Mosby's Complementary and Alternative Medicine: A Research-Based Approach,* 3rd ed. St. Louis: Mosby-Elsevier, 2009.

Friedman, Meyer, and Ray H. Rosenman, *Type A Behavior and Your Heart.* New York: Fawcett Crest, 1974.

Furlong, David. *Working with Earth Energies: How to Tap into the Healing Powers of the Natural World.* London: Piatkus, 2003.

Gardner, Martin. *Did Adam and Eve Have Navels?: Debunking Pseudoscience.* New York: W. W. Norton & Co., 2000.

Gawande, Atul. *Being Mortal: Medicine and What Matters in the End.* New York: Metropolitan Books/Henry Holt and Company, 2014.

Gore, Al. *An Inconvenient Truth: The Planetary Emergency of Global Warming and What We Can Do About It.* New York: Rodale, 2006.

Harrington, Anne. *The Cure Within: A History of Mind-Body Medicine.* New York: W. W. Norton, 2008.

Hoekema, David. "African Politics and Moral Vision," *Soundings* 96:2 (Spring 2013), 121-144.

_____. "Faith and Freedom in Post-Colonial African Politics," in *Jesus and Ubuntu: Exploring the Social Impact of Christianity in Africa,* ed. Mwenda Ntarangwi (Trenton, NJ: Africa World Press, 2011), 26-45.

_____. "African Personhood: Morality and identity in the 'Bush of Ghosts.'" *Soundings*   91:3-4 (Fall/Winter 2008).

Hougham, Paul. *The Atlas of Mind Body and Spirit.* London: Gaia, Octopus Publishing Group, 2006.

"The Hunger Games." Directed by Francis Lawrence. 2012.

I Ching or Book of Changes. Numerous editions available. There's an online version.

Jahnke, Roger, Linda Larkey, et al. A Comprehensive Review of Health Benefits of Qigong and Tai Chi. *American Journal of Health Promotion* Jul-Aug 2010; 24 (6): e1-325. Accessed March 12, 2016.

_____. *The Healing Promise of Qi: Creating Extraordinary Wellness Through Qigong and Tai Chi.* New York: McGraw Hill, 2002.

Johnson, B. H., and Abraham, M. R. *Partnering with Patients, Residents, and Families: A Resource for Leaders of Hospitals, Ambulatory Care Settings, and*

Long-Term Care Communities. Bethesda, Maryland: Institute for Patient- and Family-Centered Care. 2012.

Jones, Therese, Delese Wear, and Lester D. Friedman, eds. *Health Humanities Reader*. New Brunswick, NJ: Rutgers Univ. Press, 2014.

Juhan, Deane. *Job's Body: A Handbook for Bodywork*. Barrytown NY: Station Hill Press, 1987.

Justman, Stewart, "Montaigne on Medicine: Insights of a 16th-Century Skeptic." *Perspectives in Biology and Medicine* 58.4 (Autumn 2015): 493-506.

Kelly, Thomas R. *A Testament of Devotion*. New York: HarperCollins, 1992.

Kesey, Ken. *One Flew Over The Cuckoo's Nest*. New York: Signet,1962.

Klein, Naomi. *This Changes Everything: Capitalism vs. the Climate*. New York: Simon & Schuster, 2014.

Koenig, Harold G. *The Healing Power of Faith: How Belief and Prayer Can Help You Triumph over Disease*. New York: Simon & Schuster/Touchstone: 1999

Kolbert, Elizabeth. *The Sixth Extinction: An Unnatural History*. New York: Henry Holt, 2014.

Kübler-Ross, Elisabeth. *On Death and Dying*. New York: Macmillian, 1969.

Kuhn, Thomas S. *The Structure of Scientific Revolutions*, 3rd ed. Chicago: Univ. of Chicago Press, 1966.

Lao-Tse. *The Way of Life* [*The Tao Te Ching*], trans. R. B. Blakney. New York: Signet, 1955.

Linden, David J. *Touch: The Science of Hand, Heart, and Mind*. New York: Viking, 2015

Lown, Bernard. *The Lost Art of Healing: Practicing Compassion in Medicine*. New York: Ballantine, 1999.

Liu, Tianjun. *Chinese Medical Qigong*, eds. Tianjun Liu, and Kevin W. Chen, trans. Tianjun Liu, 3rd ed. London: Singing Dragon Press, 2010.

MacDonald, Gayle. *Medicine Hands: Massage Therapy for People with Cancer*. Forres, Scotland: Findhorn Press, 1999.

_____. *Massage for the Hospital Patient and Medically Frail Client*. Philadelphia: Lippincott Williams & Williams, 2004.

Maciocia, Giovanni. *The Channels of Acupuncture*. Philadelphia: Churchill Livingstone, 2006.

Merculieff, Larry ((Ilarion) and Libby Roderick. *Stop Talking: Indigenous*

Ways of Teaching and Learning and Difficult Dialogues in Higher Education.
Anchorage: The Univ. of Alaska-Anchorage, 2013.

Metzger, Deena. *The Woman Who Slept with Men to Take the War out of Them & Tree,* Culver City, Calif.: Peace Press, 1981.

Montagu, Ashley. *Touching: The Human Significance of the Skin,* 2nd ed. New York: Harper and Row, 1958.

Montaigne, Michel Eyquem. *Essays and Selected Writings; a Biligual Edition.*
Trans. and ed. Donald M. Frame. New York: St. Martin's Press, 1963.

_____. "Apology for Raymond Sebond." In Frame, pp. 198-253.

_____. "Of Experience." In Frame, pp. 401-461.

_____. "Of Practice." In Frame, pp. 163-195.

Montgomery, Sy. *The Soul of an Octopus: A Surprising Exploration in the Wonder of Consciousness.* New York: Atria Paperback, 2015.

Moody, Raymond. *Life after Life: The Investigation of A Phenomenon-survival of Bodily Death,* 2nd. ed. San Francisco: HarperSanFrancisco, 2001.

Mukherjee, Siddhartha. *The Emperor of All Maladies: A Biography of Cancer.*
New York: Scribner, 2010.

Myers, Thomas. *Anatomy Trains: Myofascial Meridians for Manual and Movement Therapists.* London: Churchill Livingston, 2001.

Myss, Caroline. *Anatomy of the Spirit: The Seven Stages of Power and Healing.*
New York: Three Rivers Press, 1996.

Nisbett, Richard E. *The Geography of Thought: How Asians and Westerners Think Differently…and Why.* New York: Free Press, 2003.

Novalis. *Aphorisms.* Available in several editions and online.

O'Brien, Tim. *The Things They Carried.* New York: Houghton Mifflin, 1990.

"One Flew Over the Cuckoo's Nest." Directed by Milos Forman, 1975.

Oschman, James L. *Energy Medicine: The Scientific Basis.* Edinburgh: Churchill Livingstone, 2000.

_____. *Energy Medicine in Therapeutics and Human Performance.* Edinburgh: Butterworth Heinemann, 2003.

Orr, David. *Dangerous Years: Climate Change, the Long Emergency, and the Way Forward.* New Haven: Yale Univ. Press, 2016.

Pearson, Rachel. *No Apparent Distress: A Doctor's Coming-of-Age on the Front Lines of American Medicine.* New York: W. W. Norton & Company, 2017.

Pirsig, Robert M. *Zen and the Art of Motorcycle Maintenance: An Inquiry into*

Values. New York: William Morrow & Company, 1974.

Qi: The Journal of Traditional Eastern Health & Fitness. A quarterly magazine, now in its 26th year, based in Temecula, California. See https://www. qi-journal.com/.

Radin, Dean. *The Conscious Universe: The Scientific Truth of Psychic Phenomena.* New York: HarperCollins, 1997.

Reb, A.M., et al. Qigong in Injured Military Service Members: A Feasibility Study. *Journal of Holistic Nursing* (Mar 27, 2016) pii: 0898010116638159. [Epub ahead of print]

Remen, Rachel Naomi. *My Grandfather's Blessings.* New York: Berkeley, 2001.

Rhoads, C. J. "Mechanism of Pain Relief through Tai Chi and Qigong." *Journal of Pain & Relief,* 2:115 (2013). doi: 10.4172/2167-0846.1000115

Rubik, Beverly. "Measurement of the Human Biofield and Other Energetic Instruments." In Lyn Freeman's book cited above.

Sacks, Oliver. *Musicophilia: Tales of Music and the Brain.* New York: Alfred A. Knopf, 2007.

Samuels, Mike, and Hal Zina Bennett. *The Well Body Book.* New York: Random House/Bookworks, 1973.

_____. *Well Body, Well Earth.* San Francisco: Sierra Club Books: 1983.

Scarry, Elaine. *The Body in Pain: The Making and Unmaking of the World.* New York: Oxford Univ. Press, 1985.

Schumacher, E. F. *Small Is Beautiful: Economics as if People Mattered.* 1973. Rpt. New York: Harper and Row, 2010

Shem, Samuel. *The House of God.* New York: Richard Marek Publishers, 1978.

Sweet, Victoria. *God's Hotel: A Doctor, A Hospital, and a Pilgrimage to the Heart of Medicine.* New York: Riverhead Books, 2012.

Sheldrake, Rupert. *The Sense of Being Stared At and Other Aspects of the Extended Mind.* London: Arrow Books, 2003.

_____. *Dogs that Know When Their Owners Are Coming Home and Other Unexplained Powers of Animals,* rev. ed. New York: Three Rivers Press, 2011.

Steichen, Edward. *The Family of Man.* New York: Ridge Press, 1955.

Stone, John. "He Makes a House Call," *In All This Rain.* Baton Rouge: Louisiana State Univ. Press, 1980, p. 4.

Swift, Jonathan. *Gulliver's Travels* (1726). Many editions available.

Tao Te Ching. Multiple editions available; also online; sometimes spelled Dao De Jing. See Lao-Tse above.

Tattersal, Ian, *The Fossil Trail: How We Know What We Think We Know about Human Evolution.* New York, Oxford: Oxford Univ. Press, 1995.

Taylor, Cory. *Dying: A Memoir.* Portland: Tin House books, 2017.

Tegmark, Max. "Consciousness as a State of Matter." https://arxiv.org/pdf/1401.1219v3.pdf, accessed May 8th, 2015.

_____. *Our Mathematical Universe: My Quest for the Ultimate Nature of Reality.* New York: Vintage Books, 2014.

Thoreau, *Walden* (1854). Many editions available.

Tolstoy, Leo. *The Death of Ivan Ilyich, trans. Richard Pevear and Larissa Volokhonsky.* New York: Vintage, 2009.

The Yellow Emperor's Classic of Internal Medicine, trans. Ilza Veith. Berkeley, Los Angeles, London: Univ. of California Press, 2002). Originally *Huangdi neijing,* also translated as *The Emperor's Inner Canon.*

Walton, Tracy. "Help for Hospital-Based Massage Therapy" See: http://www.massagetoday.com/mpacms/mt/article.php?id=14693, accessed March 5, 2016.

_____. *Medical Conditions and Massage Therapy: A Decision Tree Approach.* Philadelphia: Lippincott Williams & Wilkins, 2004.

Weil, Andrews. *Spontaneous Healing: How to Discover and Enhance Your Body's Natural Ability to Maintain and Heal Itself.* New York: Alfred A. Knopf. 1996.

Whitman, Walt. *Leaves of Grass* (1855), many editions, also online.

Wilson, Edward O. *Biophilia.* Cambridge, London: Harvard Univ. Press, 1984.

_____. *Half-Earth: Our Planet's Fight for Life.* New York: Liveright Pub. Corp., 2016.

Wishart, Adam. *One in Three: A Son's Journey into the History and Science of Cancer.* New York: Grove Press, 2007.

Wohlleben, Peter. *The Hidden Life of Trees: What They Feel, How They Communicate,* trans. Jane Billinghurst. Vancouver: Greystone Books Ltd., 2016.

Yang, Zwing-Ming. *Qigong for Health and Martial Arts,* 2nd ed. Boston, YMAA Pubs., 1998.

Zukav, Gary. *The Dancing Wu Li Masters: An Overview of the New Physics.* New York: Bantam Books, 1979.

Web Sources

Bender, Tom. "The Physics of Qi." tombender.org/energeticsarticles/ qi_physics.pdf. accessed March 9, 2016. This is a "brief summary" of his DVD "The Physics of Qi (2006).

Berry, Sarah L., Erin Gentry Lamb, and Therese Jones. *Health Humanities Baccalaureate Programs in the United States.* Hiram, OH: Center for Literature and Medicine, Hiram College. December 2016. http:// www.hiram.edu/wpcontent/uploads/2016/11/HHBP_12_2_16.pdf, accessed May 20, 2017.

Berry, Wendell. "Health is Membership." https://home2.btconnect.com/ tipiglen/berryhealth.html, accessed August 21, 2017.

Center for Integrative Medicine, University of Arizona,https:// integrativemedicine.arizona.edu/about/definition.html, accessed July 4th, 2016.

Centers for Disease Control and Prevention, https://www.cdc.gov/chronicdisease/about/prevention.htm, accessed Aug. 12, 2017.

Chaucer, Geoffrey. *Troilus and Criseyde.* Multiple editions available and online.

Dreaming, http://www.australia.gov.au/about-australia/australian-story/ dreaming, accessed Aug. 3, 2016.

Eisenberg, David, Ronald C. Kessler, Cindy Foster, Frances E. Norlock, David R. Calkins, and Delbanco, Thomas L. "Unconventional Medicine in the United States—Prevalence, Costs, and Patterns of Use." New England Journal of Medicine 28 January1993.328: 246-252, http://www.nejm.org/toc/nejm/328/4/DOI:10.1056/ NEJM199301283280406

Environmental Protection Agency, "10 warmest years on record worldwide have occurred since 1998," https://www.epa.gov/climate-indicators/weather-climate, accessed January 22, 2019.

Facts about Climate Change and Global Warming. GOOGLE this topic for Websites by NASA, The National Wildlife Federation, Conserve Energy Future, Friends of Science, National Geographic, The New York Times, BBC News, and others.

Fields, https://en.wikipedia.org/wiki/Field_(physics), accessed Sept. 14, 2016.

Future risks to the earth, https://en.wikipedia.org/wiki/Future_of_the_ Earth, accessed July 30, 2016.

Gould, Stephen Jay. "The Median Isn't the Message." Readily available on the Web,https://people.umass.edu/biep540w/pdf/Stephen%20 Jay%20Gould.pdf, accessed March 1, 2016.

Health Humanities Journal (Univ. of North Carolina-Chapel Hill) on line at http://hhj.web.unc.edu.

Institute for Patient- and Family-Centered Care, (http://www.ipfcc.org/ about/pfcc.htmlk, accessed June 12, 2018).

The Institute for Research on Unlimited Love, http:// unlimitedloveinstitute.org/.

Keats, John. "Ode to a Grecian Urn." Numerous editions available online.

Lemoine, Patrick. "The Placebo Effect: History, Biology, and Ethics," http://www.medscape.com/viewarticle/852144_5, accessed Sept. 16, 2017.

Measuring human biofield, http://www.faim.org/measurement-of-the-human-biofield-and-other- energetic-instruments, accessed Feb. 18, 2016.)

"Monkey mind" https://en.wikipedia.org/wiki/Mind_monkey, accessed April 20, 2016.

Moon rock. See https://airandspace.si.edu/stories/editorial/long-journey- our-lunar-touchrock.

National Institute on Aging. Comfort Care, https://www.nia.nih.gov/ health/publication/end-life-helping-comfort-and-care/providing- comfort-end-life, accessed March 5, 20016.

Panpsychism, https://en.wikipedia.org/wiki/Panpsychism#Quantum_ physics, accessed Aug. 3, 2016.

Qi: The Journal of Traditional Eastern Health & Fitness. A quarterly magazine, now in its 26th year, based in Temecula, California, https:// www.qi-journal.com/.

Sea-Bands, http://www.sea-band.com/for-medical-professionals, accessed July 14, 2016).

"Tai Chi and Qigong," The National Center for Complementary and Integrative Health, https://nccih.nih.gov/health/taichi, accessed May 17, 2016

Tegmark, Max. "Consciousness as a State of Matter." March 18, 2015.

https://arxiv.org/pdf/1401.1219v3.pdf, accessed May 8th, 2015.

Walton, Tracy. "Help for Hospital-Based Massage Therapy, http://www.massagetoday.com/mpacms/mt/article.php?id=14693, accessed March 5, 2016.

Whitman, Walt. "When I Heard the Learn'd Astronomer," https://www.poetryfoundation.org/poems/45479/when-i-heard-the-learnd-astronomer, accessed July 10, 2017.

World Health Organization, http://www.who.int/about/en, accessed December 5, 2017.

Index

www.ingramcontent.com/pod-product-compliance
Lightning Source LLC
Chambersburg PA
CBHW021546210326
41599CB00010B/333